William Henry Fox Talbot

Studies in British Art 23

William Henry Fox Talbot
Beyond Photography

Edited by Mirjam Brusius, Katrina Dean, and Chitra Ramalingam

Published by
The Yale Center for British Art
The Paul Mellon Centre for Studies in British Art

Distributed by
Yale University Press, New Haven & London

Published by

The Yale Center for British Art
P.O. Box 208280
1080 Chapel Street
New Haven, CT 06520-8280
britishart.yale.edu

and

The Paul Mellon Centre for Studies in British Art
16 Bedford Square
London WC1B 3JA
United Kingdom
www.paul-mellon-centre.ac.uk

Distributed by

Yale University Press
P.O. Box 209040
302 Temple Street
New Haven, CT 06520-9040
www.yalebooks.com/art

Volume edited by Mirjam Brusius, Katrina Dean, and Chitra Ramalingam
Publication coordinated by Eleanor Hughes at the Yale Center for British Art
Copyedited by Richard Slovak
Index by Nancy Wolff

Designed by Anjali Pala, Miko McGinty Inc.
Set in Garamond by Tina Henderson
Separations by Professional Graphics, Inc., Rockford, IL; printed in Italy by Conti Tipocolor S.p.A

Library of Congress Cataloging-in-Publication Data

William Henry Fox Talbot : beyond photography / edited by Mirjam Brusius, Katrina Dean, and Chitra Ramalingam.
pages cm—(Studies in British art ; 23)
Includes bibliographical references and index.
ISBN-13: 978-0-300-17934-7 (hardback)
ISBN-10: 0-300-17934-0 (hardback)
1. Talbot, William Henry Fox, 1800-1877. 2. Photographers—Great Britain—Biography. 3. Science—Great Britain—History—19th century. 4. Photography—Great Britain—History—19th century. I. Brusius, Mirjam. II. Dean, Katrina. III. Ramalingam, Chitra, 1979-
TR140.T3W55 2013
770.92--dc23
[B]

2013019906

A catalogue record for this book is available from the British Library.
10 9 8 7 6 5 4 3 2 1

This is the twenty-third in a series of occasional volumes devoted to studies in British art, published by the Yale Center for British Art and the Paul Mellon Centre for Studies in British Art, London.

Jacket illustrations: (*front*) William Henry Fox Talbot, "I can't recognize half of these characters!" (detail), drawing in ink on tracing paper of two tablets from the Kouyunjik Collection, date unknown. Talbot Collection, British Library; (*back*) William Henry Fox Talbot, insect wing (lantern fly) (detail), 1839, photomicrographs, negative. National Media Museum, Bradford.

Contents

INSTITUTIONS AND NETWORKS

Acknowledgments

The editors thank the contributors to this volume, which draws in part on papers for a conference in June 2010 at the Centre for Research in the Arts, Social Sciences, and Humanities (CRASSH), University of Cambridge, then host institution to Chitra Ramalingam. Special thanks are due to the CRASSH scholars and administrators who helped us transform our ambitions for the event from a modest workshop into a major interdisciplinary conference and whose conference program promotes and provides a space for experiments such as this one: Mary Jacobus, Andrew Webber, Catherine Hurley, and Michelle Maciejewska. We acknowledge additional support for the conference from the British Academy, the Gerda Henkel Stiftung, the Paul Mellon Centre for Studies in British Art, and the Master and Fellows of Trinity College, Cambridge. In addition to those represented in this volume, contributions to the conference by James Elkins, Tony Crilly, David Gange, Frank James, Robin Kelsey, John van Wyhe, and Anatoly Liberman shaped our thinking on Talbot's work in many fields. A small exhibition to accompany the conference was hosted by the Wren Library, Trinity College, Cambridge, courtesy of David McKitterick.

The editors would also like to thank the host institutions of Mirjam Brusius and Katrina Dean at the time this volume was developed. The Department of History and Philosophy of Science at the University of Cambridge hosted Mirjam's PhD research, during the course of which the project leading to this volume took shape. Heartfelt thanks are due to Simon Schaffer and Eleanor Robson for their extensive knowledge and thoughtful encouragement. Mirjam's doctoral research was funded by a United Kingdom Arts and Humanities Research Council collaborative doctoral award, the Gerda Henkel Stiftung, the Rausing Studentship, the Cambridge Trust, and the Kurt Hahn Trust.

The partner institution the British Library holds the key archival sources underpinning this work and hosted the project, of which this volume is one aspect, to bring them to a wider scholarly audience. Throughout the course of the project and the production of this volume, staff members at the British Library have generously contributed their curatorial and bibliographical expertise and assisted us in obtaining images for this publication. Special thanks are due to John Falconer, Frances Harris, Scot McKendrick, William Frame, Christopher Beckett, and Jonathan Pledge. Reproduction of works from the Talbot archive at the British Library is courtesy of Petronella Burnett-Brown and the British Library Board.

Oliver Flebbe, Thomas Gerhardt, Christian Gérini, Arnold Hunt, Hans P. Kraus, Jr., Graeme Mitchison, Ursula Sims-Williams, and Roger Watson (National Trust, Fox Talbot Museum) all assisted Mirjam in aspects of the research for her catalogue of Talbot's notebooks, which is now available online. Additional thanks are due to Jim Secord, Jennifer Tucker, and Kelley Wilder for their intellectual engagement and support.

Every aspect of this endeavor has benefited from the advice of the historian of photography and Talbot expert Larry Schaaf. We also are grateful to Richard Slovak for copyediting the volume and Anjali Pala at Miko McGinty Inc. for her beautiful design. Finally, we thank our collaborators at the Yale Center for British Art for their encouragement, professional support, and hard work on this project, namely, Amy Meyers, Eleanor Hughes, Robin Hoffman, Imogen Hart, Erika Cohn, Shaunee Cole, Christopher Lotis, Sarah Rapuano, and Craig Canfield.

Mirjam Brusius, Katrina Dean, and Chitra Ramalingam

NOTE TO THE READER

The letters to and from Talbot quoted in this volume can be found online in the Correspondence of William Henry Fox Talbot, http://foxtalbot.dmu.ac.uk. Most notebooks by Talbot are now catalogued at the British Library and can be found in its online catalogue. If an item at the British Library is labeled "partially catalogued," that means it is still in the process of being catalogued and has not received a full Add. MS reference number, the primary way to search for items in the online catalogue there. In this case a former catalogue number, such as an accession number (acc. no.), will be given where available, in addition to the overall collection number (Add. MS 88942), to help the reader trace the item with the help of the Manuscripts Reading Room staff.

INTRODUCTION
TALBOT AND HIS ARCHIVE

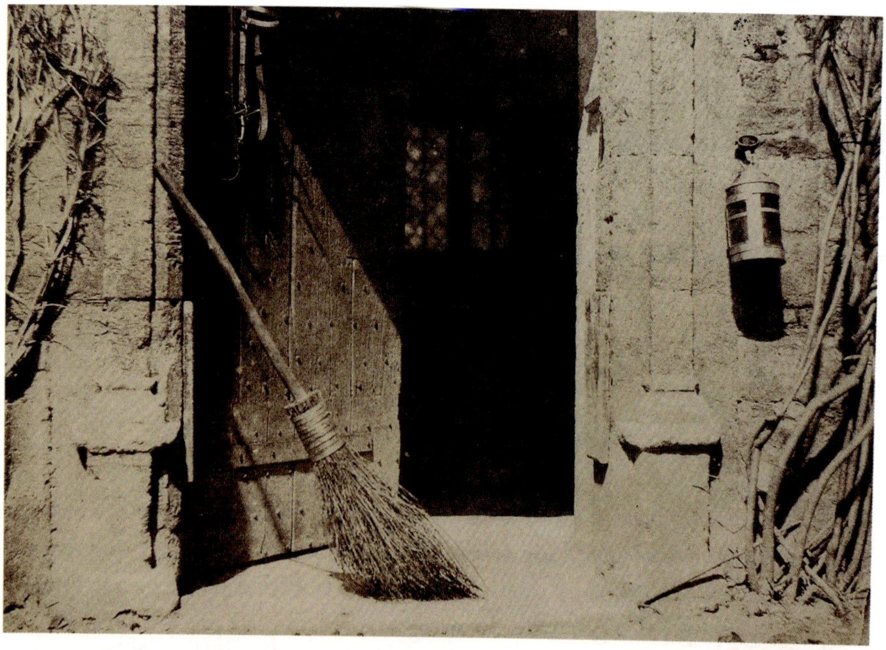

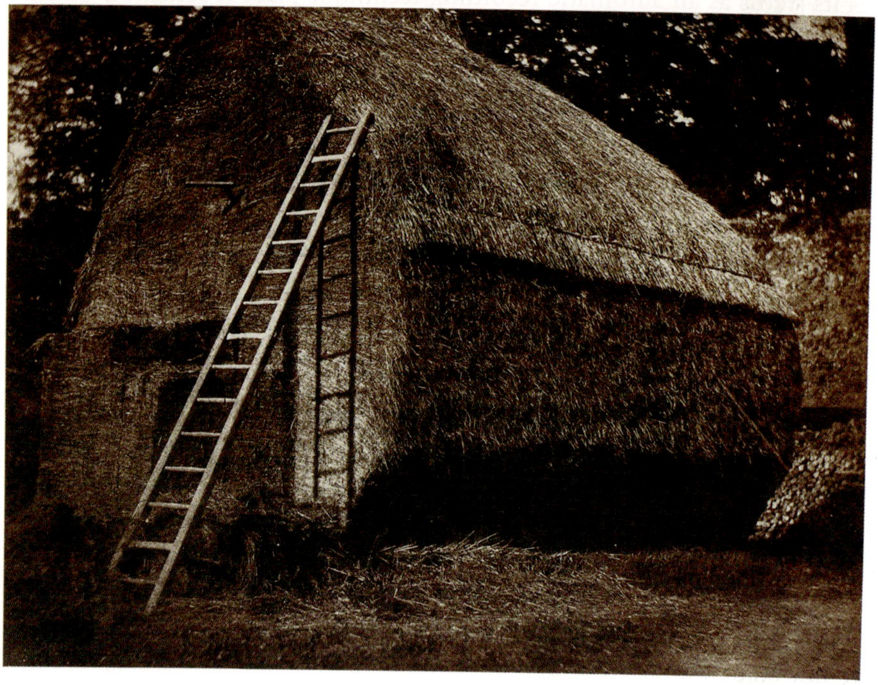

1. William Henry Fox Talbot, *The Open Door*, before 1844, salted paper print, 5⅝ x 7⅝ in. (14.4 x 19.4 cm), in Talbot, *The Pencil of Nature* (London: Longman, Brown, Green and Longmans, 1844), plate VI. British Library.

2. William Henry Fox Talbot, *The Haystack*, ca. 1841, salted paper print, 6⅜ x 8¼ in. (16.2 x 20.9 cm), in Talbot, *The Pencil of Nature* (London: Longman, Brown, Green and Longmans, 1844), plate X. British Library.

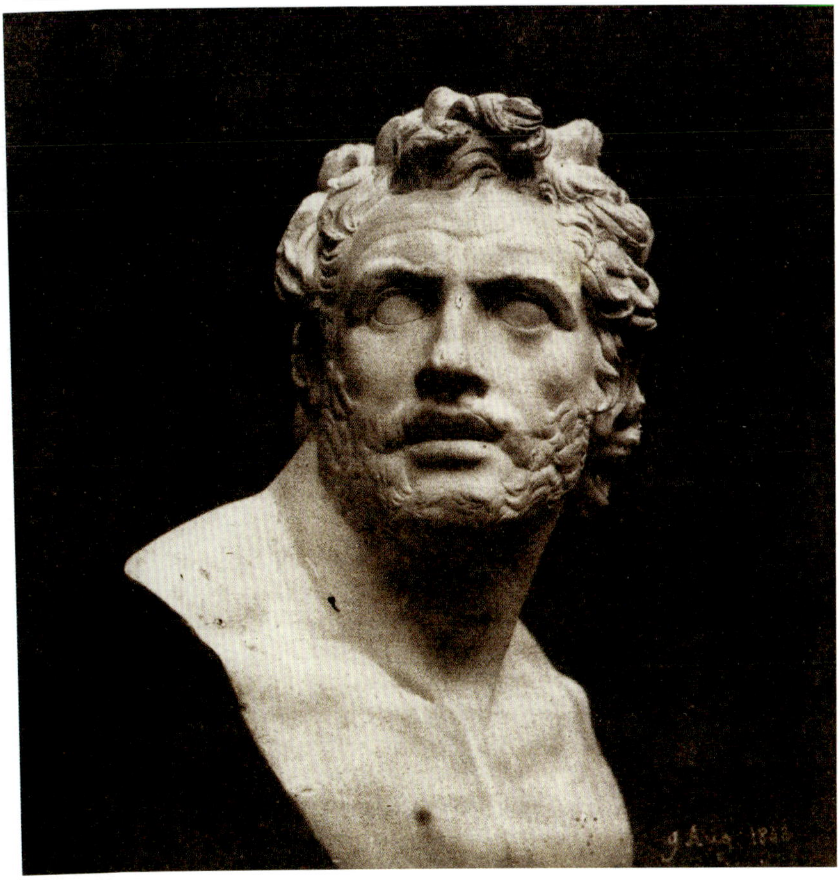

3. William Henry Fox Talbot, *Patroclus*, Sept. 8, 1841, salted paper print from calotype negative, 7 x 6⅞ in. (17.7 x 17.6 cm). Photo 2 (17), Talbot Collection, British Library.

H. J. P. Arnold first addressed the full range of Talbot's interests in his biography of 1977.[2] Many scholars have pointed to Talbot's prodigious interests and pursuits, but usually in terms of how they energized and influenced his achievements in photography, rather than on their own terms.[3] The frameworks of many of these studies allow for readings of Talbot's photographs only as works of art, enhanced by a contemporary perception of the images as valuable trade goods on the art market.[4] Scholars thus have been fascinated by his scholarly range, but most have not taken this as an opportunity to develop a more pragmatic approach to his photography. It may be that the frequent references to Talbot as a polymath or a "universal scholar" obscure rather than illuminate the real meaning of his constellation of activities. A complete contextualization of his work—one that encompasses the full range of Talbot's scientific and scholarly activities by taking into account the abundant archival record and fleshing out the networks of material and intellectual exchange that sustained his work—remains to be done.

Such an approach is more possible today than ever before, because of a recent dramatic advance in the accessibility of Talbot's archival collections. In 2003 a comprehensive online edition of nearly ten thousand letters to and from Talbot appeared as a result of the Correspondence of William Henry Fox Talbot project, directed by Larry Schaaf.[5] This excellent database helped to show that Talbot was connected to a remarkable network of nineteenth-century savants. The letters offer a point of entry into this network that can be traced through other sources, including his notebooks and the archives of his contemporaries.[6] Schaaf himself had already used the correspondence to investigate the involvement of the British scientist John Frederick William Herschel, a significant friend and correspondent during the period when Talbot made his photographic experiments.[7] In addition, the Talbot family donated a large collection of his notebooks and other manuscripts, along with photographs, to the British Library in 2006. This collection is significant in both scope and scholarly integrity. The notebooks have now been catalogued for the first time and are accessible to all scholars.[8] The time is thus ripe for new narratives and contextualized accounts of Talbot's life and work.

An interdisciplinary conference, entitled "William Henry Fox Talbot: Beyond Photography" and held at the Centre for Research in the Arts, Social Sciences, and Humanities at the University of Cambridge in 2010, attempted to approach Talbot's work with this wider perspective in mind.[9] The event brought together art historians, curators, historians of science, and practitioners in the many scholarly fields to which he contributed, including many of the contributors to this volume. Like the conference, this volume is intended to be a starting point for future work. Its aim is to situate Talbot amid the networks and institutions of Victorian intellectual enterprise, while raising basic questions about the relation between photography and these other fields, as well as pointing toward new directions of research.

The notebook collection in the British Library reflects Talbot's many and long-standing interests beyond photography, and it allows new narratives about his life and work to emerge. In about 1833 Talbot began the photographic experiments that would make him famous after he presented his invention to the world in 1839. He was most active as a photographer during this decade and early in the following one; by the mid-1840s, perhaps surprisingly, he had given up working behind the camera and turned his attention almost entirely to other activities.[10] This was the most intensely creative period in his life, and it was characterized by widely ranging pursuits.[11] Many art historians might be forgiven for assuming that Talbot spent 1839—the year when photography was presented to the public by his rival in France, Louis-Jacques-Mandé Daguerre—entirely consumed by publicizing and perfecting his technique for photogenic drawings. Yet 1839 was also when

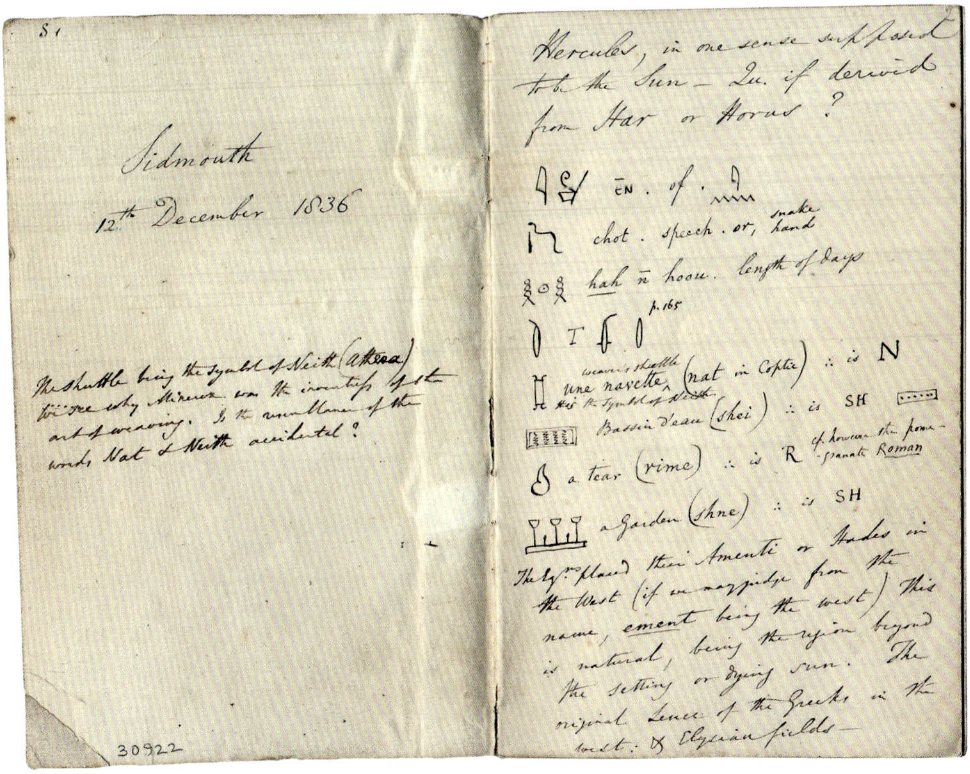

4. William Henry Fox Talbot, Notebook S.1, 1836. Add. MS 88942/1/13, Talbot Collection, British Library.

Talbot published a slim volume entitled *The Antiquity of the Book of Genesis: Illustrated by Some New Arguments*, a contribution to a common literary genre that sought to identify traces of the influence of the book of Genesis on the religions of preclassical civilizations; in the same year he published the second volume of *Hermes, or Classical and Antiquarian Researches*, an exploration of questions about the antiquities and classical traditions of Greece, Italy, and Egypt.[12] An exclusive focus on photography makes his work on these antiquarian projects at this time seem puzzling.

In fact, from the very beginning Talbot's interests ranged widely, and one of the most enduring continuities in his life was his enthusiasm for classical, antiquarian, philological, and Assyriological studies. His early scholarly successes as a student at Cambridge included the Porson Prize for Greek Verse (1820); his first publications were contributions to classical scholarship. Talbot's archive contains many notebooks on etymology and philology from the 1830s and 1840s, again during the period in which he was most active in photography (fig. 4). From the 1850s to the end of his life, Assyriology was one of the subjects on which he spent the most time, as documented by the more than one hundred Assyriological notebooks that form the majority of the notebooks in the British Library collection

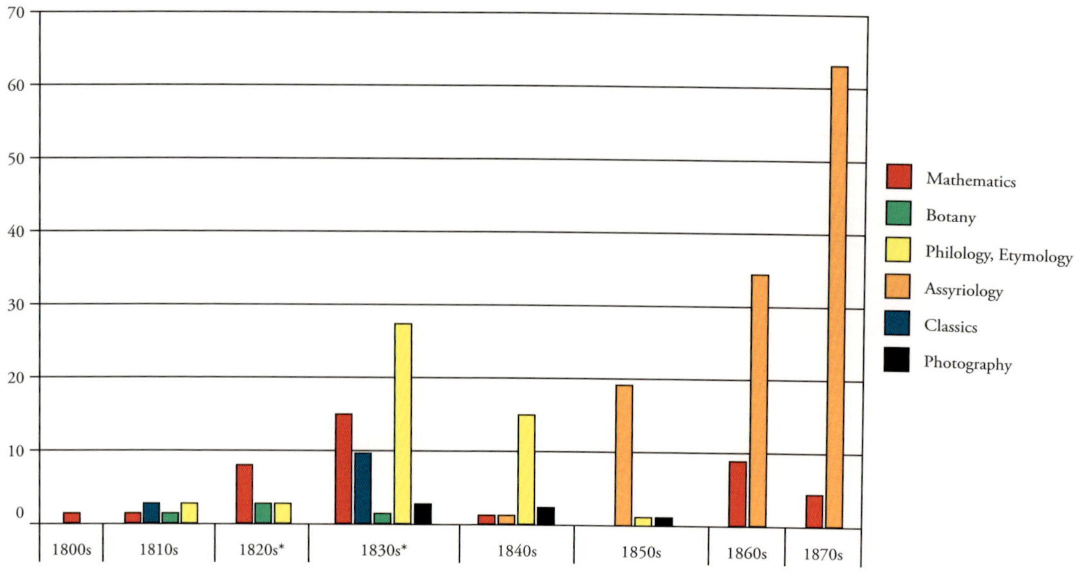

**In the 1820s and 1830s the mathematical notebooks also included notes on the physical sciences.*

5. Subject and number of Talbot's notebooks by decade.

(fig. 5).[13] He also became a fierce (but not always successful) advocate of the use of photography as a field research tool in archaeology and Assyriology, particularly for the settling of philological disputes. An abiding interest in origins and a persistent fascination with script and inscription, languages and decipherment, appear continually throughout Talbot's notebooks until the end of his life.

As a student at Cambridge Talbot also excelled in mathematics, ranking as "twelfth wrangler" in the rigorously competitive system of examinations known as the Mathematical Tripos, which in his day was understood not only to be a training ground for the best mathematicians and mathematical physicists of the Victorian era but also to serve as a key element of liberal education for gentlemen.[14] His lifelong study of mathematics is documented in his publications and notebooks on scientific and mathematical topics dating from the 1820s through the 1870s. As for the natural sciences, the notebooks show that Talbot was active in botany from an early age, and that he worked at the forefront of chemical and optical research in the 1820s and 1830s. He was a member of the scientific elite and an active member of the Royal Society, the principal learned society for natural science, during a period now seen as critical to its reorganization and professionalization.

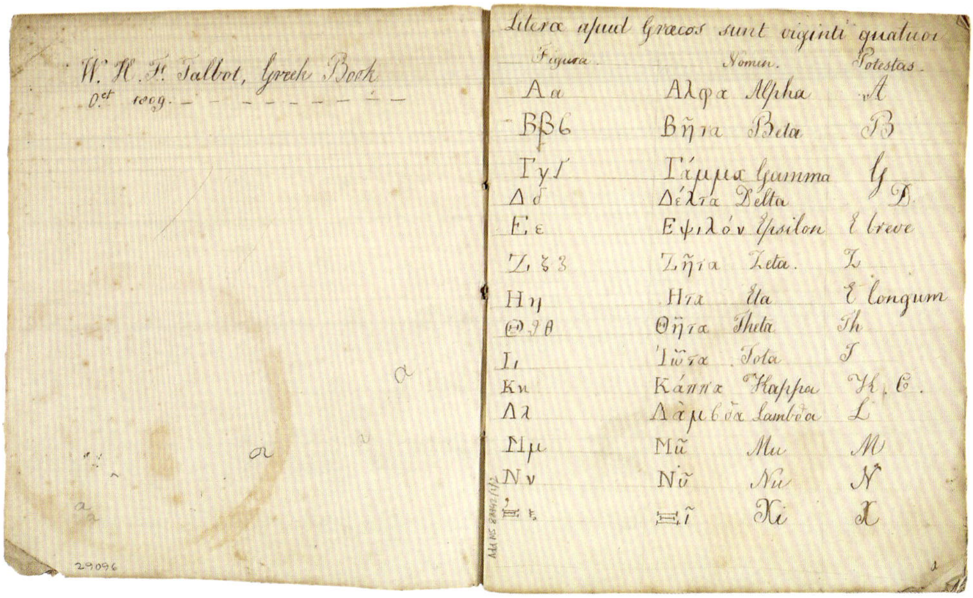

6. William Henry Fox Talbot, Exercise Book, 1809. Add. MS 88942/1/2, Talbot Collection, British Library.

NOTE-TAKING, KNOWLEDGE-MAKING

These and many other aspects of Talbot's life can be traced through the notebook collection at the British Library, which includes about 360 items.[15] The first notebooks date from when he was about ten years old (fig. 6). He continued the practice of keeping notebooks until his death in 1877 (fig. 7). Although several items are in other collections, the large assortment of notebooks held in the British Library gives a representative overall impression of Talbot's scholarship. Of these 360 items, 124 are accounts, miscellaneous notebooks, and calendars, many of which were kept by his father, his mother's servant, or other staff at Lacock Abbey in Wiltshire.[16]

Apart from notebooks the Talbot archive at the British Library contains unbound Assyriological and mathematical folios, natural specimens, photographs, diaries, letters, offprints of his articles, patents, artifacts, and a small selection of books, presumably from his library at Lacock.[17] This diversity of objects and documents invites us to look more closely at the materiality of Victorian scientific and intellectual life. In recent decades there has been a move toward the historical study of science and scholarship as practical activities, located for its practitioners in the routines of everyday life, and to an understanding of scientific knowledge as made

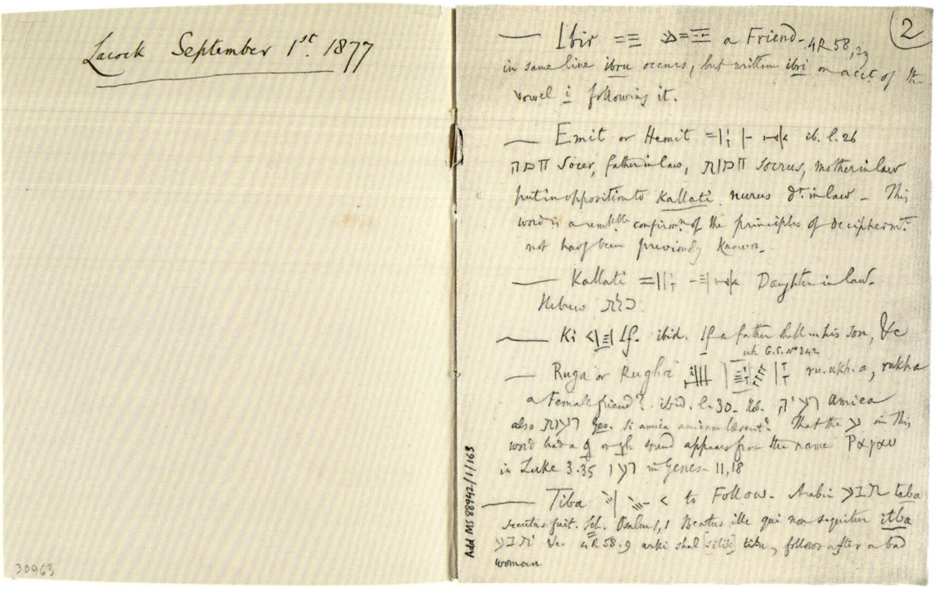

7. William Henry Fox Talbot, Notebook T, 1877. Add. MS 88942/1/163, Talbot Collection, British Library.

through human and social action rather than simply acquired. This has opened up a view of science as a process, suggesting new ways to think about old categories and dichotomies such as experiment and theory; laboratory, museum, and field; body and mind; doing and thinking. It has also encouraged attention to the materiality of scientific practices: to the instrumentation and embodied techniques that make up the day-to-day activities of science; and to the imbrication of thinking and knowing, on the one hand, and making and doing, on the other. This move has been characterized as one that treats the visual and material culture of science as "materialized epistemology," relating sensual with ideational knowing.[18]

It is this approach that we suggest can be taken with Talbot's archive. The essays in this volume highlight the variety of documents, objects, and photographs in that archive and treat them as the material records of an intellectual culture and way of life that both enabled and constrained his activities.[19] We treat practices as various as note-taking, collecting, decipherment, and experimentation, and the various material traces they leave behind, as records of the way that knowledge of different kinds was made, maintained, and circulated in the period. When situated within the Talbot archive, photography itself might be characterized as a form of "materialized epistemology," a technique for making pictures that is also a method for generating or recording knowledge, though this is certainly not the only way to understand the medium's meaning. Talbot's botanical notebooks and collection of plant specimens, for example, can be read as records of practical scientific habits—particularly the cultivation of observational skill through microscopic investigation and counting—that he then applied to the study and evaluation of his earliest photographs.[20] When read against entries in his scientific notebooks on the optical illusion of visual persistence, the instrumentation used in Talbot's proposal in 1851 for instantaneous photography by spark illumination reveals his dependence on devices already in use for decades to establish reliable visual knowledge about rapid movement through acts of immediate perception with the eye.[21] Talbot's Assyriological notebooks and his other images and objects related to decipherment, meanwhile, allow us to trace the intellectual and material resources required for decipherment in the period. As records of practice in an evolving discipline, they can help us to understand the numerous difficulties that scholars of past civilizations found when they evaluated photographs of ancient inscriptions as objects for research.[22]

In conjunction with the letters, the notebooks are thus the key to understanding Talbot's intellectual mind-set, his interests beyond photography, his exchanges with other scholars, and his publications.[23] The notebooks were tools for self-observation; he used the act of writing to stimulate the thinking process. It is here that the transition from the private toward the public takes place, for many rudimentary notebook entries resulted in letters or drafts for publications.[24] As a register of what was worth recording, the notebooks are a manifestation of the very origins of Talbot's scholarship, providing further avenues to understanding his way of acquiring, assembling, selecting, organizing, classifying, memorizing, revising, and establishing knowledge. That is, his note-taking practices are also a mirror of his methods of knowledge production. For Talbot, who traveled widely in Europe but never to remote places such as Mesopotamia, notebooks were the field; they were the place where he collected and re-collected information over a long period of time. We might understand the diaries and research notebooks as a technology that allowed for the gathering, recording, and preserving of information, a means to memorize, understand, and reflect on what one had perceived.[25]

TALBOT'S ARCHIVE AND PHOTOGRAPHY

In many cases, we suggest, Talbot's photographs might also be seen in such a light. One of the arguments between the lines of this volume is that the meaning of his photographs cannot not be reduced purely to iconography but must also be understood as a witness of scientific and scholarly concerns, just as the notebooks are.[26] By considering his photographs and his notebooks as complementary tools, we find that many of Talbot's early uses of photography concern new methods of inscription, recording, classification, visual display, collection, and, above all, reproduction.[27] For him photography's appeal had much more to do with these concerns than has previously been recognized.

Although Talbot's photographs clearly have aesthetic value, many of them were inspired by his scientific and scholarly activities, rather than produced as works of art for their own sake. In his notebooks there is little record of an interest in images in the narrow sense of pictures. On the other hand, concerns with script as a visual medium, geometry, pattern finding and decoding, ocular acuity, and more general visual problems related to observation and discernment are present throughout his archive. Talbot's understanding of photography as a means to making appealing pictures certainly was guided by the cultural and aesthetic values of his class. But we will fail to understand these values if we remain focused solely on how he engaged with the aesthetics of pictorial production in the fine arts. In a small handful of publications on photography, Talbot did adopt rhetoric suggestive of picturesque or romantic aesthetics. But there is more to the story. Even where his manuscripts do not explicitly take up photography—which is the case for all but a handful of notebooks in the collection—they are a fertile resource for exploring photography's meaning against the backdrop of scholarly and experimental practice in many fields. The variety of media in Talbot's archive enables us to make sense of the place of photography both within and against these practices.

Throughout his life, for example, Talbot continued to describe and evaluate photographs according to terminology and criteria borrowed from ways of engaging with script and printed text. This tendency can also be traced in the wider early discourse around the new medium, which incorporated ideas about inscribing, drawing, and printing.[28] For Talbot this was true even in his final decades, when decipherment and "photoglyphic engraving"—a name he explicitly modeled on words such as "hieroglyph," hinting at a suggestive analogy between photomechanically produced images and ancient scripts, both of which provoked problems of legibility—were his primary concerns.[29] Art historians have previously noted how Talbot explored linkages and divergences between text and image in his first photographic book, *The Pencil of Nature* (1844); they have also pointed out that in *The Pencil of Nature* he foregrounded the criterion of legibility in his discussion of the image entitled *Entrance Gateway, Queen's College, Oxford*.[30] But more attention must be paid to the prevalence and particularity of such concerns throughout his archive, in pursuits as various as etymology, Assyriology, optical research, and photomechanical reproduction.[31]

We also call attention to photography's place in networks of scientific and intellectual exchange. In certain contexts photographs were valued as proxies for what they represented. A photograph might stand in as a reproducible and mobile version of its more intractable referent.[32] It might also be criticized as an inadequate proxy, an imperfect copy, less informative or evocative than its original. Such criteria for evaluating photographs placed them within a varied economy of media through which distant or inaccessible objects were reproduced and circulated: drawings, engravings, etchings, lithographs, and more. The correspondence between Talbot and many other scientists and antiquarians highlights the importance of the exchange of information in nineteenth-century knowledge systems.

More generally we hope future researchers will be inspired by the contents of this volume to scrutinize conventional ideas of iconicity with respect to Talbot's photography. *The Open Door* or *The Haystack* (see figs. 1 and 2) from *The Pencil of Nature* may indeed allude to a familiar iconography, as some art historians have argued. In doing so, however, these pictures also engage with the concern for detail and accuracy that distinguished, for example, seventeenth-century Dutch painting.[33] Talbot experimented with visual challenges, using familiar iconographies and traditions to shape and foster a new image-generating device. Other Talbot photographs valued for their composition and aesthetic appeal can likewise be read as rigorous experiments to negotiate how photographers could tackle specific technical challenges, such as the depiction of scale. *The Open Door* and *The Ladder* (fig. 8), for example, explore how distance, shadows, and associative relationships can be depicted through chemistry. The iconicity of Talbot's photographs is thus determined by experimental trials of how things are depicted *in a photograph*, and by their information content. Even when Talbot used works of art as subjects of photography, he did so by turning them into objects of study and experiment. In doing so he both diminished and increased their status as works of art. Aesthetic values could provide useful avenues to communicate knowledge, while knowledge might generate aesthetic visual effects.[34]

The diversity of functions carried out by photography has long prevented us—along with Talbot himself—from clearly defining its nature. As Vered Maimon has shown, the ambivalence prevalent throughout the text of *The Pencil of Nature* suggests that Talbot most likely was undecided himself about what the medium was and what it might be.[35] This was also apparent in the early history of photographic terminology and the naming of the medium.[36] Even so, throughout *The Pencil of Nature* he introduced photography as a tool directed by a confident specific guiding interest. Talbot's gaze was that of a gentleman scientist and antiquarian who knew about the objects he photographed. For him objects were registered on paper just as thoughts were registered in letters and notebooks. Perhaps in photography, as in note-taking, we may find reflected the entire breadth of Talbot's methods of knowledge production: collecting, reading, recording, classifying, studying, deciphering, memorizing, displaying, and circulating.

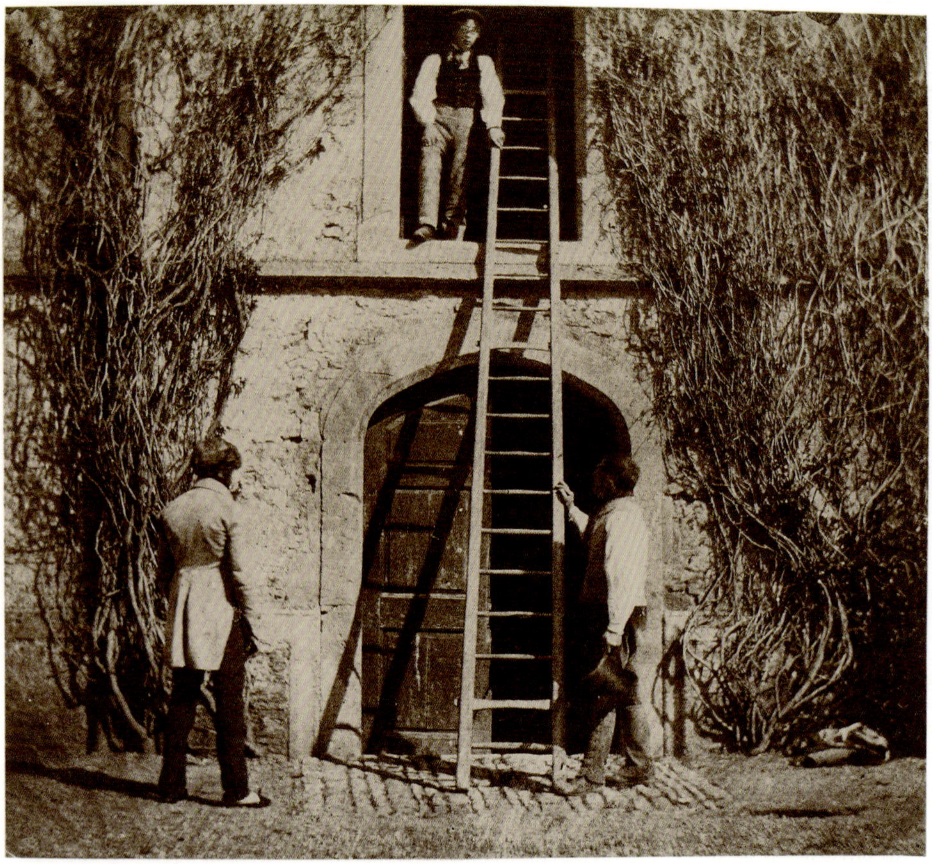

8. William Henry Fox Talbot, *The Ladder*, before 1845, salted paper print, 6⅝ x 7⅛ in. (16.9 x 18.2 cm), in Talbot, *The Pencil of Nature* (London: Longman, Brown, Green and Longmans, 1844), plate XIV. British Library.

ART AND SCIENCE

For those readers who may be familiar only with the photographs for which Talbot became famous, the many images in this volume are intended to reveal the visual and material variety of his archive.[37] They bring into view the rich aesthetics at the heart of Victorian scientific and intellectual culture, particularly in antiquarian pursuits and the natural sciences. They also complicate our understanding of the aesthetics of early photographic production by situating photography against practices, techniques, and activities that have rarely been connected to the medium's history.

Talbot's wide range of activities and interests reflects a crucial period in the intellectual development and social formation of Victorian knowledge systems, when the meanings of even such seemingly basic terms as "art" and "science," not to mention "discovery," "invention," "discipline," "professional," "amateur," and "aesthetics," were very much in flux. It is our contention that the modern distinction between art and science cannot be deployed uncritically in an analysis of Talbot's life and work, or in the early history of photography more generally. Talbot himself certainly does not seem to have seen any sort of binary opposition of this kind, referring to his new invention frequently as "a new Art" while focusing his efforts on publicizing it to the "scientific public."[38]

Talbot's failure to distinguish sharply between the medium's scientific and aesthetic value is less surprising than it might seem. In fact, neither term had a stable meaning in this period. The very notion of a unified "science" was consolidated only in mid-century. The term "scientist," notably, did not exist in English until William Whewell coined it in 1833 and proposed that it replace the broader term "natural philosopher."[39] Meanwhile, discourses on art that previously emphasized links to craftsmanship and technical skill began to take on aesthetic ideologies that linked it to imagination, creativity, and beauty.[40] The appearance of photography exactly at the time when these shifts were taking place is no coincidence.[41] As an image-making process photography shared something with pictorial art, but as a mechanical and chemical process as well as a recording tool it also shared something with science and its aims, as well as with art as craftsmanship and technical skill.[42] It was exactly this ambiguous relationship between skilled craftsmanship and self-inscription that made the nature of photography difficult to grasp.

Instead of assuming the separateness of these two categories, the essays in this volume open up a richer history of techniques and practices that are not so easily designated as belonging exclusively either to art or to science. They do so by calling attention to the materiality of record-keeping and imaging practices for nineteenth-century scientists and intellectuals, by examining values having to do with the proper or enjoyable exercise of visual judgment and observation, and by exploring how photography was compared to or combined with other recording practices in scientific and scholarly practice. In doing so we mean to open up broader questions about how certain practices, images, or objects come to be designated as belonging to either of these categories in the first place. For the nineteenth century—a period during which the modern scientific disciplines first emerged only gradually, the very notion of a unified "science" was contested, and the cultural position of art and its relation to artisanal and technical craft were also realigned—this seems to be a much more historically sensitive and fruitful way to probe the changing economies of art and science, as well as the techniques of representation that lie between them at interesting boundary sites such as the early history of photography.[43]

VICTORIAN KNOWLEDGE NETWORKS

In France early photography emerged in the context of visual art and showmanship; Daguerre, Talbot's main rival, was a set designer and panorama painter, a master of the art of theatrical illusion.[44] By contrast Talbot and Herschel, the British innovators who contributed to the invention of photography, were gentlemen of science: gentle-born men of leisure, means, and honor for whom science and scholarship were amateur activities.[45] One of the aims of this volume is to contextualize Talbot with respect to the extensive literature on the social and cultural history of Victorian science. Insights gained from the study of science and scholarship in its historical context can complement and shed new light on art historical approaches not only to Talbot's life and work but also to the early history of photographic invention. His most active years as a scientist, scholar, and photographer took place during a period of massive social and cultural transformation in Britain. The agrarian, rural, aristocratic, and Christian society of his youth would become, by the end of his life, an industrial, urban, middle-class, and secular society. These changes were also reflected in the social and epistemic economies of science of the period, which were characterized by dramatic institutional, disciplinary, and methodological innovations. The transition between the "natural philosophy" of the eighteenth century and the emergence of the more recognizable modern scientific disciplinary order was accompanied by a social shift in the scientific and intellectual elite from gentlemen of science to professionals, an institutional shift from privileged, private sites of knowledge-making to public institutions and universities, and a decline in the prominence of natural theology in favor of the more secular, expertise-oriented outlook of scientific naturalism.[46] These shifts were also reflected in early photography itself, which was at first dominated by genteel amateurs in a relatively homogeneous community. But class dynamics in photography changed just as they did in science in general.[47] By the late 1850s people from many social backgrounds identified themselves as photographers, and photography had become a commercial practice. Photography appeared on the scene just as new systems of authority, truth, and scientific expertise were consolidated. The discourse involving the new medium responded in part to the tensions surrounding these new regimes and debates over what scientific practice was.[48]

Yet none of these changes occurred uniformly or inexorably, and the 1830s and 1840s were a period of maximum flux. Natural science was still commonly assumed to be a gentleman's pursuit, but in fact the community of scientific practitioners was heterogeneous, encompassing not only genteel amateurs like Talbot but also instrument makers and craftsmen, members of such professions as law and medicine, and itinerant lecturers and showmen. Antiquarian scholarship, on the other hand, was more fully dominated by genteel enthusiasts with backgrounds and education similar to Talbot's. In both areas, however, ideals of gentlemanly, amateur comportment coexisted with more professional, expertise-based norms and a strong ethos of moral self-improvement. Talbot's genteel social position and his wide range of scholarly and scientific activities make him a useful subject for illuminating the shifting and socially heterogeneous landscape of nineteenth-century knowledge networks, as well as their geographically dispersed character and the cultures of communication and exchange that sustained them. The essays in this volume, in different ways, explore Talbot's movements through this variegated landscape. They find his activities shaped and constrained by discourses of class, skill, expertise, training, and discipline.

One way to begin to think about such matters is by highlighting the often-overlooked metropolitan aspect of Talbot's scientific and intellectual life.[49] He was in London, away from Lacock, for a good part of every year when he was not traveling abroad, particularly in the most productive years of his life. In the 1830s and 1840s he usually spent much of the traditional London "season" in town. For a gentleman of science like Talbot, unlike the provincial scientific elites of the eighteenth century, the concentrations of wealth, population, and technical skill in the metropolis were crucial resources.[50] The social diversity of British science was matched by a variety of urban places of scientific experiment, display, and exchange, which were also crucial sites around which the early photographic community would form. Talbot made frequent use of the city's artisan instrument makers and chemists to obtain supplies for his optical and chemical experiments; visited the city's public "galleries of science," machine displays, and museums; participated in activities at the Royal Society, Royal Institution, and other elite scientific societies; and relaxed at the Athenaeum, a gentlemen's club for men of literary and philosophical achievements and an important network for those with scientific or scholarly interests. Several of the essays in this volume invite us to pay closer attention to the variety of sites in which Talbot was active and the connections and tensions between them. For example, we can learn something of the shifting norms for participation in Victorian networks of science and scholarship by contrasting Talbot's relationship to two sites in London: the Royal Institution and the British Museum.

The Royal Institution, London's premier venue for experimental science, was an important focus for Talbot's scientific activities in the capital. With the famous experimental philosopher Michael Faraday at its head, it was the best-equipped chemical and physical laboratory in the country.[51] On numerous occasions in the 1820s and 1830s, Talbot collaborated with Faraday on chemical and optical experiments, and his well-regarded research on the spectral analysis of flames was conducted in part at Faraday's laboratory in 1826.[52] Faraday encouraged and advised Talbot, while fending off more time-consuming requests, and frequently had his assistant at the laboratory perform experiments or prepare materials for Talbot's use.[53] Faraday was also one of the signatories on the certificate of recommendation confirming Talbot's election as a fellow of the Royal Society.[54] The productive relationship between the two men reflects the social landscape of early nineteenth-century science in London. Talbot's wealth and effortless social standing contrasted sharply with Faraday's humble origins and carefully negotiated rise in status; the son of a blacksmith, Faraday had labored long and hard to establish his reputation and independence as a "man of science."[55] Their apparently close working relationship is complicated somewhat by the cautious tone of several of Faraday's extant letters to Talbot, which suggests that he felt their social difference keenly. His wary deference in response to Talbot's requests recalls an earlier period in Faraday's life, when as a laboratory assistant his experimental services were available to other higher-status men of science, his own status occasionally overlapping uncomfortably with that of a servant. Despite their stark differences in background, Talbot frequently found himself relying on the resources of the Royal Institution's laboratory and on Faraday's expertise, encouragement, and respect. He also depended on Faraday's well-established procedures at the Royal Institution for establishing priority in scientific discoveries through public performance and display. It was at the Royal Institution Library that Talbot first displayed his photogenic drawings to the world in early 1839, in a presentation overseen by Faraday.[56]

Talbot's family connections and social standing brought him easy access to membership in the scientific elite as a genteel amateur; however, when he turned more toward his antiquarian and philological interests later in life, he found that access was not so easy. He was particularly interested in the ancient objects excavated in Mesopotamia by the British explorer and antiquarian Austen Henry Layard between 1845 and 1851, and brought to the British Museum. The first cargo of Layard's finds had arrived in London in October 1848, including colossal statues of winged bulls and lions as well as a collection of clay tablets with cuneiform inscriptions in an as-yet-undeciphered script. The following year Talbot started working to decipher this script. But despite his successes in decipherment, Talbot struggled to be taken seriously as a full-fledged member of the scholarly community of Assyriologists, particularly when he began to advocate for the use of photography both in the field and at the museum. When a project to photograph cuneiform inscriptions in the British Museum's collections in the 1850s finally began, with the intention of promoting the progress of decipherment by making the fragile objects carrying the inscriptions more mobile, Talbot found his access to these photographs impeded by the highly controlled and hierarchical systems for distributing Assyriological information from the museum. It was initially restricted to scholars holding positions at prestigious institutions, even while public access to the museum's collections was increasingly improved for all social classes. Talbot thus occasionally found himself dismissed as an amateur in fields that were increasingly, but unevenly, professionalizing. His proposals for how photography could be used as a pragmatic tool in fields where his expertise was not taken for granted—such as in botany, archaeology, or Assyriology—were sometimes disregarded.[57]

Our attention to Talbot's full range of activities as an early Victorian scholar and gentleman of science provides a new outlook on his life and work. The essays in this volume invite us to consider the specificity of London, Lacock, and other sites in his life as settings for scientific and intellectual activity, as well as the class dynamics that shaped them and the networks of communication and exchange that linked them. Talbot's voluminous notebooks and correspondence offer a rich resource for tracing how nineteenth-century knowledge was situated in specific places, how it traveled between sites, and how it was maintained or modified through translation and circulation.

Attention to these concerns can also enhance our understanding of photography itself. A memoranda notebook from 1839, for example, includes lists of individuals to whom Talbot sent photographs in his efforts to establish his priority of discovery as well as to convey to those colleagues the visual possibilities of the new medium. Most of these were scientific colleagues in a variety of fields. One typical entry, sent to his friend the physicist and telegraphic pioneer Charles Wheatstone, reads: "May 10 — Wheatstone. 4 in all. Tower of Lacock. Lace. Transp. No. 1" (fig. 9).[58] These records show how Talbot's efforts to publicize his own role in defining the new medium were pursued through his existing scientific networks. The place of photographs in these networks also suggests new ways of thinking about their aesthetics; sending one of his many images of the bust of Patroclus (fig. 10) to his colleagues was as much about the mobility of objects of study, the domestication of knowledge, and the photographic experiment as it was about rendering an art object. Highly staged and artful photographs such as *Patroclus* undermine claims that photography was mere mechanical self-inscription. Thus the question to ask is not whether Talbot sought more objective pictures in opposition to graphic art.[59] Instead, how did representation function for Talbot? What happened when various forms of representation, inscription, and reproduction interacted with one another? Above all, what was the scholarly motivation behind these pursuits in the early history of photography?

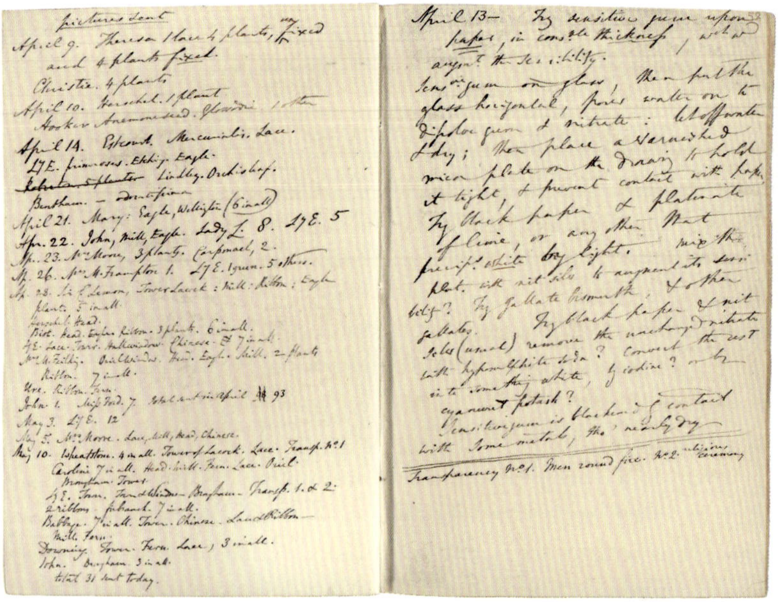

9. William Henry Fox Talbot, opening of Notebook "Memoranda," 1838–1840. Add. MS 88942/1/272, Talbot Collection, British Library.

THE STRUCTURE OF THIS VOLUME

The essays in this volume, contributed by art historians and historians of science, offer a variety of ways to approach these questions. The main body of the volume is divided into three parts: "Models for Investigation," "Invention and Discovery," and "Institutions and Networks." Our intention here is to challenge the assumption that Talbot's work can be easily divided into two categories: the photographic and the nonphotographic. Instead, early uses for and ideas about the new medium appear in most of the essays in this volume, interwoven—as they were in his activities—with other methods of copying, recording, transmitting, and displaying information.

Part 1, "Models for Investigation," introduces three of Talbot's lifelong passions: botany, mathematics, and the literary works of Sir Walter Scott. In her essay on Talbot's interest in botany, Anne Secord shows that botanical study—particularly the visually demanding study of mosses, Talbot's favored specialty—provided a model for both observation and the exercise of visual and mental judgment that he would later apply to his photogenic drawings. Although Talbot excelled in mathematics early in life, June Barrow-Green argues that his lifelong habit of working and publishing in mathematics grew out of its place in a liberal education. Mathematics was a training ground for the mind rather than simply a pursuit in itself. Much has been made of Talbot's activities in both botany and mathematics as evi-

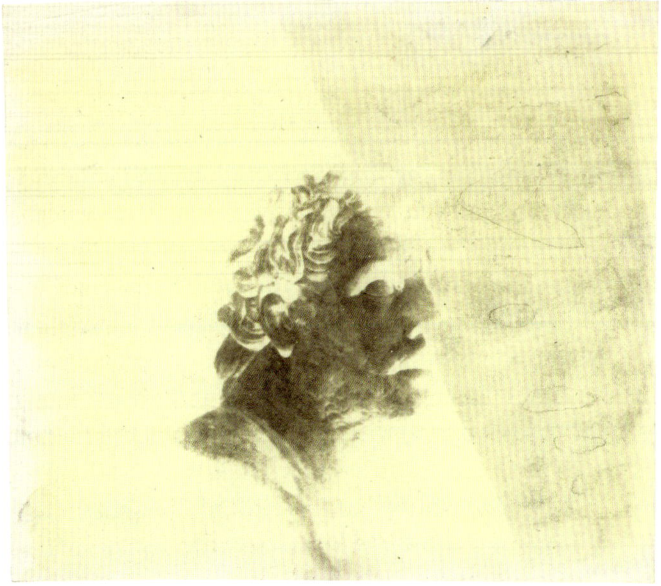

10. William Henry Fox Talbot, *Patroclus*, 1840, 7 x 7 in. (17.7 x 17.6 cm), salted paper print. Photo 3 (13), Talbot Collection, British Library.

dence of his exceptionally prodigious expertise. These two essays offer a different perspective: a view into a world in which mathematics and botany could serve as leisure pursuits and as models for intellectual activity in other areas, and in which Talbot may have been a typical rather than exceptional contributor to those fields. In addition, Graham Smith explores Talbot's enjoyment of Scott's novels and romantic verse, finding that his experience of these works can be traced through his activities relating to travel, history, architecture, and photography. Talbot's *Sun Pictures in Scotland* of 1845, a remarkable photographically illustrated book very different from the more miscellaneous *Pencil of Nature*, thus becomes a profound record of the intertwining of reading, experience, and memory.

Part 2, "Invention and Discovery," foregrounds the status of photography as Talbot's invention, but notions of invention or discovery are invoked very differently by the three authors in this section. This is fitting for a period in which the proper method for establishing claims to priority was contested, and the distinctions between scientific discovery and technical invention were being produced.[60] Herta Wolf and Vered Maimon treat these as discursive rather than technical achievements. Wolf's essay closely examines the emergence of the trope of "Nature as a drawing mistress" in the immediate press coverage of the first public announcement about photography in London in 1839. Maimon, on the other hand, analyzes the discursive production of photography in two of Talbot's canonical discovery accounts, finding that his early conceptions of photography should be read against methodological frameworks specific to English science in this period. The third essay, by Larry Schaaf, takes up Talbot's decades-long quest to perfect a practical means for book illustration by translating photography into printer's ink, a technique he referred to first as an "Improvement in the Art of Engraving." After 1850 Talbot largely abandoned making photographs, a fact often understood to mark a break in his interests. But Schaaf's essay makes clear the persistence of Talbot's concern with the mundane but fundamental problems of copying and reproduction, suggesting a continuation of his desire to "perfect" photography.

Part 3, "Institutions and Networks," moves outward to explore Talbot's participation in these aspects of Victorian intellectual enterprise. The essays by Eleanor Robson and Mirjam Brusius examine Talbot's efforts to establish himself among the community of British Assyriologists during the years when he worked to decipher Assyrian inscriptions. Robson's essay looks at the context surrounding the heroic "decipherment event" of 1857, which led to widespread acceptance that cuneiform could really be read. Robson also reveals the many mediations that were required to access ancient Assyrian texts: in order to make their way into English, inscriptions on clay tablets were first reproduced in various media and then interpreted through multiple linguistic traditions. Interestingly, most Assyriologists—including, in certain cases, Talbot himself—considered photographs of cuneiform tablets to be inadequate as working objects for decipherment. The perceived difficulties with using photographs of antique objects for research purposes are also taken up by Brusius, whose essay examines Talbot's efforts to convince the British Museum and other scholars studying the antique to integrate photography into research and collection practices in the field and at the museum. Chitra Ramalingam's essay traces Talbot's pursuit of photographic instantaneity, particularly his experiments with spark-illuminated photography, through a series of encounters with optical illusions and the investigation of human vision both enabled and constrained by London's demonstration-centered culture of scientific experiment. The volume ends with a commentary by Simon Schaffer that offers an incisive, critical reflection on the essays.

This introductory essay has explored how the contents of the Talbot archive at the British Library can be used to illuminate his identity as a nineteenth-century scholar and scientist, as well as to inspire new ways of thinking about the meaning of early photography and its role in nineteenth-century intellectual networks. We have dwelled particularly on tracing links between photography and the note-taking and knowledge-making practices documented in the notebook collection. But the Talbot archive is part of an even more diverse collection originally held at Lacock Abbey, a collection that extends not only "beyond photography" but indeed beyond Talbot himself. The following introductory essay, by Katrina Dean, suggests another perspective on photography by examining the significance of Lacock Abbey as a site of preservation and heritage.

1. See the essay by Katrina Dean in this volume.
2. H. J. P. Arnold, *William Henry Fox Talbot: Pioneer of Photography and Man of Science* (London: Hutchinson Benham, 1977).
3. For example, see John Hannavy, *Fox Talbot: An Illustrated Life of William Henry Fox Talbot, "Father of Modern Photography," 1800–1877* (Princes Risborough, UK: Shire Publications, 1976); Arnold, *William Henry Fox Talbot*; Gail Buckland, *Fox Talbot and the Invention of Photography* (Boston: David R. Godine, 1980); Hubertus von Amelunxen, *Die aufgehobene Zeit: Die Erfindung der Photographie durch William Henry Fox Talbot* (Berlin: Verlag Dirk Nishen, 1988); Larry J. Schaaf, *Out of the Shadows: Herschel, Talbot & the Invention of Photography* (New Haven and London: Yale Univ. Press, 1992); Vered Maimon, "Displaced 'Origins': William Henry Fox Talbot's *The Pencil of Nature*," *History of Photography* 32, no. 4 (2008): 314–25; and Graham Smith, "Talbot's Epigraph in 'The Pencil of Nature,'" *History of Photography* 34, no. 1 (2010): 90–95.
4. Critical alternative frameworks, for example, in relation to formations of labor and social class, are tested in Steve Edwards, *The Making of English Photography: Allegories* (University Park, PA: Pennsylvania State Univ. Press, 2006). Studies treating photography as a phenomenon depending on technological, social, scientific, and cultural changes in the nineteenth century include Bernd Busch, *Belichtete Welt: Eine Wahrnehmungsgeschichte der Fotografie* (Munich: Hanser Verlag, 1989); and Mary Warner Marien, *Photography: A Cultural History* (London: Laurence King, 2002).
5. The Correspondence of William Henry Fox Talbot, http://foxtalbot.dmu.ac.uk (accessed Jan. 30, 2012), hereafter "Correspondence."
6. See, for example, the Babbage Papers at the British Library and the Herschel Papers held by the Royal Society Library.
7. Schaaf, *Out of the Shadows*.
8. For the notebook catalogue, see searcharchives.bl.uk.
9. "William Henry Fox Talbot: Beyond Photography," conference at the Centre for Research in the Arts, Social Sciences, and Humanities (hereafter CRASSH), Univ. of Cambridge, June 24–26, 2010.
10. After this period, however, he did continue to advocate the practical use of photography in a variety of fields and to develop techniques for photomechanical reproduction; see the essays by Larry Schaaf, Mirjam Brusius, and Chitra Ramalingam in this volume.
11. Schaaf, *Out of the Shadows*.
12. William Henry Fox Talbot, *The Antiquity of the Book of Genesis: Illustrated by Some New Arguments* (London: Longman, Orme, Brown, Green and Longmans, 1839); and Talbot, *Hermes, or Classical and Antiquarian Researches*, vol. 2 (London: Longman, Orme, Brown, Green and Longmans, 1839). For more on the circumstances surrounding these publications, see Arnold, *William Henry Fox Talbot*, 88–92. For the context of Talbot's work on the antiquity of Genesis, see David Gange, "The Antiquity of Genesis and the Origins of Civilisation" (paper presented at "William Henry Fox Talbot: Beyond Photography," conference at CRASSH, Univ. of Cambridge, June 24–26, 2010).
13. See Notebooks on Assyriology and Decipherment, mid-1850s–1877, Add. MSS 88942/1/68–182, Talbot Collection (hereafter "TC"), British Library (hereafter "BL").
14. See the essay by June Barrow-Green in this volume; see also Andrew Warwick, *Masters of Theory: Cambridge and the Rise of Mathematical Physics* (Chicago: Univ. of Chicago Press, 2003).
15. For the notebook collection, see Mirjam Brusius, "Beyond Photography: An Introduction to William Henry Fox Talbot's Notebooks in the Talbot Collection at the British Library," *Electronic British Library Journal* (2010), http://www.bl.uk/eblj/2010articles/pdf/ebljarticle142010.pdf. The photographs are in the Jerwood Charitable Foundation catalogue of photographs at the British Library, http://www.bl.uk/catalogues/photographs/. Several photographic notebooks and thousands of important prints (originating in a donation to the Science Museum by Talbot's granddaughter Matilda in the 1930s) are held at the National Media Museum, Bradford, and a smaller number of items are in other public and private collections.

16. The notebooks vary in size, scope, and the manner in which they were held. While the account books are usually smaller, the scholarly notebooks are typically twice as large, some hardcover and some softcover, with differing length and design.

17. Add. MS 88942, TC, BL (partially catalogued). At time of publication, these items were still being catalogued; see Note to the Reader, p. viii.

18. M. Norton Wise, "Making Visible," *Isis* 97 (2006): 75–82. Wise's term is a generalization of Pamela Smith's notion of "artisanal epistemology," in which she treats image making as a form of knowledge production and argues that in cases in which images are imitations of the processes of nature, visual art and science should be understood to share the same goals. Pamela H. Smith, *The Body of the Artisan: Art and Experience in the Scientific Revolution* (Chicago: Univ. of Chicago Press, 2006).

19. On photographs as material objects and as elements of an archive, see Elizabeth Edwards and Janice Hart, eds., *Photographs Objects Histories: On the Materiality of Images* (New York and London: Routledge, 2004); and Elizabeth Edwards, *The Camera as Historian: Amateur Photographers and Historical Imagination, 1885–1918* (Durham: Duke Univ. Press, 2012).

20. See the essay by Anne Secord in this volume.

21. See the essay by Chitra Ramalingam in this volume.

22. See the essay by Eleanor Robson in this volume.

23. Comparably comprehensive archival collections for Victorian scientists include those of Charles Darwin and Michael Faraday. The Darwin archive of letters and notebooks is accessible in various electronic resources: http://www.darwinproject.ac.uk/darwins-letters and http://darwin-online.org.uk/EditorialIntroductions/vanWyhe_notebooks.html (both accessed Jan. 30, 2012). The bulk of the Faraday archive of research notebooks, manuscripts, and correspondence is held at the Royal Institution, London, and the complete research notebooks and correspondence have been published: *Faraday's Diary: Being the Various Philosophical Notes of Experimental Investigation Made by Michael Faraday during the Years 1820–1862*, ed. Thomas Martin, 8 vols. (London: G. Bell and Sons, 1932); and *The Correspondence of Michael Faraday*, ed. Frank A. J. L. James, 6 vols. (London: Institution of Electrical Engineers, 1991–2011).

24. For the private and the public in nineteenth-century scholarly archives, see, for example (in connection with Darwin), Martin Rudwick, "Charles Darwin in London: The Integration of Public and Private Science," *Isis* 73 (1982): 186–206.

25. Mirjam Brusius, "Preserving the Forgotten: William Henry Fox Talbot, Photography and the Antique" (PhD diss., Univ. of Cambridge, 2011; Akademie Verlag, forthcoming, and English edition in preparation). More perspectives on note-taking and scientific research notebooks can be found in Frederic Lawrence Holmes, "Scientific Writing and Scientific Discovery," *Isis* 78 (1987): 220–35; Frederic Lawrence Holmes, Jürgen Renn, and Hans-Jörg Rheinberger, eds., *Reworking the Bench: Research Notebooks in the History of Science* (London: Kluwer Academic Publishers, 2003); Ann Blair, "Note Taking as an Art of Transmission," *Critical Inquiry* 31 (2004): 85–107; Lorraine Daston, "Taking Note(s)," *Isis* 95 (2004): 443–48; and Anke Te Heesen, "The Notebook: A Paper-Technology," in *Making Things Public: Atmospheres of Democracy*, ed. Bruno Latour and Peter Weibel (Cambridge, MA: MIT Press / ZKM Karlsruhe, 2005), 582–89.

26. For iconographic and iconological approaches to Talbot's photographs, see Mike Weaver, "Talbot's Broom and Swift's Broomstick," *History of Photography* 15 (1991): 242–43; Weaver, *Henry Fox Talbot: Selected Texts and Bibliography* (Boston: G. K. Hall, 1993); Lars Kiel Bertelsen, "Interpreting Talbot's 'The Ladder,'" *History of Photography* 24 (2000): 184–85; Larry J. Schaaf, *The Photographic Art of William Henry Fox Talbot* (Princeton, NJ: Princeton Univ. Press, 2000); Geoffrey Batchen, "A Philosophical Window," *History of Photography* 26 (2002): 100–112; and Anne McCauley, "Talbot's Rouen Window: Romanticism, Naturphilosophie and the Invention of Photography," *History of Photography* 26, no. 2 (2002): 124–31.

27. Brusius, "Beyond Photography."

28. For the metaphor of "Nature as a drawing mistress," see the essay by Herta Wolf in this volume.

29. On decipherment see the essays by Eleanor Robson and Mirjam Brusius in this volume; on photogravure and Talbot's choice of the term "photoglyphic engraving," see the essay by Larry Schaaf in this volume.

30. William Henry Fox Talbot, *The Pencil of Nature* (London: Longman, Brown, Green and Longmans, 1844). For *The Pencil of Nature*, see the detailed introduction by Larry Schaaf in Talbot, *The Pencil of Nature: Anniversary Facsimile with an Introduction by Larry J. Schaaf* (New York: Hans P. Kraus, Jr., 1989). The range of secondary literature on the book is vast. Some detailed and specific studies include Hubertus von Amelunxen, "William Henry Fox Talbot: 'The Pencil of Nature' (1). Ein kleines Plädoyer für eine neue 'Lektüre,'" *Fotogeschichte* 4 (1984): 17–28; Larry J. Schaaf, "Henry Fox Talbot's 'The Pencil of Nature': A Revised Census of Original Copies," *History of Photography* 17 (1993): 388–96; Michael Gray, "Notes about W. H. F. Talbot's Texts and Drafts for 'The Pencil of Nature,'" in *Per Paolo Costantini: Fotografia e raccolte fotografiche*, ed. Tiziana Serena (Pisa: Scuola Normale Superiore, 1999), 63–97; Catherine Rogers, "W. H. F. Talbot, the Pencil of Nature and the Camera," *History of Photography* 26 (2002): 141–50; Robin Kelsey, "Photography, Chance, and *The Pencil of Nature*," in *The Meaning of Photography*, ed. Robin Kelsey and Blake Stimson (Williamstown, MA: Sterling and Francine Clark Art Institute, 2008), 15–33; and Maimon, "Displaced 'Origins.'" On *Queen's College Oxford* and legibility, see Talbot, *Pencil of Nature*, plate XIII; and Carol Armstrong, *Scenes in a Library: Reading the Photograph in the Book, 1843–1875* (Cambridge, MA: MIT Press, 1998), 160 n. 19.
31. The legibility of inscriptions as an important criterion for evaluating the success of photographic images is a theme in the essays by Mirjam Brusius, Chitra Ramalingam, Eleanor Robson, and Larry Schaaf in this volume.
32. See the discussion of "immutable mobiles" in Bruno Latour, "Drawing Things Together," in *Representation in Scientific Practice*, ed. Michael Lynch and Steve Woolgar (Cambridge, MA, and London: MIT Press, 1990), 19–68.
33. See Svetlana Alpers, *The Art of Describing: Dutch Art in the Seventeenth Century* (Chicago: Univ. of Chicago Press, 1983).
34. Brusius, "Preserving the Forgotten," 48–49.
35. Maimon, "Displaced 'Origins,'" 314.
36. Schaaf, *Out of the Shadows*, 54, 62, 112; and Geoffrey Batchen, "The Naming of Photography: 'A Mass of Metaphor,'" *History of Photography* 17 (1993): 22–32.
37. For this reason all images of notebooks in this introduction have been produced in order to convey their materiality as three-dimensional objects with physical presence, rather than disembodied, two-dimensional pages. In the following essays, where individual authors refer to a specific piece of information on a notebook page, notebook images sometimes have been cropped to emphasize that information. Our argument remains, however, that the notebooks, along with Talbot's photographs, should be understood not as abstracted texts but as elements of a rich material archive.
38. William Henry Fox Talbot to John Frederick William Herschel, Jan. 25, 1839, Doc. No. 3774, Correspondence (original: Royal Society, London); and Talbot to William Jerdan, Jan. 30, 1839, Doc. No. 3782 (the original letter has not been located; this is reproduced from the printed version in the section on fine arts, "The New Art," *Literary Gazette; and Journal of the Belles Lettres, Arts, Sciences, &c.* [hereafter *Literary Gazette*], Feb. 2, 1839, 73).
39. Simon Schaffer, "The History and Geography of the Intellectual World: Whewell's Politics of Language," in *William Whewell: A Composite Portrait*, ed. Menachem Fisch and Simon Schaffer (Oxford: Clarendon Press, 1991), 201; and Richard R. Yeo, *Defining Science: William Whewell, Natural Knowledge, and Public Debate in Early Victorian Britain* (Cambridge: Cambridge Univ. Press, 1993).
40. Raymond Williams, *Culture and Society, 1780–1950* (Hardmondsworth, UK: Penguin Books, 1961), 15–16.
41. An influential recent discussion of this relationship is Peter Galison and Lorraine Daston, *Objectivity* (Boston: Zone Books, 2007).
42. Ann Bermingham, *Learning to Draw: Studies in the Cultural History of a Polite and Useful Art* (New Haven and London: Yale Univ. Press, 2000), 238.
43. For useful methodological discussion of the relations between art and science, see Caroline A. Jones and Peter Galison, eds., *Picturing Science, Producing Art* (New York: Routledge, 1998); and Smith, *Body of the Artisan*. On the evolving public personae of the artist and the scientist, see Galison and Daston, *Objectivity*.
44. Stephen C. Pinson, *Speculating Daguerre: Art and Enterprise in the Work of L. J. M. Daguerre* (Chicago: Univ. of Chicago Press, 2012).

45. The classic study is Jack Morrell and Arnold Thackray, *Gentlemen of Science: Early Years of the British Association for the Advancement of Science* (Oxford: Clarendon Press, 1981). See also Roy Porter, "Gentlemen and Geology: The Emergence of a Scientific Career, 1660–1920," *Historical Journal* 21, no. 4 (1978): 809–36; Martin Rudwick, *The Great Devonian Controversy: The Shaping of Scientific Knowledge among Gentlemanly Specialists* (Chicago: Univ. of Chicago Press, 1985); and Marie Boas Hall, *All Scientists Now: The Royal Society in the Nineteenth Century* (Cambridge: Cambridge Univ. Press, 2002).
46. Susan Faye Cannon, *Science in Culture: The Early Victorian Period* (New York: Dawson and Science History Publications, 1978); Morrell and Thackray, *Gentlemen of Science*; Ian Inkster and Jack Morrell, *Metropolis and Province: Science in British Culture, 1780–1850* (Philadelphia: Univ. of Pennsylvania Press, 1983); Frank M. Turner, *Contesting Cultural Authority: Essays in Victorian Intellectual Life* (Cambridge: Cambridge Univ. Press, 1993); Bernard Lightman, ed., *Victorian Science in Context* (Chicago: Univ. of Chicago Press, 1997); and Richard R. Yeo, *Science in the Public Sphere: Natural Knowledge in British Culture, 1800–1860* (Aldershot, UK: Ashgate, 2001).
47. Jennifer Tucker, *Nature Exposed: Photography as Eyewitness in Victorian Science* (Baltimore: Johns Hopkins Univ. Press, 2005).
48. Tucker, *Nature Exposed*; and Galison and Daston, *Objectivity*.
49. For methodological reflections on understanding science as an urban activity, see Sven Dierig, Jens Lachmund, and J. Andrew Mendelsohn, "Introduction: Toward an Urban History of Science," *Osiris*, 2nd ser., 18 (2003): 1–19.
50. Iwan Morus, Jim Secord, and Simon Schaffer, "Scientific London," in *London—World City, 1800–1840*, ed. Celina Fox (New Haven: Yale Univ. Press in association with the Museum of London, 1992), 129–42; and Bernard Lightman, "Refashioning the Spaces of London Science: Elite Epistemes in the Nineteenth Century," in *Geographies of Nineteenth-Century Science*, ed. David N. Livingstone and Charles W. J. Withers (Chicago: Univ. of Chicago Press, 2011), 25–50.
51. For the history of the Royal Institution, see Morris Berman, *Social Change and Scientific Organization: The Royal Institution, 1799–1844* (Ithaca, NY: Cornell Univ. Press, 1978). For details on the running of the laboratory, see Iwan Rhys Morus, "Different Experimental Lives: Michael Faraday and William Sturgeon," *History of Science* 30 (1992): 14.
52. William Henry Fox Talbot, "Some Experiments on Coloured Flames," *Edinburgh Journal of Science* 5 (1826): 77–82. See also letters between Talbot and Herschel, Brewster, and Faraday for this period, in Correspondence.
53. For example, see Talbot to Michael Faraday, March 29, 1833, Doc. No. 2672, Correspondence (original: McGill Univ. Libraries Rare Book and Special Collections).
54. See the essay by June Barrow-Green in this volume.
55. L. Pearce Williams, *Michael Faraday: A Biography* (London: Chapman & Hall, 1965); and Iwan Rhys Morus, *Frankenstein's Children: Electricity, Exhibition, and Experiment in Early-Nineteenth-Century London* (Princeton, NJ: Princeton Univ. Press, 1998), 13–42.
56. "New Art," 74–75. For more on the Royal Institution as a site for Talbot's photographic work, see the essay by Chitra Ramalingam in this volume.
57. See the essay by Mirjam Brusius in this volume. On the uneven professionalization of science in this period, see Ruth Barton, "'Men of Science': Language, Identity and Professionalization in the Mid-Victorian Scientific Community," *History of Science* 41 (2003): 73–119. For comparable problems between amateurs and forms of professional expertise in late nineteenth-century visual sciences, see Thomas F. Gieryn, *Cultural Boundaries of Science: Credibility on the Line* (Chicago: Univ. of Chicago Press, 1999); and Bernard Lightman, *Victorian Popularizers of Science: Designing Nature for New Audiences* (Chicago: Univ. of Chicago Press, 2007).
58. Notebook "Memoranda," 1838–1840, Add. MS 88942/1/272, TC, BL.
59. See Kelley Wilder, "William Henry Fox Talbot und 'The Picture Which Makes Itself,'" in *Von Selbst: Autopoietische Verfahren in der Ästhetik des 19. Jahrhunderts*, ed. Friedrich Weltzien (Berlin: Reimer, 2006), 189.
60. On debates in this period over the nature of intellectual property, see the essay by Simon Schaffer in this volume.

Lacock Abbey and the Country House Archive

Katrina Dean

THE HERITAGE OF LACOCK ABBEY

A pageant in 1932 to celebrate the seven hundredth anniversary of Lacock Abbey featured its then owner, Matilda Talbot, dressed up as its thirteenth-century founder, Ela Longespée, third Countess of Salisbury (fig. 11). Having consulted a medieval historian in the planning of the event, she explained, "We had come to the conclusion that all those taking part . . . should be shown engaged in his own trade, the blacksmith should be a blacksmith, the shepherd a shepherd and so forth," emphasizing continuity between past and present. Some 10,000 visitors enjoyed the spectacle.[1] The heritage interpretation of Lacock continued following its donation to the National Trust in 1944, with the number of visitors approaching 150,000 in 2010–11 (fig. 12).[2] On the site is the Fox Talbot Museum of photography, recognizing the contribution to this field of Matilda Talbot's grandfather William Henry Fox Talbot. Not least through Talbot's archive, his role in the history of photography has become a feature in the legacy of Lacock Abbey and its historical reconstructions.

Writers in the field of heritage studies have periodically reminded us of both the distinctions and the connections between heritage and history.[3] Most credible heritage projects, as in the case of the Lacock pageant in 1932, draw to some extent on historical scholarship, whereas many projects in support of historical scholarship, including the public acquisition of archives, are justified in part by their heritage value: the often-intangible cultural, political, or social benefits of engaging with heritage connected to present-day interests and concerns. Conversely, the evidence of history generally lies in the material culture that embodies heritage. For these and other reasons, not least the professional requirement for historians to produce relevant scholarship, the subjects and sources of history are often shaped by contemporary interests. Recognizing the modern perspective that takes for granted the importance of photography, this volume aims to explore links

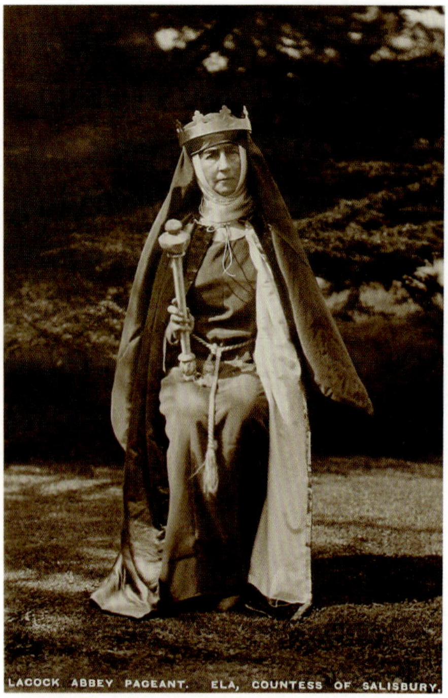

11. Matilda Talbot as Ela Longespée, third Countess of Salisbury, at the Lacock Pageant, Lacock Abbey, Wiltshire, 1932. Fox Talbot Museum, the National Trust.

between photography, knowledge, and art around the middle of the nineteenth century before the importance of photography was established.

The disposition of Talbot's archive and his reputation as a pioneer of photography are intertwined. From the mid-1930s Matilda Talbot championed her grandfather's role as an inventor of photography, donating Notebook P and Notebook Q—which encompass, among other subjects, Talbot's photographic experiments—to the Science Museum, London, along with a large collection of artifacts connected to the art. This was followed by the donation of important prints representing Talbot's photographic oeuvre.[4] Besides his photographic work, the "photographic" notebooks are among the best-known Talbot sources.[5] His works are also represented in other art collections and continue to attract strong commercial valuations and the interest of private collectors.[6]

In 2006 the British Library acquired much of the remainder of Talbot's archive from his descendants and heirs who until recently continued to live at Lacock, where the National Trust managed and provided access to the archive. The valua-

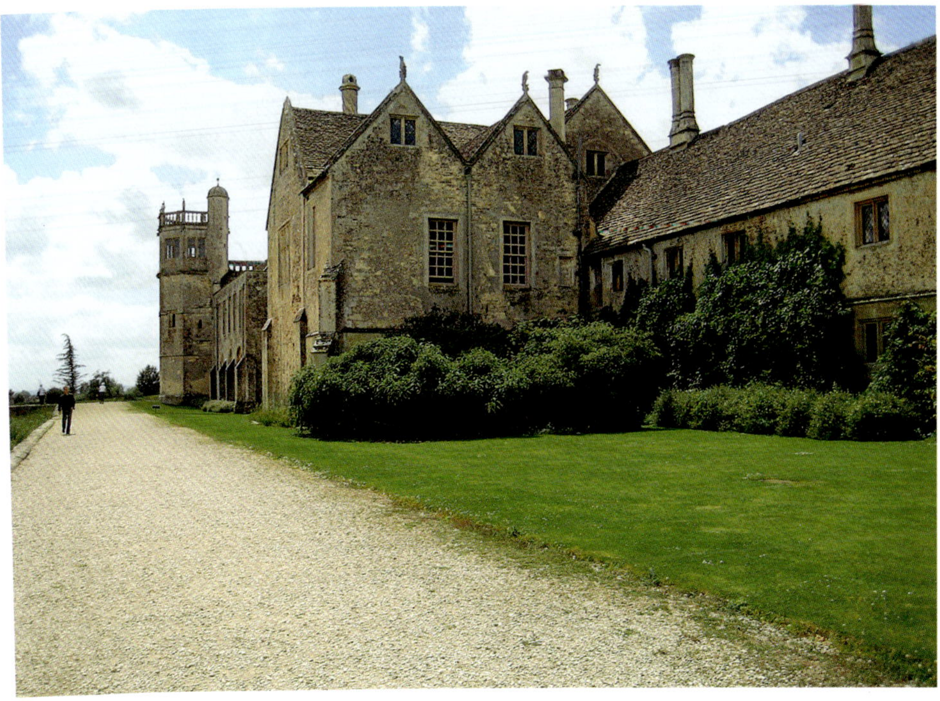

12. Lacock Abbey, May 2010. Photograph courtesy of Mirjam Brusius.

tion and addition of this collection to the national collection of historical manuscripts at the British Library reflected the richness of the archive as well as Talbot's status as a pioneer of photography. The photographic works were separated from the rest of the archive for specialized cataloguing. Many of the photographic prints were featured in an exhibition at the British Library in 2009 on photography in the nineteenth century.[7]

Yet most of Talbot's papers at the British Library concern his nonphotographic interests: many of them document fields of endeavor now belonging to the natural sciences. The papers also represent a strong interest in the history of languages, archaeology, and the study of classical and religious texts. Talbot's wide and voluminous correspondence on a myriad of subjects is easily accessible through the online Correspondence of William Henry Fox Talbot project.[8] Lesser known is the remainder of the Talbot archive, including more than two hundred notebooks documenting Talbot's scholarly interests, some of which form part of the same series as Notebook P and Notebook Q.[9] Much of the archive also concerns Talbot's management of the estate at Lacock. Other aspects document his travels and his social and family life.

The diversity of Talbot's archive, as well as his limited profile in the mainstream historiography of nineteenth-century science and other disciplines, shaped the formulation of the historical question leading to this volume.[10] What is the relationship between Talbot, gentleman of science—or, more widely, Victorian scholar—and his role in the invention of photography? Larry Schaaf and others have documented Talbot's chemical and optical experiments connected with photography.[11] Reference to the newly catalogued notebooks at the British Library and the consideration of well-known sources from new perspectives allow a more general discussion to emerge through this volume. The effect is to characterize Talbot's experiments in photography as the iteration of a wider ambition to cultivate knowledge across many fields. The purpose of this ambition was in part to promote discovery and invention, concepts undergoing radical redefinition at this time.[12] Talbot, for example, was much criticized for his successful attempts to patent his inventions of the calotype (1841 and 1842) and photogravure (1852) as potentially harmful to the progress of the art, perhaps with the implication that pecuniary interest in his inventions was also ungentlemanly.[13] On the other hand, cultivation of knowledge across several fields would not only safeguard against the pitfalls of failure in one but, more importantly, also build foundations for the proper conduct of the sciences, polite society, and even good government.[14] In particular, Talbot's tenure as a reform member of Parliament for Chippenham during the 1830s, his most fruitful period for photography and several other pursuits, suggests a common purpose and structure underlying his intellectual, political, and social activities.[15]

Perhaps nowhere is the confluence of Talbot's archive and his opportunities, duties, ambitions, interests, and achievements better understood than by reference to the context of his ancestral home, Lacock Abbey. While periods at Cambridge, in London and Edinburgh, and on the Continent furthered one or several of Talbot's endeavors, it was at home in the country that things came together, or at least were juxtaposed. The importance of Lacock to Talbot's photographic efforts is well documented, not least in many of his most famous photographs. Yet if we take Talbot's archive as a record of what mattered to him, the canvas is much wider. Lacock provided opportunities not only to pursue his many activities but also to record and preserve them. Here Lacock joins a tradition of country houses that remain as heritage sites in their own right, as places for the preservation of historical documents and historical sites for science.

LACOCK ABBEY AS A SITE OF PRESERVATION

In Britain great houses are major sites for the preservation of private archives, which sometimes also stand in for public archives retained by the holders of public office. Private archives provide unique information that is not available in public sources, and they enhance the interpretation of heritage. The interests of scholars and antiquarians have in the past strained the integrity of these private collections. The Students of Antiquity, founded by the seventeenth-century MP and antiquary Sir Edward Dering and others, decreed in part: "That every one do endeavor to borrow of other strangers with whom he hath interest all such books, notes, rolls, deeds etc as he can obtain."[16] In a regime established and monitored by the Historical Manuscripts Commission since 1869, the arrangements for private archives have been regularized through support for private management, public custodianship, and tax benefits resulting in the public acquisition of major collections of private archives.[17] Archives documenting several generations of families with major roles in public life feature in the national collection of historical manuscripts at the British Library, and these are often named for the great houses in which they were preserved.[18] Among these are, for example, the papers of the Evelyn family of Wotton, Surrey, primarily containing the papers of the diarist, gardener, numismatist, and antiquarian John Evelyn, a founding fellow of the Royal Society who wrote on forestry and compiled information on technical crafts (fig. 13).[19] While the Evelyn papers at the British Library also contain the records of John Evelyn's family and descendants, Talbot's comprise only his individual collection, the remainder of the Lacock Abbey archive being held on deposit at Wiltshire and Swindon Archives, Chippenham.[20]

What seems important in most cases, including the ones discussed above, is the role of the house in preserving the archive. For example, while Evelyn lived much of his life away from Wotton until the end of the seventeenth century, when he inherited the estate and the seat from his elder brother, the archive was preserved at Wotton and added to from that time until 1945. The Talbot family's connection with Lacock naturally involved many changes of tenure, inheritance often being preserved through the female line, with the effect that the estate and family archive is an eclectic mix of documents incorporating new influences across the generations.[21] However, Lacock has served not only as the backdrop to these historical influences but crucially as the window through which an otherwise fragmented past is preserved.

In the case of the Talbot family archive, this preservationist context has a long and distinctive provenance. Lacock Abbey has its origins as a religious house—among the key medieval institutions for documentary preservation—founded by

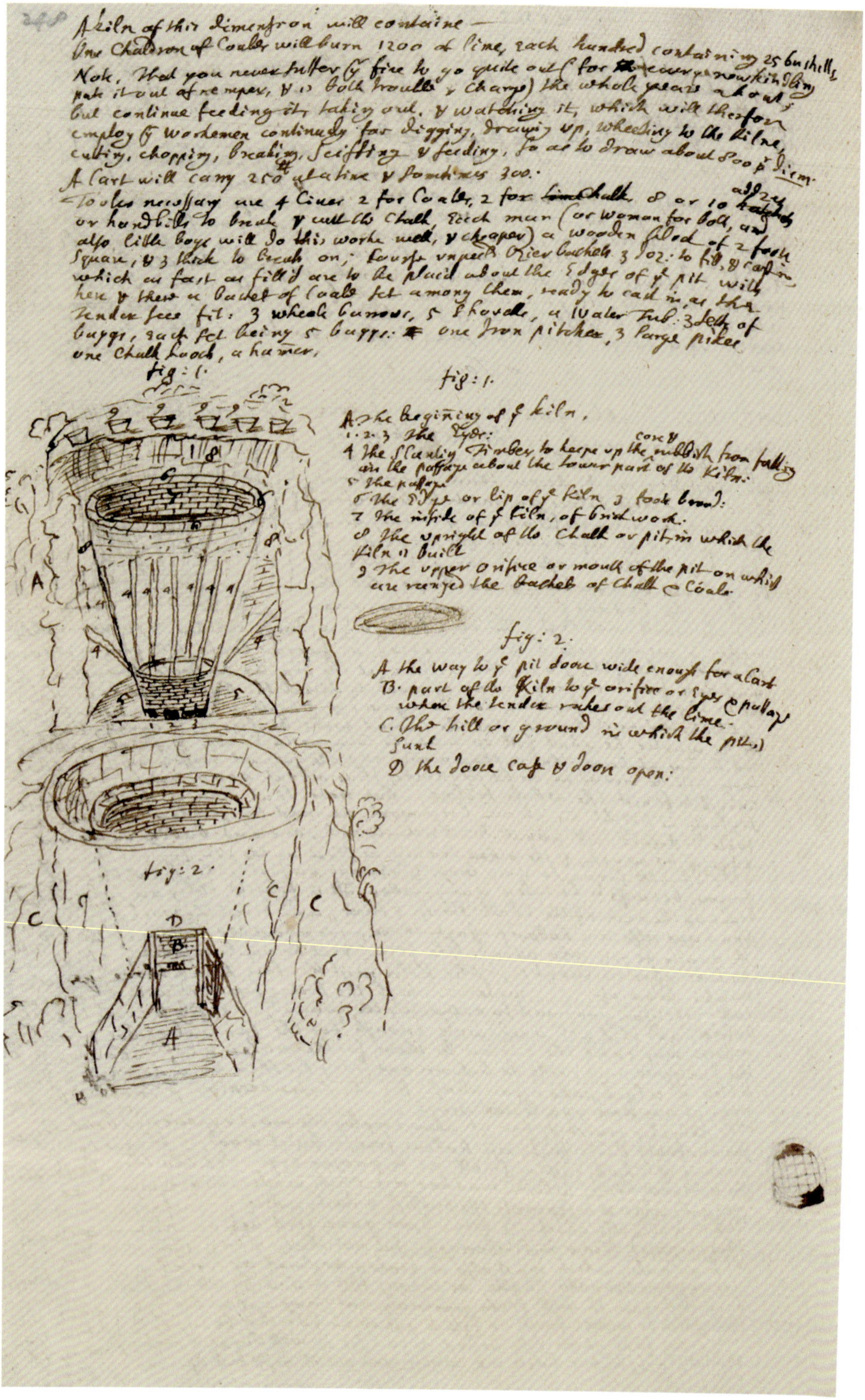
A kiln of this dimension will containe —
One Chaldron of Coales will burn 1200 of lime, each hundred containing 25 bushills
Note, that you never suffer ye fire to go quite out (for every new kindling puts it out of temper, & is both troublesome & chargeable) the whole yeare about, but continue feeding it, taking out, & watching it, which will therefore employ ye workemen continually for digging, drawing up, wheeling to the kilne, cutting, chopping, breaking, sifting & feeding, so as to draw about 800 p diem.
A Cart will carry 250 at a time & sometimes 300.
Tooles necessary are 4 Lines 2 for Coales, 2 for chalk 8 or 10 adzes or handbills to break & cutt the Chalk, Each man (or woman for both, and also little boys will do this worke well, & cheaper) a wooden block of 2 foote square, & 3 thick to break on; course unpeeld Ozier buckets 3 doz: to fill, & carry which as fast as fill'd are to be plac'd about the Edges of ye pit with here & there a Bucket of Coals set among them, ready to cast in, as the tender sees fit: 3 wheele barrows, 5 Shovells, a Water tub. 3 setts of buggs, each set being 5 buggs: one Iron pitcher, 3 large pikes one Chalk hooke, a hamer.

fig: 1.
A the beginning of ye kiln,
1.2.3 the Sydes:
4 the Scaling Timber to keepe up the rubbish from falling in the passage about the lower part of the kiln:
5 the passage
6 the Edge or lip of ye kiln a foote broad:
7 the inside of ye kiln, of brickwork.
8 the upright of the Chalk or pit in which the kiln is built
9 the upper orifice or mouth of the pit on which are ranged the buckets of Chalk & Coals.

fig: 2.
A the way to ye pit doore wide enough for a Cart
B. part of the kiln to ye orifice or eye & pullay where the tender rakes out the lime.
C the hill or ground in which the pit is sunk
D the doore cast & doors open:

13. John Evelyn, "Trades, Secrets and Receipts mechanical as they came casually to hand," autograph manuscript, ca. 1650, Evelyn Papers, vol. 172, fol. 126. Add. MS 78339, British Library.

Ela, Countess of Salisbury, in 1232. Her husband, William Longespée, third Earl of Salisbury, was one of the nobles who advised King John to grant the Magna Carta and served as sheriff of Wiltshire, an office held by Ela after her husband's death. She joined the order of nuns at Lacock in 1238, becoming abbess the following year. The Lacock Magna Carta of 1225 was preserved there until Matilda Talbot donated it to the British Museum in 1944 (fig. 14). An interesting example of document as heritage, it was displayed at the Library of Congress in Washington, DC, in 1946, at an installation opened by the British ambassador, referring perhaps to the restorative nature of Anglo-American political liberty at the end of World War II.[22] The Lacock Magna Carta is one famous item in a significant local collection of medieval charters that formed the basis for the collections built by the subsequent owners of Lacock House, the Talbot family.[23]

The acquisitive and preservationist activities of the subsequent inhabitants of Lacock can be bracketed by two representatives: the first private individual to inhabit Lacock Abbey, William Sharington (whose niece Olive married Sir John Talbot of Salwasp, Worcestershire, establishing the connection with the Talbot family); and Charles Henry Talbot, the son of William Henry Fox Talbot and the last to pass the legacy within the Talbot family (to his niece Matilda).

Following the dissolution of the monasteries, in 1540, the abbey and its contents were purchased by Sharington, who was later accused of clipping coins in his role as under-treasurer of the Bristol Mint. He built a tower room, in the tradition of the *studiolo* (small study), to view the abbey's riches, including the Magna Carta, privately. In contrast, the last private owner of Lacock at the time of his death was Charles Talbot, an antiquarian of the county of Wiltshire who was described in a guidebook of 1930 as "an enthusiastic archaeologist, one who was recognised in the county as a leading authority on ancient buildings. With the utmost patience he had elucidated the history of the Abbey, the changes it has undergone and the genealogical facts connected with the ramifications of the different families who have been associated with it."[24] While the aims of Charles Talbot seem to us nobler than those of Sharington, each valued the muniments of Lacock Abbey as signs of a vanishing past or as symbols of an active power, respectively.

The collection of estate and family papers accumulated in the intervening period, most held on deposit at Wiltshire and Swindon Archives, has been added to by each generation at Lacock. As mentioned earlier, it was only from the late 1930s that this substantial religious and country house archive began to be subject to a process of dispersal connected with a series of discrete—if interconnected in their overall ambition—photographic, architectural, local, and national heritage projects. From this time onward, significant parts of the collection were cared for and made accessible in public repositories under various arrangements of gift or loan.

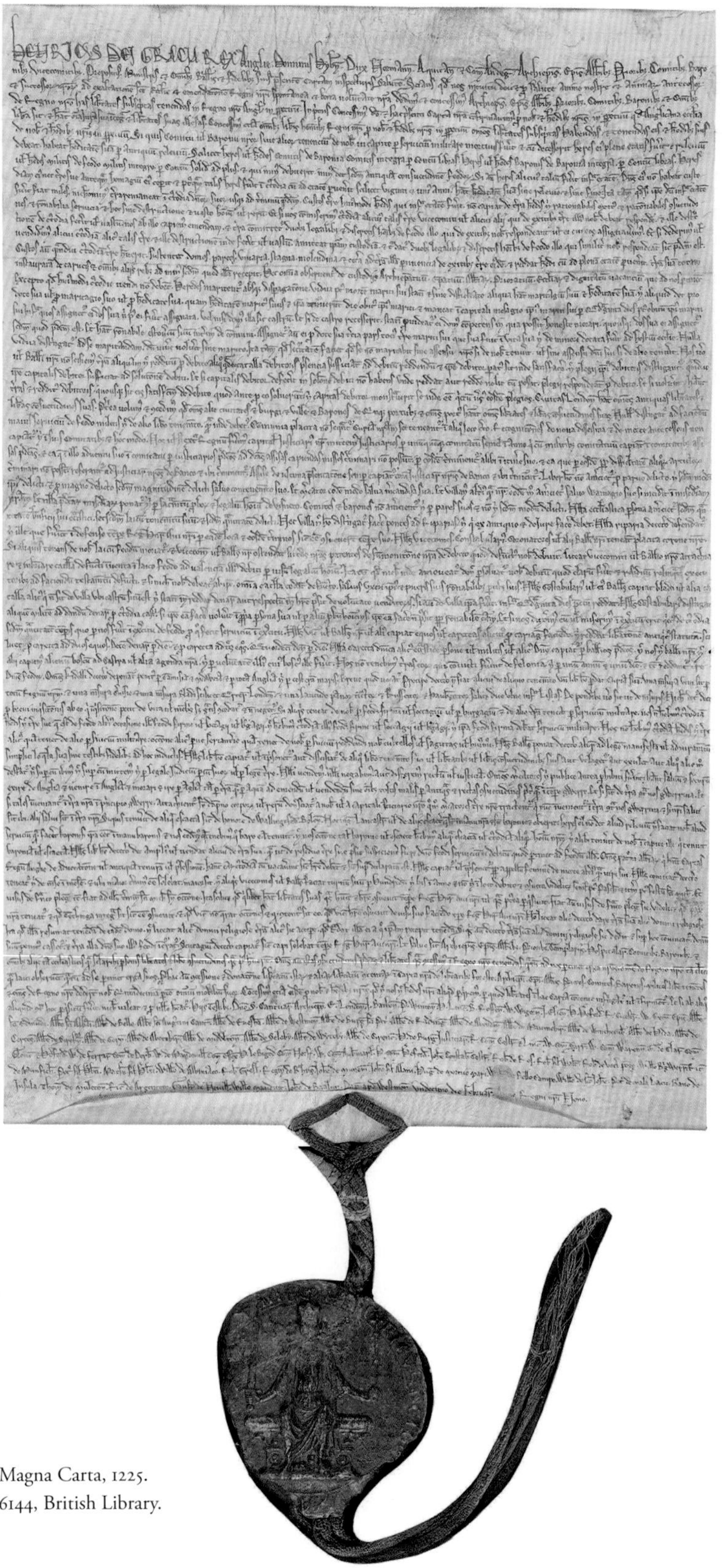

14. Lacock Magna Carta, 1225.
Add. MS 46144, British Library.

William Henry Fox Talbot's archive was destined to become an installment in this wider collection of estate and family papers, although he may not have known this in the earliest days of his childhood correspondence. In light of his own statements (for example, as a boy he asked his family not to destroy any of his letters) and later scholarly activities, however, it is possible that he conceived of his enterprise in a preservationist tradition.[25] This seems especially relevant to Talbot's scholarly career, which, as shown in this volume, displayed a strong interest in the origins of words, texts, traditions, and disciplines as well as the sources for delineating them. Yet as we see, too, in Talbot's work these attempts at preserving the past were also wellsprings of innovation.

COUNTRY HOUSE SCIENCE

Historians' accounts of country house science have focused on the seventeenth and latter half of the nineteenth centuries in reference to wider arguments about the emergence and professionalization of science. Occurring at the same time as the emergence of public scientific institutions, the important discoveries of country house scientists, such as Robert Boyle in the earlier period and Frank Balfour, Charles Darwin, and John William Strutt, third Baron Rayleigh, in the later period, have been documented to show continuities with traditions of gentlemanly science and its interconnections with the emergence of modern science and its professionalization.[26] In comparison there seems to be relatively little discussion of country house science in the first half of the nineteenth century—the exceptions including Talbot's correspondent William Parsons, third Earl of Rosse—possibly because its ubiquity is assumed under the label of "gentlemanly science."[27] Undoubtedly Talbot's experience at Lacock and in the family homes of his childhood suggests some common themes with other periods.

These themes might be summarized as follows:

- Country house science (and society) is aristocratic, Anglican, and politically conservative.
- Country houses are scientifically significant sites through interdependence with metropolitan political and scientific institutions.
- Country houses are associated with a bucolic tradition of intellectual solitude and productivity.
- Country houses are sites of technological change.
- The domestic context of country houses is important, including the role of women, servants, and others as researchers, educators, and technicians.

Talbot's experience attests to the relevance of these themes in the first half of the nineteenth century, with some exceptions, notably in politics. (His mother, Lady Elisabeth Feilding, was a Whig supporter, and Talbot, as previously mentioned, was himself a reform MP for Chippenham in the early 1830s; on the other hand, Talbot was a unionist on the Irish question.) Tying these themes together is the notion of the importance of the country house in gaining acceptance for scientific and technological change, because it is at the same time a powerful institution where established traditions are preserved. As discussed in several essays in this volume, Talbot's work used technological innovation both as a tool for preserving the past and as a means of establishing new knowledge.

TALBOT'S PHILOSOPHY OF SCIENCE

But how typical was Talbot among his contemporaries? Plentiful archives such as Talbot's can skew our understanding of historical experience as much as scant ones, because they encourage us to emphasize the lives and work of those who are most documented.[28] It might be argued that the series of essays culminating in this volume have confirmed that Talbot's key innovation was photography and that his well-documented work in other fields is undoubtedly interesting from a historical point of view but not remarkable. The purpose of this volume, however, is not to argue that Talbot's contributions to various fields were the key ones historically, but rather to explore the extent to which his archive and pursuits can illuminate an emergent approach to knowledge in the first half of the nineteenth century.

Talbot fits poorly into the historiography of nineteenth-century science.[29] An advocate for neither the Anglo-Tory natural theological establishment of the first half of the nineteenth century nor the professionalizing radicals of the latter part, Talbot was nonetheless a gentleman of science in a wider landscape of natural philosophers from bourgeois or modest backgrounds who ministered, wrote, and lectured to support intellectual pursuits: William Whewell, David Brewster, and Michael Faraday, among others. Talbot's friend John Frederick William Herschel required employment to support his research, and even the wealthier Charles Babbage did not manage a landed estate. In 1838 Talbot wrote to his mother, Lady Elisabeth: "There is a sad dearth of scientific aristocracy; & in a country, the mainsprings of whose wealth and commerce are so intimately connected with the improvements of science!"[30] Rather than being representative of a ubiquitous species, Talbot perhaps considered himself prodigious in his class for his scientific accomplishments and the degree of seriousness with which he approached his research.

At the same time, in stark contrast to his contemporaries, Talbot wrote little—other than his accounts of the discovery of photography that are discussed in the essay by Vered Maimon in this volume—on the general problems of science; its relation to religion, education, and research; or the need for professional reform.[31] In Maimon's essay Talbot's approach to epistemology in one account concurred with that expounded in the well-known writings of Whewell, whose theory of fundamental ideas emphasized the constructive role of the mind in science and the historical growth of knowledge. Sensations, Whewell argued, are not sufficient as the basic element of knowledge, and their apprehension cannot be separated from the active mind.[32] Elsewhere Talbot endorsed a more cumulative, one might say mechanical, interpretation, following Herschel. Some of the proposed uses of Talbot's invention of photography, such as in the fields of cuneiform decipherment, archaeology, and botany—as discussed in essays by Mirjam Brusius and Anne Secord in this volume—can be reviewed in light of these more philosophical concerns. In this sense Talbot's efforts to accumulate and preserve his archive might be seen as something more than a natural desire to preserve the traditions of Lacock Abbey and its country house archive. His desire to develop photography as a technique for preserving scientific and cultural objects displays a more systematic approach to the discovery and establishment of new knowledge. It is intended that this volume will situate the study of Talbot's photography in the context of his wider archive and contribute to the history of Victorian cultures of knowledge.

1. Oliver Garnett, *Lacock Abbey, Wiltshire* (Swindon, UK: National Trust, 2003), 46.
2. National Trust (Great Britain), *Annual Report, 2010–11* (Swindon, UK: National Trust, 2011), http://92.52.118.192/annualreport2011.pdf (accessed Jan. 30, 2012), 91.
3. See Michael Oakeshott, "The Activity of Being an Historian," in *Rationalism in Politics and Other Essays* (London: Methuen, 1962), 137–67; Oakeshott, *On History and Other Essays* (London: Basil Blackwell, 1983); and David Lowenthal, *The Heritage Crusade and the Spoils of History* (Cambridge: Cambridge Univ. Press, 1998).
4. The notebooks and prints are held by what is now the National Media Museum, Bradford, part of the National Museum of Science and Industry, following the transfer of items in 1989 from the Science Museum, London, to the newly instituted (and since renamed) National Museum of Photography, Film and Television. See Larry J. Schaaf, *Records of the Dawn of Photography: Talbot's Notebooks P & Q* (Cambridge: Cambridge Univ. Press, 1996), xv.
5. Schaaf, *Records of the Dawn of Photography*.
6. Searches of the lot finders on the websites of Christie's and Sotheby's yield several examples. One, originating in a gift by Matilda Talbot in about 1944, is William Henry Fox Talbot, *A Bush of Hydrangea in Flower*, ca. 1842, Christie's, London, sale 7040, May 18, 2005, lot 4, www.christies.com/lotfinder (accessed Jan. 30, 2012). See also the Sun Pictures series of catalogues; for example, Larry J. Schaaf, *Sun Pictures Catalogue Ten: British Paper Negatives, 1839–1864* (New York: Hans P. Kraus, Jr., 2001), http://www.sunpictures.com (accessed Jan. 30, 2012).

7. John Falconer and Louise Hide, *Points of View: Capturing the 19th Century in Photographs* (London: British Library, 2009).
8. The Correspondence of William Henry Fox Talbot, http://foxtalbot.dmu.ac.uk (accessed Jan. 30, 2012), hereafter "Correspondence."
9. See the introductory essay by Mirjam Brusius and Chitra Ramalingam in this volume. For details, see Mirjam Brusius, "Beyond Photography: An Introduction to William Henry Fox Talbot's Notebooks in the Talbot Collection at the British Library," *Electronic British Library Journal* (2010), http://www.bl.uk/eblj/2010articles/pdf/ebljarticle142010.pdf.
10. A prolific literature since the 1960s has discussed the social and cultural as well as intellectual aspects of science in this period but still tends to focus on figures associated with major developments in scientific theory (such as Charles Darwin on evolution) and related experimental techniques. A contrast with the treatment of Michael Faraday is pertinent given Talbot's connection with him, discussed in the essay by Chitra Ramalingam in this volume. See, for example, *The Correspondence of Michael Faraday*, ed. Frank A. J. L. James, 6 vols. (London: Institution of Electrical Engineers, 1991–2011). There is, however, an entry on Talbot in Bernard Lightman, ed., *The Dictionary of Nineteenth-Century British Scientists*, 4 vols. (Chicago: Univ. of Chicago Press, 2004). On the other hand, Talbot is also overshadowed in the historiography of technology in this period, which primarily concentrates on the Industrial Revolution and its heroes. See, for example, Christine MacLeod, *Heroes of Invention: Technology, Liberalism and British Identity, 1750–1914* (Cambridge and New York: Cambridge Univ. Press, 2007). A lack of scholarship on Talbot is also evident in other disciplines where he is now seen to be less successful, such as Assyriology, discussed in the essay by Eleanor Robson in this volume, and etymology, discussed in Anatoly Liberman, "Talbot as a Student of Word Origins" (paper presented at "William Henry Fox Talbot: Beyond Photography," conference at the Centre for Research in the Arts, Social Sciences, and Humanities, Univ. of Cambridge, June 24–26, 2010).
11. Some of these aspects are discussed in Larry J. Schaaf, *Out of the Shadows: Herschel, Talbot & the Invention of Photography* (New Haven and London: Yale Univ. Press, 1992). For optical experiments, see also the essay by Chitra Ramalingam in this volume.
12. For a discussion of Talbot's deployment of the meaning of discovery, see the essay by Vered Maimon in this volume. For invention, see MacLeod, *Heroes of Invention*.
13. Keith Adamson, "Early British Patents in Photography," *History of Photography* 15 (1991): 313–23, esp. nn. 52, 122, 155. William Parsons, third Earl of Rosse, for one, attempted to persuade Talbot to relinquish his patent rights in their correspondence.
14. See the essay by Anne Secord in this volume, including the important reference to J. W. Barrow, *Whigs and Liberals: Continuity and Change in English Political Thought* (Oxford: Clarendon Press, 1988).
15. See the essay by Chitra Ramalingam in this volume on the importance of the London season's coinciding with the sitting of Parliament, both to Talbot's scientific experiments and to a schedule for the move between town (London) and country (Lacock).
16. Quoted in Christopher Kitching, "Public Interest or Private Property? In Celebration of Private Archives," Maurice Bond Memorial Lecture, 2004, *Archives* 30 (2005): 3–4. Indeed it remains common for biographers to borrow archival documents held in private hands, and it seems that this occurred with regard to Talbot's archive at Lacock. See Schaaf, *Records of the Dawn of Photography*, xxvii n. 7.
17. Kitching, "Public Interest or Private Property?"
18. See "Modern Historical Manuscripts," British Library, http://www.bl.uk/reshelp/findhelprestype/manuscripts/modernmanuscripts/index.html (accessed Jan. 23, 2012). I also would like to acknowledge Frances Harris for her personal reflections on private papers based on experience and expertise.
19. For a discussion of this archive and its history, see Theodore Hofmann et al., "John Evelyn's Archive at the British Library," *Book Collector* 44 (1995).

20. Lacock Abbey archive, deposit no. 2664, Wiltshire and Swindon Archives, Wiltshire and Swindon History Centre, Chippenham. Here there are further parallels with the Evelyn Papers in that the Wotton estate archive is split between the British Library and the local county record office, with the main sequence of manorial records and deeds on deposit at the Surrey History Centre, Woking.

21. The Lacock Abbey archive includes early deeds and charters, manorial court books and rolls, surveys, estate papers, legal papers, personal and household accounts, family papers and correspondence for the Talbot, Feilding, and Davenport families, and other manuscripts, including seventeenth- and eighteenth-century recipe books.

22. *Program Incident to the Installation Held in the Coolidge Auditorium of the Library of Congress, Sunday, December 15, 1946, at 4 p.m.* (Washington, DC: Library of Congress, 1947).

23. Kenneth Rogers, ed., *Lacock Abbey Charters* (Devizes, UK: Wiltshire Record Society, 1979).

24. *Lacock, the Abbey, Church and Village: Thirteen Illustrations* (Trowbridge, UK: B. Lansdowne and Sons, 1930).

25. H. J. P. Arnold, *William Henry Fox Talbot: Pioneer of Photography and Man of Science* (London: Hutchinson Benham, 1977).

26. Steven Shapin and Simon Schaffer, *Leviathan and the Air Pump: Hobbes, Boyle and the Experimental Life* (Princeton, NJ: Princeton Univ. Press, 1985); Steven Shapin, "The House of Experiment in Seventeenth-Century England," *Isis* 79 (1988): 373–404; Soraya de Chadarevian, "Laboratory Science versus Country-House Experiments: The Controversy between Julius Sachs and Charles Darwin," *British Journal for the History of Science* 29 (1996): 17–41; Simon Schaffer, "Physics Laboratories and the Victorian Country House," in *Making Space for Science: Territorial Themes in the Shaping of Knowledge*, ed. Crosbie Smith and Jon Agar (Basingstoke, UK: Macmillan, 1998); and Donald L. Opitz, "Aristocrats and Professionals: Country-House Science in Late Victorian Britain" (PhD diss., Univ. of Minnesota, 2004).

27. Simon Schaffer, "The Leviathan of Parsonstown: Literary Technology and Scientific Representation," in *Inscribing Science: Scientific Texts and the Materiality of Communication*, ed. Timothy Lenoir (Stanford: Stanford Univ. Press, 1998), 182–222.

28. Kitching, "Public Interest or Private Property?" For comparison, the notebooks of Leonardo da Vinci are discussed in George Sarton, "The Literary Works of Leonardo da Vinci; The Notebooks of Leonardo da Vinci," *Isis* 35 (1944): 184–87, esp. 185.

29. An early discussion is Susan Faye Cannon, *Science in Culture: The Early Victorian Period* (New York: Dawson and Science History Publications, 1978). For subsequent literature, see the introductory essay by Mirjam Brusius and Chitra Ramalingam in this volume, notes 45 and 46.

30. William Henry Fox Talbot to Lady Elisabeth Feilding, Nov. 17, 1838, Doc. No. 3756, Correspondence (original: Talbot Collection, British Library).

31. Arnold, *William Henry Fox Talbot*, 79.

32. Richard R. Yeo, *Defining Science: William Whewell, Natural Knowledge, and Public Debate in Early Victorian Britain* (Cambridge: Cambridge Univ. Press, 1993).

MODELS FOR INVESTIGATION

Talbot's First Lens: Botanical Vision as an Exact Science

Anne Secord

It was in the "brilliant summer" of 1835 that William Henry Fox Talbot discovered that his photogenic drawings, produced by long exposure to the sun, acquired greater detail the smaller they were. He accordingly "had several small boxes made, in which I fixed lenses of shorter focus, and with these I obtained very perfect but extremely small pictures: such as without great stretch of imagination might be supposed to be the work of some Lilliputian artist. They require indeed examination with a lens to discover all their minutiae."[1] The tiny photographs revealed for the first time a "camera vision,"[2] but one that required magnification to be fully appreciated. Until Talbot refined the light sensitivity of his photographic paper, the potential of his discovery of photography was best assessed by the combination of these two ways of seeing, both of which required the use of lenses.

Talbot's minute pictures were just a stage in his experimentation with photographic images. The combination of photography and microscopy became a Victorian craze, however, with the production of microphotographs from the 1840s by the scientific and optical instrument maker John Benjamin Dancer. By the 1850s the subjects of Dancer's commercially produced sets of images ranged from magnified fleas to the moon, and from buildings to composite group portraits. The appeal of these microphotographs was that they played with scale not only in the subjects they depicted but also by requiring a microscope to view them. Victorians marveled at photography, but their delight in photographic miniatures owed more to the pleasures of microscopy.[3]

With the increasing availability of affordable microscopes from the 1830s, microscopic investigation in natural history became hugely popular. In order to understand what was being seen when the microscope revealed the structure of things not readily visible to the naked eye, observers were encouraged to engage in a process of calibration by measuring and counting. Dancer drew on the natural historian's preoccupation with counting to confer credibility on his microphotographs, inviting viewers to add up the number of heads in a composite portrait or the number of letters in a text. In this way Dancer conferred the authority of natural history (in which the minuscule had to be measured in order to be believed) on what was little more than a popular visual spectacle. This mimicry of scientific practice also provided a way of measuring the technical skills of microphotographers, as shown by claims of being able to capture the image of a "giant cathedral" that allowed the viewer "to count every stone."[4]

In similar fashion, back in the summer of 1835, Talbot had also equated counting with credibility and value. Among the Lilliputian photographs he produced in that year is the darkened, postage-stamp-sized one of the latticed oriel window of his ancestral home, Lacock Abbey. He stuck the small image on a piece of paper and made the following note: "When first made, the squares of glass about 200 in number could be counted, with help of a lens."[5]

Talbot's emphasis on precision and verisimilitude, as measured by the exactness of counting with the help of a magnifying lens, was born not out of his photographic practice but out of his early botanical interests. In particular his fascination with mosses (a difficult, little-studied, and unpopular family of plants among early nineteenth-century botanists) meant that Talbot was well versed in the process of microscopic vision long before there was a popular culture of microscopy. Magnification was required not only because counting was essential in the identification of mosses but also because it allowed observers to exercise their judgment in establishing a classification of these plants. The use of the microscope came after the initial discovery of a moss in nature; it was used to assess and judge the nature of the plant. In like manner Talbot used magnification to confirm the quality of his early photograph of the oriel window. Beyond honing his observational skills, the study of mosses gave him a more extensive training of the mind, as he himself recognized.

Talbot's early photographs—or photogenic drawings—of plants frequently have been attributed to his interest in botany; it is also assumed that pictures of plants must be botanical. By exploring his practice of botany, however, I will argue that the relationship between Talbot's botanical and photographic interests is more complex and telling for understanding his intellectual development and his discovery of photography. Rather than photography being the means by which to illustrate Talbot's botanical interests, it was botany that provided a model for both observation and the exercise of judgment. He employed the tools and practices essential for the study of mosses to assess his earliest "Lilliputian" photogenic drawings. Moreover, it was Talbot's early experience of observational science that provided him with a precedent for what he later recounted as the application of "the inductive method of modern science" to his discovery of photography.[6]

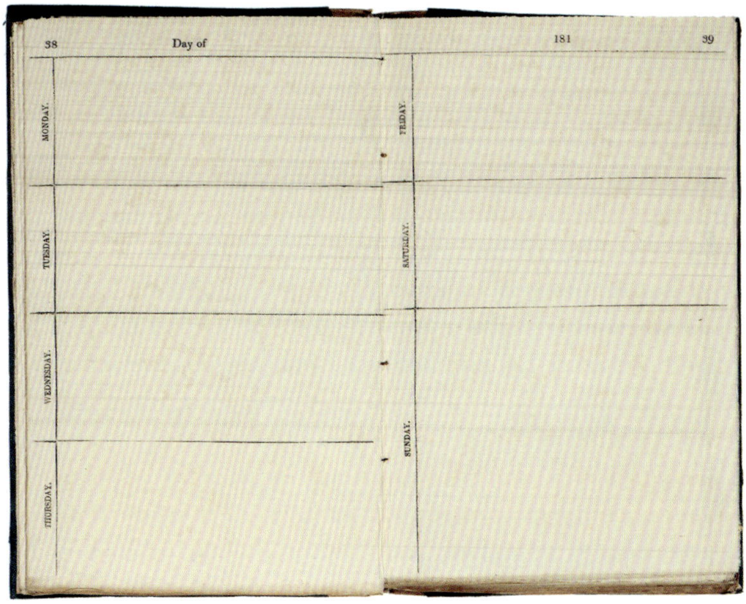

15. *The Student's Journal: Arranged, Printed, and Ruled, for Receiving an Account of Every Day's Employment, for the Space of One Year*, 4th ed. (London: Taylor and Hessey, [1815]), 38–39. Cambridge University Library.

THE TRAINING OF THE MIND

In 1815 Talbot's mother, Lady Elisabeth Feilding, sent her son *The Student's Journal: Arranged, Printed, and Ruled, for Receiving an Account of Every Day's Employment, for the Space of One Year* (two pages of which are shown in fig. 15). In the early nineteenth century, British record keeping in the form of notebooks, diaries, and journals became almost obsessive. This monitoring of one's own progress was seen as an essential step toward an active and useful life in society; by 1822 the "art of employing time to the greatest advantage" could be proclaimed the "true source of happiness." This happiness boiled down to consolidating whatever course of self-improvement had been embarked upon by keeping a journal in which an "analytical summary" of the employment of each day was to be made.[7] This advice was avidly taken up by the middle classes and the gentry, and although the specific reasons for keeping a journal might not be the same in all cases, there was a shared preoccupation with not squandering time.[8] For example, the middle-class Thomas Hirst of Halifax, apprenticed from the age of fifteen to an engineering surveyor, kept a journal that served as an aid to and record of a strenuous program of self-education and spiritual progress; so did the upper-class Anne Lister of Shibden Hall, Halifax, who, already a compulsive diary and journal writer, was glad to glean "several hints for journal-keeping" when she read *The Art of Employing Time* in 1823.[9]

The sequence of fifty-two blank openings in *The Student's Journal* did not appeal to the fifteen-year-old Talbot, striking him as being "too methodical."[10] And yet, by this time, with the encouragement of his aunt Lady Mary Lucy Cole (previously Mary Talbot), herself an ardent gardener and botanist, Talbot had already opened a botanical notebook in which he kept a scrupulous record of the rare plants he had observed.[11] This, however, was a record of pleasure, not duty, and a register of only the most important of his botanical observations. *The Student's Journal* resembled his botanical notebook only in its declaration that its "province is Action, not Speculation."[12]

Even though Talbot's letters to his mother—written from Harrow School and from his aunt's home, Penrice Castle on the Gower Peninsula in Wales—are littered with references to his various studies and wide-ranging reading, Lady Elisabeth evidently came to fret about the limitations of botany as a mere science of naming, just as she later worried that Cambridge would foster in her son an intense and too-narrow preoccupation with mathematics.[13] In the early nineteenth century a gentleman—especially one with the aspirations Lady Elisabeth hoped to inculcate in her young son—needed to show that he could manage a wide range of knowledge. It was even more important that Talbot could display many and varied interests in the 1830s, when issues of gentility and knowledge were at the core of the reforming Whig program with which Talbot was politically involved.[14] As well as avoiding the dangers of too-narrow and ungentlemanly specialization, Talbot, on leaving Cambridge in 1821, also acknowledged that there "is nothing like having two or three different pursuits for se délasser [relaxation] when anything goes wrong."[15]

In 1814, in response to his mother's doubts about the benefits of botanical study, Talbot put up a spirited defense that showed that he, too, believed in spending time well: botany, he declared, was a "great resource" with which "to fill up the time, which would otherwise be spent in ennui."[16] More importantly, however, he also claimed that botany was not only a pleasurable pursuit but also a training of the mind. Writing from his aunt's home in Penrice Castle, with its magnificent garden and isolated location, so appropriate for the study of plants, Talbot explained that botany

> is a science which extends pretty far, & which by no means consists entirely of nomenclature— It affords excellent exercise to the powers of discrimination, & practises the memory very much. I am sure that I shall find Euclid much easier, after having accustomed myself as I do here, to the attentive examination of plants; in the descriptions of which, every term & expression must be well weighed in the mind, & thoroughly understood. Far from there

> being no *mind* in it, I think that if you . . . ever read Smiths Introduction to Botany, you must confess that there *is* something more in Botany than to know every plant when you see it. Aunt Mary says there is a difference between a philosophical, & a *stupid* botanist.[17]

Talbot's linking of Euclid and botany is perhaps more telling than he could have realized, given the importance he would later place on inductive reasoning in his discovery of photography. Although he did not articulate the nature of the link, his enthusiasm for these subjects and his awareness of the value of botany in comprehending geometry recall the philosopher John Stuart Mill. Seven years Talbot's junior, Mill claimed that his intellectual development and arguments concerning inductive reasoning were founded on the natural sciences and mathematics. In 1819 "Master John Mill, London, (Under thirteen years of age.)" provided solutions to two geometric problems in the *New Series of the Mathematical Repository*.[18] Just a year later Mill's journal—reflecting his father's insistence that he record all his activities—shows that he was botanizing (collecting and drying plants) as well as studying mathematics and chemistry during a visit to France. Botany would become a lifelong avocation for Mill following his first vivid experiences of botanizing, which, under the tutelage of George Bentham, with whose family he was staying, included three weeks in the Pyrenees.[19]

It was Mill's direct experience of the empirical sciences such as botany that led him to argue in 1843 that mathematics, hitherto taken as the archetypal deductive science, had the same origin as the rest of our knowledge. David Bloor has explained how this view was justified by Mill's claim that when we learn geometry and arithmetic, we draw on our experiences about the properties and behavior of material objects and the ways in which they can be ordered and sorted. Experience tells us that some facts apply only to limited ranges of objects, and this knowledge makes up the empirical sciences; some facts, however, apply to a wide range of things that can be arranged in patterns and arrays in many different ways. Mill believed that these wide-ranging truths about the ordering and patterning of objects underlay mathematics, and that, as Bloor put it, "the processes of reasoning in mathematics are just pale shadows of physical operations with objects." We are not aware of referring back to experiences of things because habit renders mathematical processes mechanical.[20]

In his autobiography Mill was explicit about the role that botany had played in his thinking: "Logic, and the dialectics of Plato . . . had given me a strong relish for accurate classification. This taste had been strengthened and enlightened by the study of botany, on the principles of what is called the Natural Method, which I had taken up with great zeal, though only as an amusement, during my stay in

France." When Mill found that he could also apply scientific classification to complex legislative subjects, he "felt taken up to an eminence" from which he could "survey a vast mental domain, and see stretching out into the distance intellectual results beyond all computation."[21] Talbot's own youthful perception of botany as a training of the mind and as a model for processes of reasoning was also born out of practice. This practice with respect to mosses combined microscopy, mathematical precision, and observational judgment in order to arrange these tiny plants in a natural classification. Talbot's inculcation of practical scientific habits can be seen in this area of study as revealed by his correspondence, his collection of dried plant specimens, and his botanical notebooks.

OBSERVING MOSSES

Along with the disappointing *Student's Journal*, Talbot's mother sent him what he really wanted: a new microscope.[22] When hinting at the start of December 1815 that this would be a fine present, Talbot had been careful to stipulate that the microscope should be of as high a magnifying power as possible.[23] A year earlier he had vented his frustration with his old microscope and the difficulties of studying mosses when comparing botanical notes with his cousin Jane Talbot, at whose family home, Penrice Castle, he spent many of his holidays. "I have found quantities [of mosses] whose names I am ignorant of," he told her, but "I am very busy, & have not time to investigate them. Moreover, the genus Hypnum is so intensely hard that it discourages me. Moreover, my microscope is not strong enough for small mosses. Moreover, they are so like each other, that it requires experience, to distinguish them."[24] Talbot thus effectively summed up the difficulties that mosses presented to botanists.

From the late eighteenth century mosses had been classified according to a system devised by the Leipzig physician and botanist Johann Hedwig, in which the genera were determined by the form and situation of the minute "fruit," or threadlike stems topped with capsules (fig. 16). Apart from the difficulties of finding mosses bearing their seasonal fruit, the process of classifying mosses demanded the use of a microscope, as is made clear in the illustration on the title page of Hedwig's monograph on mosses (fig. 17). Hedwig claimed that the structure of the peristome, or fringe divided into teeth, on the inside of the capsule was an essential character. One of the remarkable features of this fringe, which varies in different genera, is that, whether it is single or double, the number of teeth, regardless of differences in size, number, or arrangement, is always a multiple of four. This feature obviously appealed to Talbot. Early in his study he told his mother that

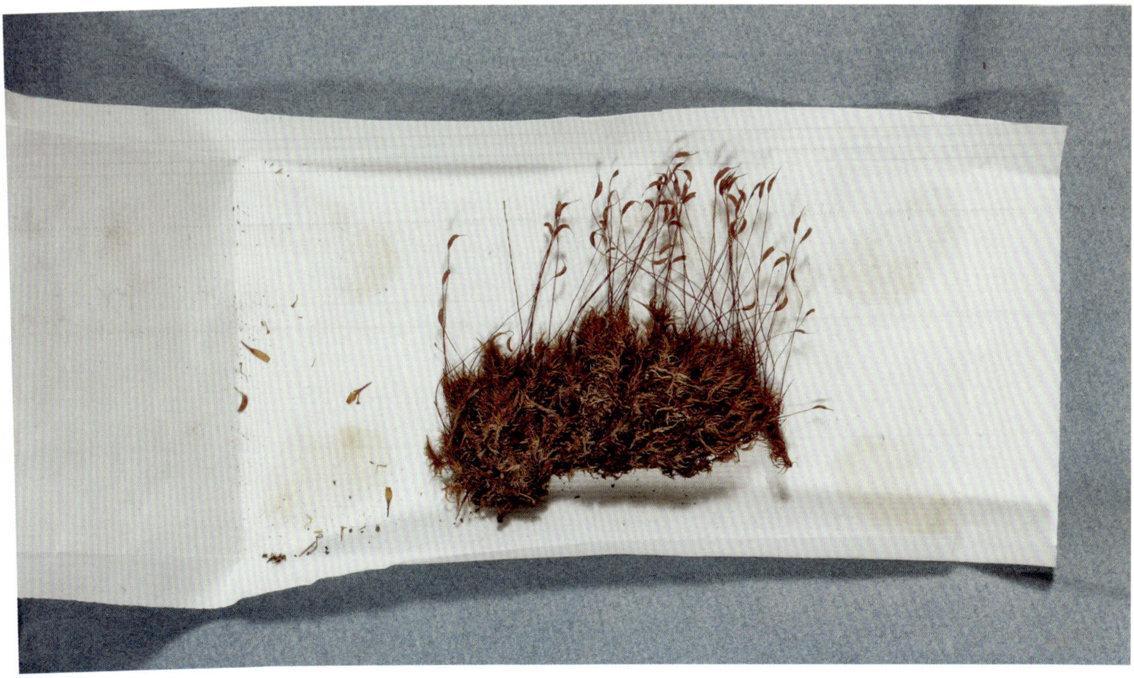

16. Specimen of *Dicranum scoparium* (detail), included in a botanical specimen album of mosses in Talbot's collection, undated. Acc. no. 36418, Add. MS 88942, Talbot Collection, British Library (partially catalogued).

"mosses are very curious if examined carefully. They have a little fringed blossom; the fingers, or *teeth* as they are called, of which, are always in number 4, 8, 16, 32, or 64: as exact as possible."[25]

While genera were defined in this exact manner (fig. 18), identifying species was more problematic, especially those distinguished by subtle variations in the shape and aspect of their leaves. The botanical classification of mosses therefore demanded precision to establish the genus and judgment to determine the species. The genus *Hypnum*, which was so intensely difficult that it discouraged Talbot, fell into this category (fig. 19). The different species of *Hypnum* were so similar to one another that without a suite of dried specimens for comparison (fig. 20), it was often impossible for any but the most experienced observers to identify what they had found. Whether the result of an explanation of the divisions of *Hypnum* and help with identification of specimens from the botanist Lewis Weston Dillwyn, or as a consequence of his own later investigations, the first page devoted to this genus in Talbot's specimen book bears witness to his difficulties: every species name bar one has been scratched out (fig. 21).

IOANNIS HEDWIG

MED. DOCT. AC PROFESSORIS BOTANICES IN ACAD. LIPSIENSI, ACADEM. IMPERIAL. NATVRAE CVRIOSORVM, REG. SCIENT. HOLM. SOCIETATISQVE LONDINENS. MEMBRI, PHYSIOPHILOR. BEROL. TVRICENS., MEDICO-CHIRVRG. TVRIC., OECONOM. LIPS. ET MARCHICO-POTSDAMENSIS, MATHEMATICO-PHYSICAE ERFORDIENSIS SODALIS

SPECIES

MVSCORVM FRONDOSORVM

DESCRIPTAE ET TABVLIS AENEIS COLORATIS ILLVSTRATAE

OPVS POSTVMVM.

SVPPLEMENTVM PRIMVM

SCRIPTVM

A

FRIDERICO SCHWAEGRICHEN

MED. ET PHILOS. DOCT., BOTANICES ET HISTORIAE NATVRALIS PROFESSORE IN ACADEMIA LIPSIENSI, SOCIETATT. REG. PHYSIC. GOETTINGENS. MEMBR., REG. MONACENS. CORRESP., PHYTOGRAPHIC. MOSCVENS., BOTANIC. RATISB., PHYSIOGRAPHIC. WETTERAVIENS., HEIDELBERGENS. OECON. LIPS. SODAL.

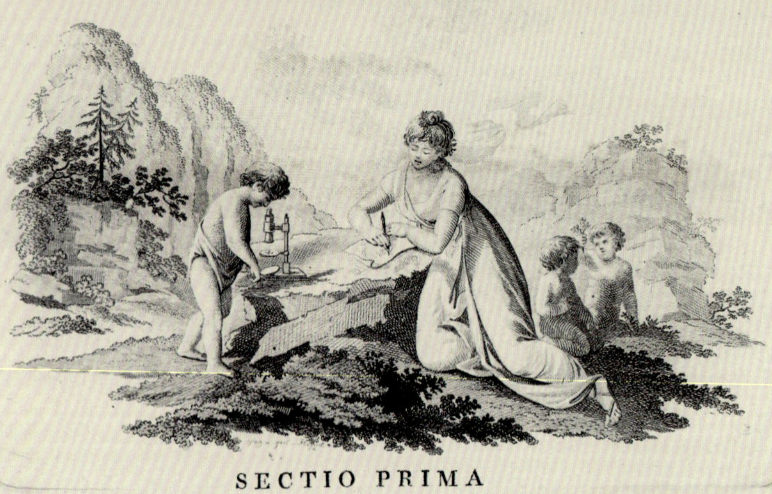

SECTIO PRIMA

TABVLIS AENEIS COLORATIS L. ILLVSTRATA.

LIPSIAE

SVMTV IOANNIS AMBROSII BARTH

MDCCCXI.

PARISIIS APVD TREVTTEL ET WÜRZ, RVE DE LILLE N°. 17.

17. Johann Hedwig, *Species Muscorum Frondosorum descriptae et tabulis aeneis*, ed. Frederic Schwaegrichen (Leipzig: Sumtu Joannis Ambrosii Barthii, 1801), title page. Manuscripts and Archives, Yale University Library.

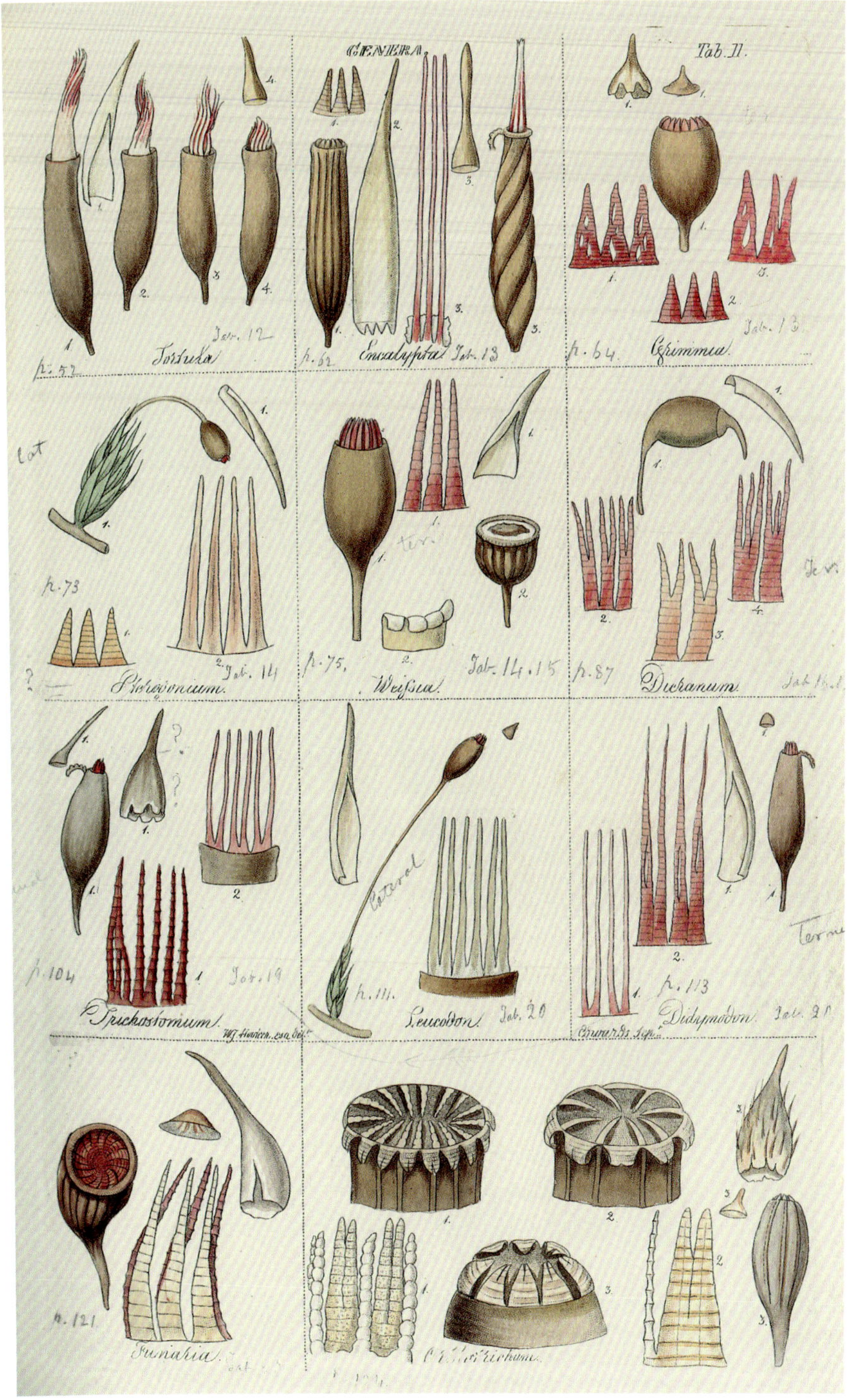

18. Illustrations of some genera of mosses, in William Jackson Hooker and Thomas Taylor, *Muscologia Britannica: Containing the Mosses of Great Britain and Ireland, Systematically Arranged and Described*, 2nd ed. (London: Longman, Rees, Orme, Brown, and Green, 1827), plate II. Farlow Library of Cryptogamic Botany, Harvard University, Cambridge, MA.

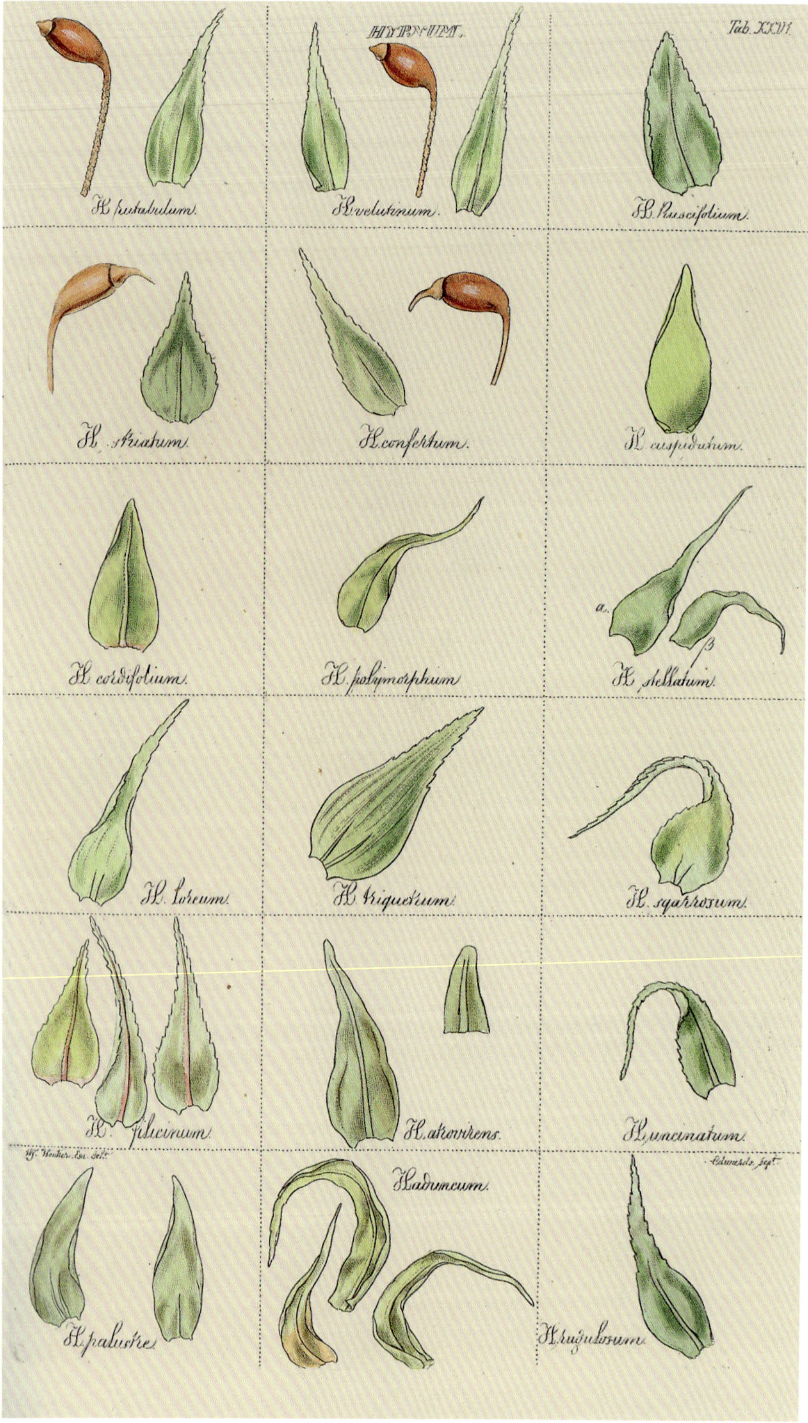

19. Illustrations of some species of *Hypnum*, in William Jackson Hooker and Thomas Taylor, *Muscologia Britannica: Containing the Mosses of Great Britain and Ireland, Systematically Arranged and Described*, 2nd ed. (London: Longman, Rees, Orme, Brown, and Green, 1827), plate XXVI. Farlow Library of Cryptogamic Botany, Harvard University, Cambridge, MA.

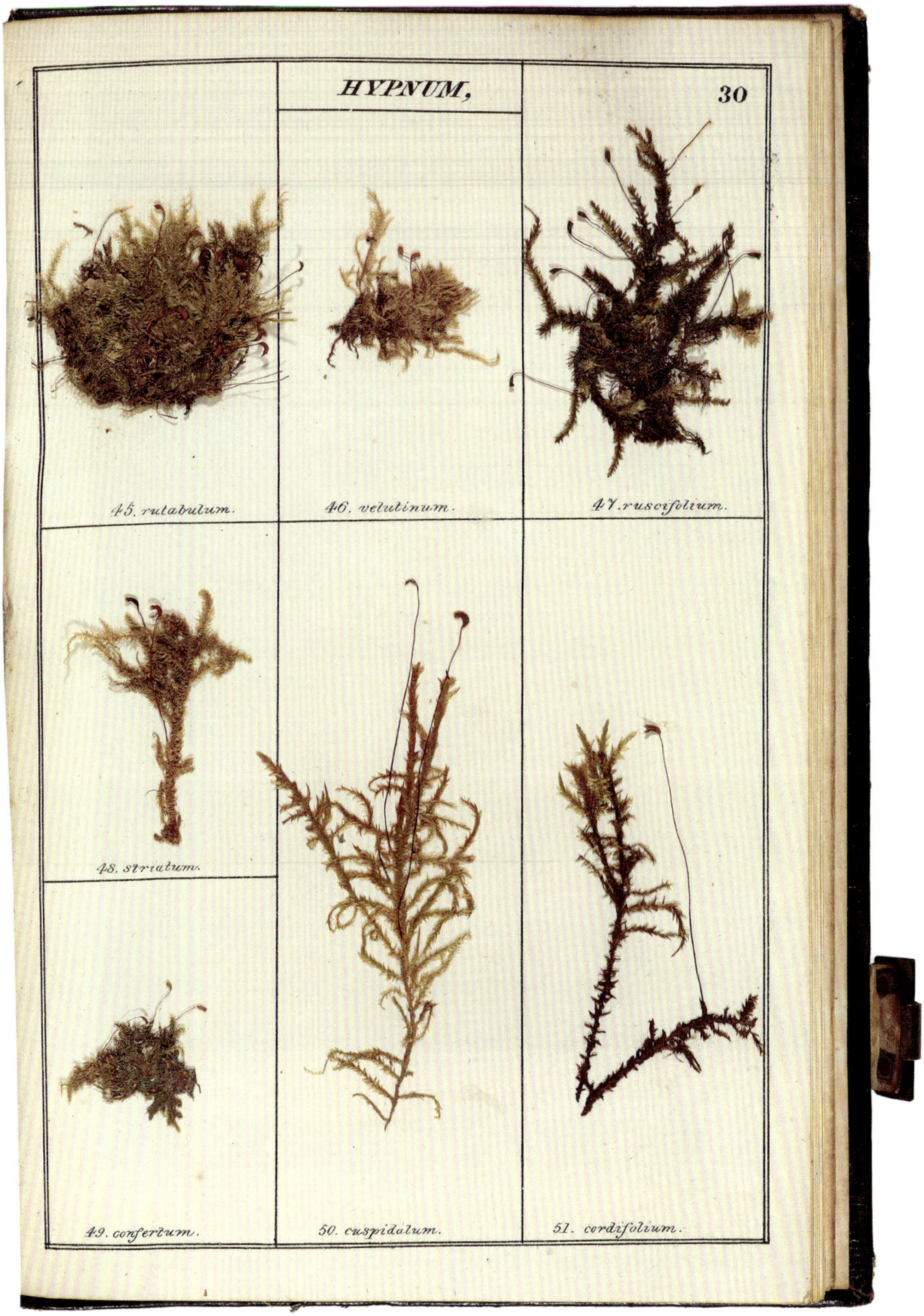

20. Specimens of different species of *Hypnum,* in George Gardner, *Musci Britannici, or Pocket Herbarium of British Mosses* (Glasgow: Allan & Ferguson, 1836), 30. Old Library Stack, Selwyn College, Cambridge.

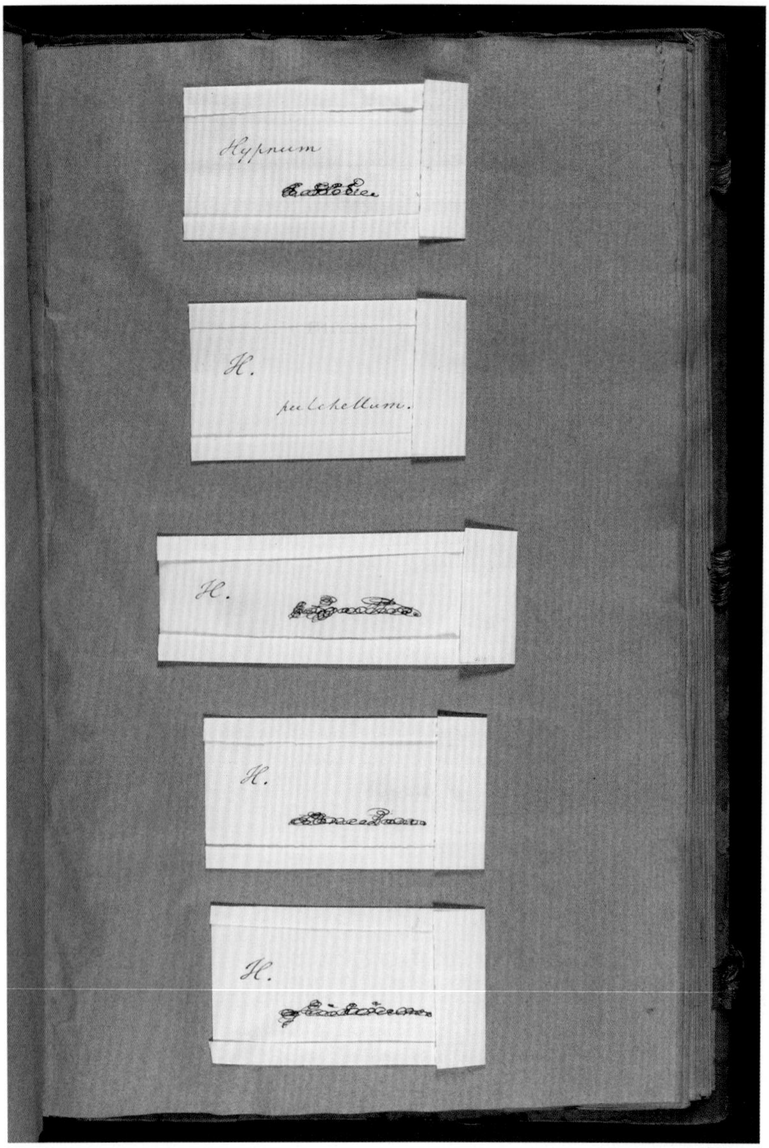

21. Envelopes containing specimens of *Hypnum*, included in a botanical specimen album of mosses in Talbot's collection, undated. Acc. no. 36419, Add. MS 88942, Talbot Collection, British Library (partially catalogued).

Much has been made of Talbot's contacts with the most eminent botanists of the day.[26] Connections of this sort, however, were not unusual in the world of British botany, which was largely dependent on private individuals of all social classes scattered across the country. Indeed, without these independently motivated enthusiasts, sciences such as botany would never have been able to accumulate a critical mass of observations. Correspondence was the glue that held together the social networks by which specialists obtained the specimens and information they required, and many publications made clear the wish of experts to obtain such information and the help they could offer local correspondents in return.[27] In the difficult and generally unpopular area of nonflowering plants such as mosses, this was even more the case. Talbot's contacts with the recognized botanical authorities of the day were therefore entirely typical. Nonetheless, his social status was undoubtedly an advantage in maintaining and developing contacts. After their correspondence began Talbot met Dillwyn, who was a visitor at Penrice, and found him to be "an exceedingly clever man" who

> entertained us in the Ev^{g} with extraordinary arithmetical puzzles, & solved with great ease, all that *we* could propose to him. . . . He is moreover an excellent magistrate, a very eminent naturalist, & as you know, famous for making good & beautiful china, which he is daily improving. There is so much to be learned from such a man: Especially when joined to a very obliging disposition.— He began his Natural History, with Botany, then Insects, &c &c. Birds, & now Shells, on which he is writing a book, & is therefore become a very early riser, that it may not interfere with his magisterial duties.[28]

The mastery of such a wide range of activities clearly impressed Talbot, and although Dillwyn was a china manufacturer rather than a man of independent wealth, he provided a model for emulation.[29]

Despite the limitations of correspondence, it was the means by which Talbot learned that botany was a communal activity. Botanists had long insisted that a free but equitable exchange of information and specimens should be the norm. Although correspondence played an essential role in making this possible, the interaction between different social classes that letter writing allowed led to the necessity of determining the trustworthiness of an unknown correspondent before parting with specimens.[30] Again, Talbot's status as a gentleman, and his family connections, gave him an advantage by indicating that he was honorable and that his word could be trusted. His status was also underlined by the fact that although botanists—eager to encourage as wide a range of participants in botany as possible—emphasized that even unknown correspondents need not stand on ceremony with one another, high social position was sometimes acknowledged by an unexpected degree of formality. Talbot's own expectation of informal and direct exchanges between botanists is revealed by his confusion with respect to etiquette on receiving a copy of Dawson Turner's book on Irish mosses with a very formal note. He sent Turner's letter to his mother, stating, "I am at a loss what answer to return: so I shall expect you to send me your advice on that knotty point: unless you can find me a parallel case in the '*Universal Letter Writer.*'"[31]

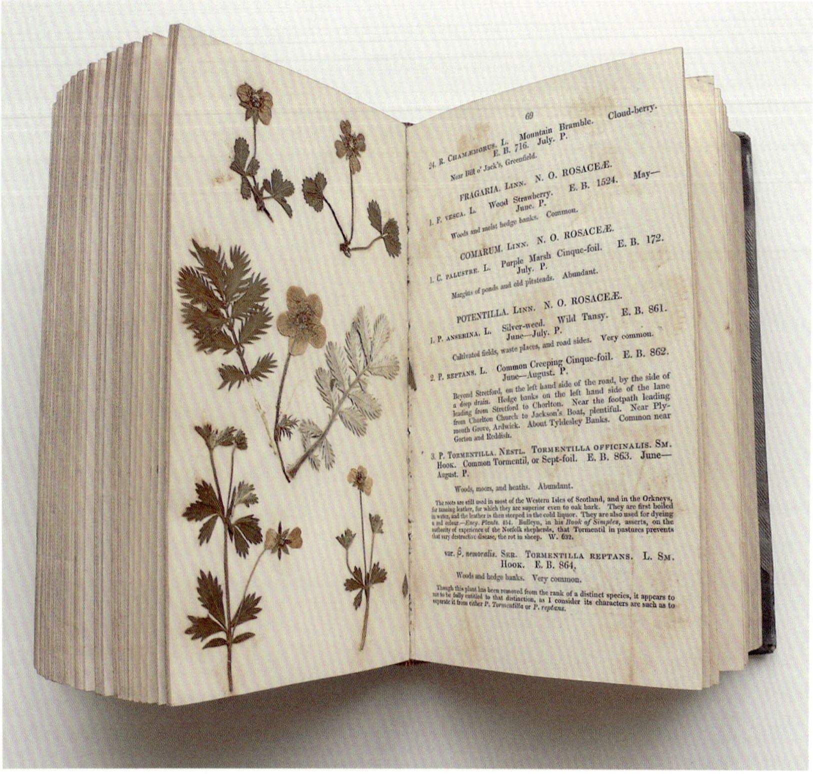

22. Specimens glued into an interleaved copy of Richard Buxton, *Botanical Guide to the Flowering Plants, Ferns, Mosses, and Algae, Found Indigenous within Sixteen Miles around Manchester* (London: Longman, 1849). Author's collection.

THE COMMUNITY OF BOTANISTS

Correspondence was also used to spread news of new and promising botanical enthusiasts, especially those interested in little-studied groups like the mosses. By this means the Reverend James Dalton of Yorkshire heard from a friend that Talbot was a "strenuous muscologist," and he, in turn, immediately wrote to Talbot, offering advice to the schoolboy.[32] It was Dalton who encouraged Talbot to make visible, through sorting and ordering according to the natural system of classification, the species that constituted the genera of mosses. Elaborating on a widely adopted practice that brought together textual descriptions and plants (fig. 22), Dalton outlined the best way to go about this:

> My plan with mosses, is, to attach a specimen to an interleaved Flor: Brit: opposite the description — & to draw magnified leaves &c, from microscopic observation. This facilitates investigation beyond conception. . . . you had best . . . wait for the work now preparing by Hooker . . . have it inter-

leaved, & proceed accordingly. He attends much to the Natural Orders, & makes daily discoveries towards the renewal of those links in the chain, which, it appears, have been obscured by rust, but not lost. Linnæus, like a niggard, carried the discoveries he had made upon the subject . . . along with him to the grave; — for which I doubt not he is now wandering without breeches amongst the brambles which border old Styx.[33]

Whether or not Talbot followed this advice to interleave a copy of William Jackson Hooker and Thomas Taylor's *Muscologia Britannica* when it was published in 1818 or to do so to James Edward Smith's *Flora Britannica*, "a delicious book" that he already owned, it is clear that he wanted to learn more about the natural orders of mosses.[34] Well before Hooker and Taylor's *Muscologia* was published, Talbot had requested and received a list of the genera that Hooker had determined to adopt in his monograph, as well as an explanation of why some were to be omitted.[35] The exactness of observation and the exercise of judgment required for determining a classificatory system for the mosses, especially when many new ones were still being discovered, led Dalton to caution Talbot with respect to the *Muscologia*: "Perfection must not be hoped for; — but the study will, doubtless, be greatly facilitated, if it be only by the rejection of the numberless plants now admitted as species, w^{h} are, certainly, no more than varieties."[36]

In this aim of at least starting to clarify the classification of mosses, Talbot was able to help Hooker. As early as 1816 Talbot had honed his observational skills sufficiently to be able to distinguish many of the tiny and particularly earthbound mosses in the genus *Phascum* (fig. 23)—and to put Hooker "right."[37] From specimens sent by Talbot, Hooker made a drawing of *Phascum curvicollum* for his monograph (fig. 24), but so minute are these mosses that, at little more than one-eighth of an inch (four millimeters) high, Talbot's own specimens of *Phascum curvicollum* are no longer even visible, being hidden in an earthy mess (fig. 25). Even if the passage of time has affected the state of these tiny plants, it has not altered the quantity of earth. If the specimens Hooker received were in this state, he, perhaps out of deference to Talbot's social status, made no complaints.[38] This contrasts starkly with Hooker's instructions to another collector at the opposite end of the social scale, the Manchester weaver Edward Hobson. Although Hooker would later state that he did "not know any Naturalist who . . . searched for Mosses more successfully than Hobson . . . nor one who . . . discriminated them more accurately,"[39] he frequently complained that Hobson's specimens were "so loaded with . . . earth . . . I can hardly distinguish the fructification nor fasten [them] down in my herbarium"; "with all mosses," Hooker stressed, "the more free from the earth they are the better & in travelling the soil does not . . . injure them."[40]

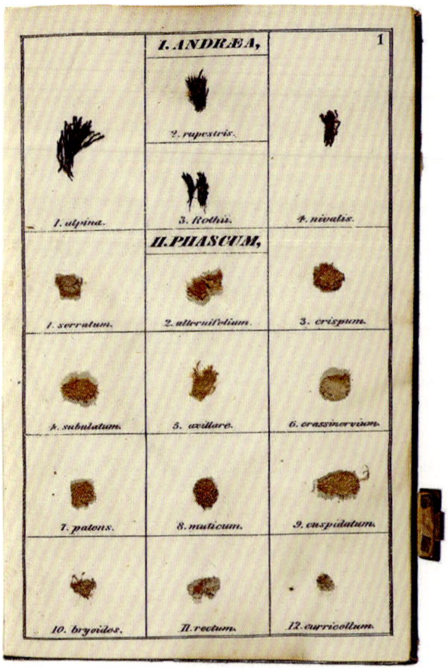

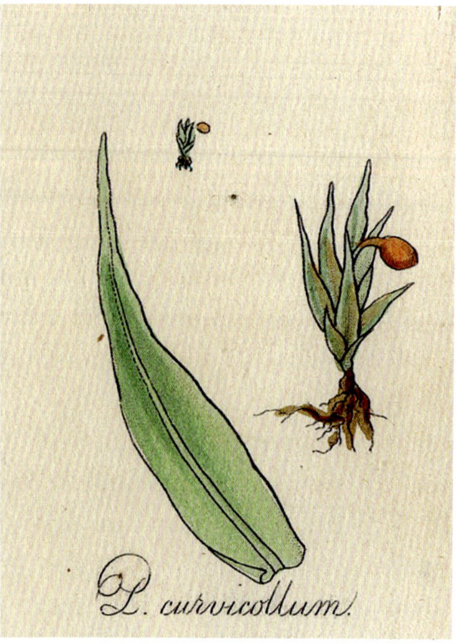

23. Specimens of *Phascum,* in George Gardner, *Musci Britannici, or Pocket Herbarium of British Mosses* (Glasgow: Allan & Ferguson, 1836), 1. Old Library Stack, Selwyn College, Cambridge.

24. Magnified drawing of *Phascum curvicollum,* in William Jackson Hooker and Thomas Taylor, *Muscologia Britannica: Containing the Mosses of Great Britain and Ireland, Systematically Arranged and Described,* 2nd ed. (London: Longman, Rees, Orme, Brown, and Green, 1827), detail of plate V. Farlow Library of Cryptogamic Botany, Harvard University, Cambridge, MA.

Nonetheless, with so many mosses yet to be discovered, Hooker's appreciation of a keen eye led him to cultivate Talbot, who, as well as his observational skill, also possessed the advantage of staying in different parts of the country.[41]

The difficulty with mosses was not confined to judging what you had found, but also to finding these tiny plants in the first place. The development of Talbot's observational skill in spotting the growing forms of mosses in nature is apparent from both his early botanical notebook and his collection of specimens. In the latter the three exclamation marks that follow the label "Bryum roseum in fruit" signaled Talbot's achievement in finding this moss with the capsules by which it could be securely identified.[42] His notebook entries are more restrained; his record of finding the "very rare" *Pterogonium smithii* at Penrice (fig. 26) contrasts with his cousin Jane's raptures when she reported finding the same moss at Kingston House, Dorset: "What Joy equals mine at this moment — *I* find *Pterogonium Smithii* — is it possible!"[43]

Exchanges such as this with Talbot's affectionate "Coz," as well as with his mother and other members of his family, especially his uncle the diplomat and botanist William Fox Strangways, were—and remained—immensely important in sustaining Talbot's interest in botany. Botany, in turn, continued to bind the family together, intertwining travels and memories, as the Talbot family albums

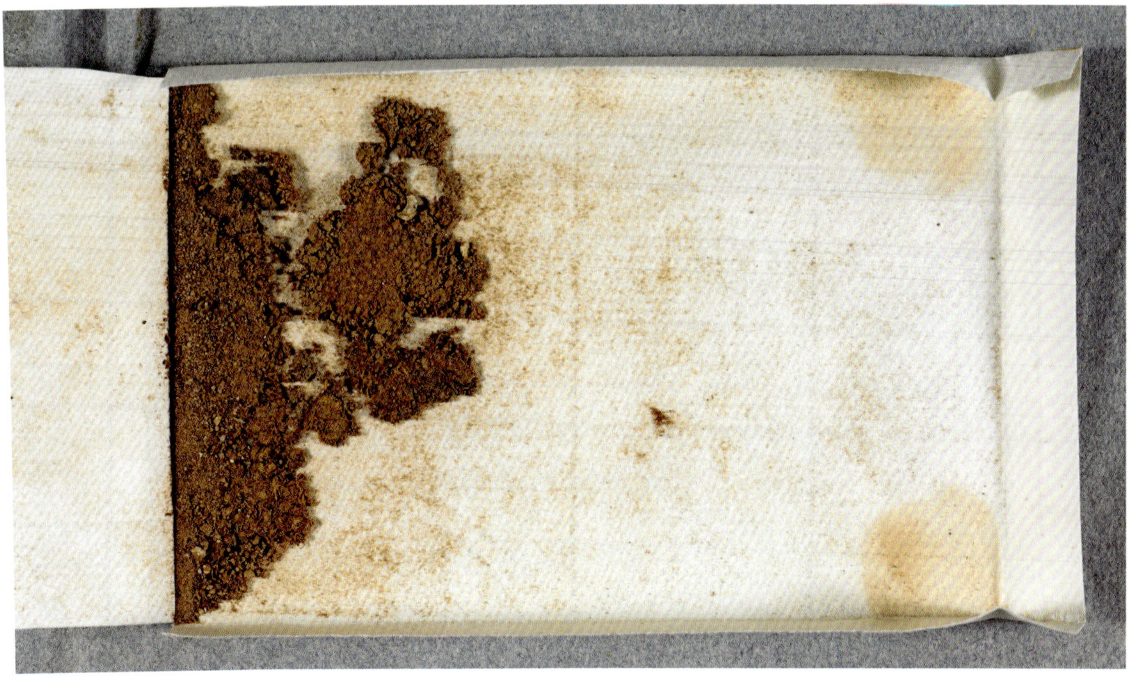

25. Specimen of *Phascum curvicollum* (detail), in a botanical album of mosses in Talbot's collection, undated. Acc. no. 36418, Add. MS 88942, Talbot Collection, British Library (partially catalogued).

show.[44] Even though Talbot's attention turned to horticulture and establishing the garden at Lacock Abbey from 1821, his interest in mosses was rekindled years later when he heard about George Gardner's *Musci Britannici, or Pocket Herbarium of British Mosses*, a book designed to hold dried specimens of mosses with every other page "ruled off into compartments, suited to the size of the species . . . & the names printed in lithography" (see figs. 20 and 23).[45] The book contained a number of specimens when purchased, and the owner was expected to fill the empty spaces. The observational skills required by this activity, and the resulting suites of dried specimens, were essential for the study of groups of plants such as mosses, seaweeds, and grasses, whose classification was difficult and not yet settled. It was this botanical need that created a market for the production and sale of books of specimens.[46] The labor of producing multiple copies of these works was prodigious, however, and no two copies were identical. They varied as to number and quality of specimens (Hooker offered Talbot a copy of Gardner's work with the best specimens and the smallest number of blank spaces to be filled).[47] Talbot's experience of this incompleteness and variability—together with his appreciation that in contested groups of plants, arrangements of dried specimens allowed readers to observe and judge for themselves—may have contributed to the potential he saw for photography in botany.

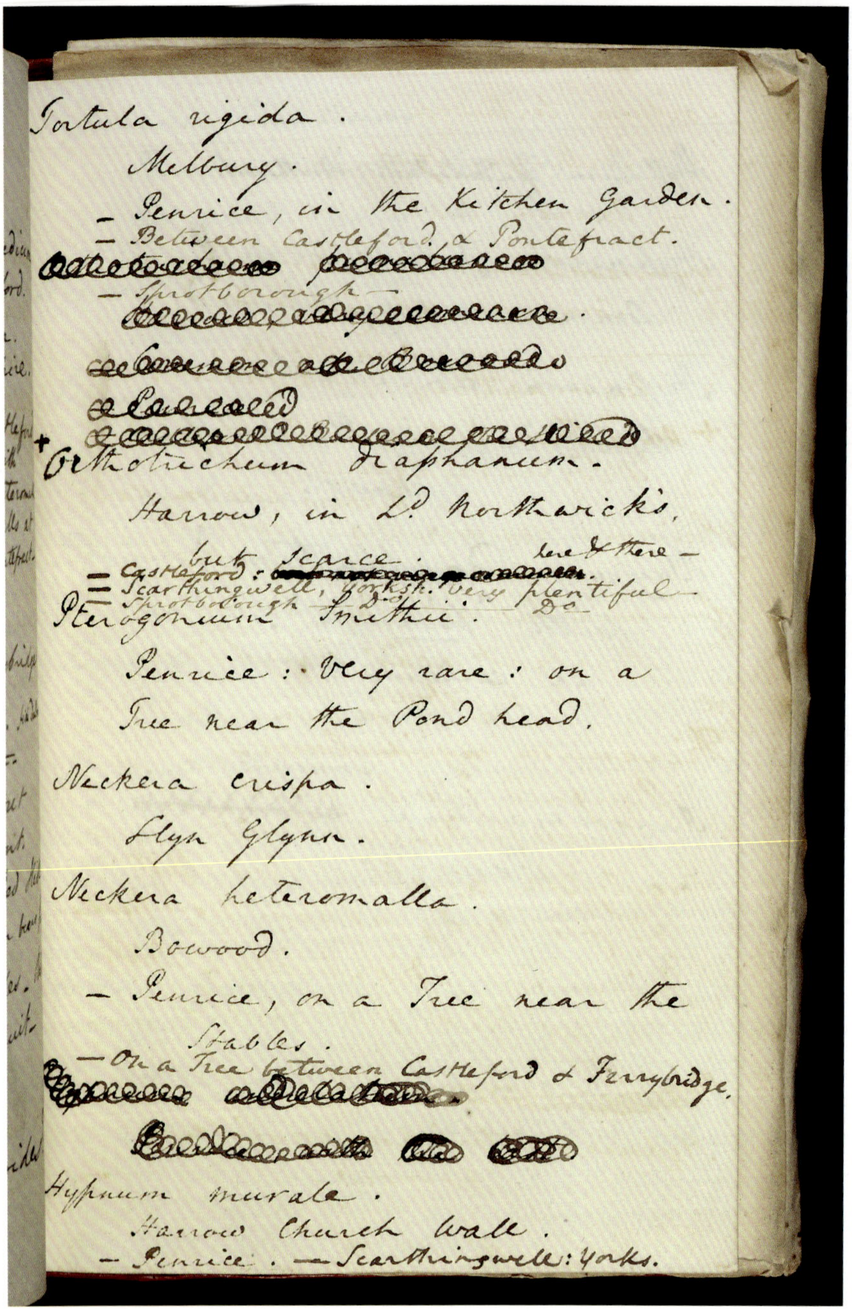
Tortula rigida.
Melbury.
— Penrice, in the Kitchen Garden.
— Between Castleford & Pontefract.
— Sprotborough —
+ Orthotrichum diaphanum.
Harrow, in Ld Northwick's,
but scarce. here & there —
= Castleford:
= Scarthingwell, Yorksh: very plentiful
= Sprotborough Do Do
Pterogonium Smithii.
Penrice: Very rare: on a
Tree near the Pond head.
Neckera crispa.
Llyn Glynn.
Neckera heteromalla.
Bowood.
— Penrice, on a Tree near the
Stables.
— On a Tree between Castleford & Ferrybridge.
Hypnum murale.
Harrow Church Wall.
— Penrice. — Scarthingwell: Yorks.

26. William Henry Fox Talbot, record of finding *Pterogonium Smithii* at Penrice, 1814–16, autograph notebook. Add. MS 88942/1/219, Talbot Collection, British Library.

NATURAL ILLUSTRATIONS

The achievement of photography, Talbot announced, was "to introduce into our pictures a multitude of minute details which add to the truth and reality of the representation."[48] In some instances photography's inability to distinguish between the essential and the nonessential could have been seen as an advantage; however, botanists rejected photographic images even in those cases when a multitude of minute details was desirable, namely, for groups such as the mosses, grasses, and seaweeds, whose classification was complex and the characteristics by which they were best ordered were not yet fully established. Striking parallels can be drawn with the rejection of photography by students of the antique (as described by Mirjam Brusius in her essay in this volume), but the differences rather than the similarities are more telling. In contrast to archaeologists' study of unique artifacts, botanists had no need to rely on photography when supplies of plants could always be replenished—albeit with difficulty from remote locations. Moreover, three-dimensional plants, once pressed into two-dimensional dried specimens, could be incorporated into books, not just as sets of specimens but also to illustrate descriptive texts.[49]

In *The Shock of the Real* Gillen D'Arcy Wood described how members of the cultural elite familiar with painterly depictions of landscape recoiled from the truth and reality of the "banal objectivism" of Talbot's *Sun Pictures in Scotland*.[50] For botanists familiar with specimen-illustrated books, however, photography was more about replacing the real with an image of the real. It is telling that Talbot's earliest photogenic drawings included not only fresh plants but also the two-dimensional forms of pressed dried plants selected from his herbarium.[51] Even though these photogenic drawings, produced by plants being laid directly on light-sensitive paper, could lay claim to the authenticity of the natural object itself, they were eclipsed by "nature printing," in which the impression made on a soft lead plate by an actual plant specimen could be used to produce multiple prints for publication.[52] By the time this process was perfected, the camera-made photograph had replaced camera-less photography and thus broken the direct association with the objects themselves as well as the notion of photography as a form of drawing. Moreover, both before and after Talbot began publishing *The Pencil of Nature* in 1844, botanists using actual specimens to illustrate their works were claiming that these were "nature's own pencil illustrating herself" and truly "natural illustrations."[53]

Photographs of plants were equally unsuccessful in replacing botanical drawings. When Hooker received some of Talbot's photographs of plants in 1839, his enthusiasm was less for the use of photography to illustrate botanical works than for the potential use of photographs as a means of copying botanical drawings, which could then be reproduced more cheaply than etchings or engravings.[54] Botanical drawings were essential because they clarified the salient features by which a plant was classified, while leaving out all other details. Novices relied on drawings to learn how to see botanically, and for experienced botanists they served to refresh the memory. Even when, later in the century, technological advances made it easier to produce sharper images of three-dimensional plants, not even amateur botanists photographed the rare plants they found in the field. Instead, the tradition of drawing plants in the margins of the most popular botanical handbook of the day continued well into the twentieth century.

When the pencil is in nature's hand rather than that of the botanist, the processes of botanical observation are separated from the production of the image itself. Making a botanical drawing, by contrast, required an observer to exercise the form of visual acuity that could detect which elements defined the object of scrutiny and thus, more importantly, which features should be ignored. Drawing helped to develop a botanist's powers of observation, whereas photography did not. Moreover, drawings established a particular way of seeing that photography could not imitate: "I wish you could contrive to mend Natures perspective," Talbot's uncle William Fox Strangways commented; "we draw objects standing up & she draws them lying down which requires a correction of the eye or mind in looking at the drawing."[55]

For botanical experts, drawings remained an essential part of the taxonomic enterprise, acting not only as virtual specimens on occasion but also as the means by which microscopic or other features of groups of plants could be compared. This "optical consistency," as Bruno Latour has shown, results from the practical means by which not just *things* can be juxtaposed but, through the use of *representations*, it becomes possible to compare things of vast or minute scale.[56] Given Talbot's success with making photomicrographs using a solar microscope in 1839, it is surprising that he did not attempt to produce magnified images of mosses for this purpose.[57]

Talbot nonetheless remained certain that photography would be a boon to botanists, if only as a means to compensate for some of the shortcomings of drawings and specimens, especially when traveling. In 1847, on hearing that William Hooker's son had set off on a botanical expedition to India, Talbot did not think it too late to propose that Joseph Hooker "send home from India a Series of photographic delineations of the plants of that country," explaining that he had devised a process so easy for travelers that "a Hindoo servant would learn it in a day." The advantages listed by Talbot included the following: "plants of delicate structure or of very complicated foliage would be pictured *while fresh*"; "thick specimens," usually not collected by travelers on account of space, could at least come home in the form of a picture; and, finally, "the pictures so obtained are at once ready for publication. They have only to be copied photographically. Thus the exact transcript of the original specimen would pass into the hands of many botanists—"[58] There are no photographs from Joseph Hooker's Indian journey.

Although the "Traveller's Camera" could capture images of three-dimensional objects, botanists were unimpressed by the resulting images and remained reluctant to give up the practices of drawing and pressing plants.[59] In 1859, despite finding a "photoglyphic engraving" of a fern to be "a most faithful representation," William Hooker was still explaining to Talbot, "What we want . . . is not only a representation of the plant, but the analysis of the fructification, venation &c. in magnified figures. . . . &, there besides, is the defect arising from inequality of surface in the plant; & where there is pubescence, & fructification — all that is blurred."[60] Even though the photoglyphic process answered Hooker's call for photomechanical reproduction using a conventional printing press, botanists continued to have reservations about the quality of the image being reproduced. Talbot's images of fresh flowers may be aesthetically striking, but they remain pictures of plants, not botanical photographs.[61]

CONCLUSION

Talbot must have been surprised by botanists' lack of appreciation of photography, just he was taken aback by the accusation that his discovery removed the need for artistic skill. He denied this charge by pointing out that in photography, "there is ample room for the exercise of skill and judgment."[62] Talbot's insistence that he had not eliminated skill depended, however, on his relocation of it from the hand to the eye. Photography possessed the virtue, in his opinion, of removing the laborious process of producing drawings, but it did not dispense with the skill of noticing picture-worthy aspects of lighting or of objects that might seem unremarkable to ordinary people.[63] As a notoriously poor artist himself, Talbot regarded photographic images, with their promise of being able to stand in for the actual object—be it a plant or an archaeological artifact—as superior types of drawings, and ones, moreover, that eliminated the idiosyncrasies of the individual observer. But how could someone as well versed in botany as Talbot suggest that photography might replace botanical drawing, the purpose of which, in recording what was seen, was less to develop artistic ability than observational skill? It would seem that he failed to recognize the importance of both the process of making a drawing and the study of drawings in teaching botanists how to observe plants. This failure, seemingly at odds with Talbot's botanical expertise, may be due to his particular interest in mosses. As we have seen, botanists placed more emphasis on specimens than on drawings for training the eye when the identification and classification of complex plants (such as mosses) were still unsettled. When confronted with specimens, the ability to notice subtle distinctions that others could not see characterized the observational skill of a botanist—and here, too, Talbot seems to have valued the eye over the hand, to judge by his mucky specimens of the minuscule *Phascum curvicollum* (see fig. 25).

If photography had not turned out to be so significant an invention, would we be considering the role Talbot thought it might play in nineteenth-century botany, or even his botanical pursuits? Scholars celebrating Talbot's botanical skill have usually assessed it from the perspective of his photographs of plants. But what does his botany look like and signify when we consider it apart from his photographic experiments? As records of observational experience, Talbot's botanical notebooks, specimen books, and letters are stereotypical and unremarkable. His attainment of sufficient skill to contribute to the enumeration of the British flora is, of course, an achievement, but one that he shared with a vast number of other botanical enthusiasts. His later horticultural activities at Lacock Abbey can be seen as part of a widespread appreciation of plants and of botany as a polite pursuit. What is remarkable, however, is Talbot's early enthusiasm for mosses, as well as the demands they made on the visual skills of the observer in their collection, identification, and ordering. Recognizing that the method of reasoning this empirical science engendered could be applied to other intellectual pursuits provided Talbot with the experience that underlay the inductive method he stressed in recounting his discovery of photography. Rather than Talbot's photographs of plants focusing our attention on his botany, his botany offers a lens through which to see his early intellectual development and the way this shaped his photographic experiments. This shift of perspective also serves to show that Talbot, far from being a single-minded photographic pioneer, was typical of his social and political class in his pursuit of a wide range of knowledge. As in the case of Mill, whose study of scientific classification helped to build a legislative picture from above, Talbot also saw his intellectual enterprise as a social and moral one.

1. William Henry Fox Talbot, "Some Account of the Art of Photogenic Drawing," *London and Edinburgh Philosophical Magazine and Journal of Science* 14 (1839): 206.
2. Larry J. Schaaf, *The Photographic Art of William Henry Fox Talbot* (Princeton, NJ: Princeton Univ. Press, 2000), 38.
3. Marina Benjamin, "Sliding Scales: Microphotography and the Victorian Obsession with the Minuscule," in *Cultural Babbage: Technology, Time and Invention*, ed. Francis Spufford and Jenny Uglow (London and Boston: Faber and Faber, 1996), 99–122.
4. Benjamin, "Sliding Scales," 120.
5. Schaaf, *Photographic Art*, frontispiece, 18.
6. Quoted in Larry J. Schaaf, *Out of the Shadows: Herschel, Talbot & the Invention of Photography* (New Haven and London: Yale Univ. Press, 1992), 6.
7. *The Art of Employing Time to the Greatest Advantage: The True Source of Happiness*, 2nd ed. (London: Henry Colburn, 1822), 193.
8. Stephen Colclough, "Recovering the Reader: Commonplace Books and Diaries as Sources of Reading Experience," *Publishing History* 44 (1998): 5–37.
9. Hirst's journal keeping is discussed in James A. Secord, *Victorian Sensation: The Extraordinary Publication, Reception, and Secret Authorship of Vestiges of the Natural History of Creation* (Chicago and London: Univ. of Chicago Press, 2000), 345–53. For Lister, see Helena Whitbread, ed., *I Know My Own Heart: The Diaries of Anne Lister, 1791–1840* (London: Virago, 1988), 315.
10. William Henry Fox Talbot to Lady Elisabeth Feilding, Dec. 16, 1815, Doc. No. 656, The Correspondence of William Henry Fox Talbot, http://foxtalbot.dmu.ac.uk (accessed Jan. 31, 2012), hereafter "Correspondence" (original: Talbot Collection, hereafter "TC," British Library, hereafter "BL").
11. Talbot's earliest extant botanical notebook in the British Library is inscribed

    Plantarum rariorum stationes,
    à W.H.F.T.
    annis 1814, 1815, 1816;
    observatæ.—

    See Notebook "Plantarum rariorum stationes," Add. MS 88942/1/219, TC, BL. Talbot may have kept earlier botanical notebooks when he and his school friend Walter Calverley Trevelyan compiled a handwritten "Flora and Fauna of Harrow" in 1811. See Katie Fretwell, "Fox Talbot's Botanic Garden," *Apollo* 17 (1993): 25.
12. *The Student's Journal: Arranged, Printed, and Ruled, for Receiving an Account of Every Day's Employment, for the Space of One Year*, 4th ed. (London: Taylor and Hessey, [1815]), iv.
13. Talbot to Lady Elisabeth Feilding, Sept. 6, 1814, Doc. No. 610, Correspondence (original: TC, BL); and Lady Elisabeth to Talbot, May 19, 1817, Doc. No. 767, Correspondence (original: TC, BL). Mary Talbot was also anxious that her own children received an education appropriate for polite society; see Joanna Martin, *Henry and the Fairy Palace: William Henry Fox Talbot and Glamorgan* (Aberystwyth: National Library of Wales, 1993), 13.
14. J. W. Barrow, *Whigs and Liberals: Continuity and Change in English Political Thought* (Oxford: Clarendon Press, 1988), 31–32. Talbot served as the Whig member of Parliament for Chippenham in 1832–37.
15. Talbot to Charles Feilding, April 8, 1821, Doc. No. 922, Correspondence (original: TC, BL).
16. Talbot to Lady Elisabeth Feilding, Sept. 6, 1814, Doc. No. 610, Correspondence (original: TC, BL).
17. Talbot to Lady Elisabeth Feilding, Sept. 6, 1814, Doc. No. 610, Correspondence (original: TC, BL). Talbot refers to Euclid's *Elements* and James Edward Smith, *An Introduction to Physiological and Systematical Botany* (London: Longman, Hurst, Rees, and Orme, 1807; 2nd ed., 1809; 3rd ed., 1814). For Penrice Castle, see Martin, *Henry and the Fairy Palace*.
18. Thomas Leybourn, ed., *New Series of the Mathematical Repository*, 6 vols. (London: W. Glendinning, 1806–35), 4:121, 128–30.

19. John M. Robson, introduction to *Journals and Debating Speeches by John Stuart Mill*, ed. John M. Robson (Toronto: Univ. of Toronto Press, 1988), xiii, xiv, xvii. For Talbot and inductive reasoning see the essay by Vered Maimon in this volume.
20. David Bloor, *Knowledge and Social Imagery*, 2nd ed. (Chicago and London: Univ. of Chicago Press, 1991), 88.
21. *Autobiography and Literary Essays by John Stuart Mill*, ed. John M. Robson and Jack Stillinger (Toronto: Univ. of Toronto Press, 1981), 69. Mill learned about the natural system of classification while botanizing in France with George Bentham.
22. Talbot to Lady Elisabeth Feilding, Dec. 16, 1815, Doc. No. 656, Correspondence (original: TC, BL). This microscope is no longer extant, but three of Talbot's later microscopes are in the National Museum of Scotland, Edinburgh; two (inv. nos. T.1936.90 and T.1936.91) were made by Andrew Ross and Co., London, and the third (inv. no. T.1936.114) by the Italian instrument maker Giovan Battista Amici.
23. Talbot to Lady Elisabeth Feilding, Dec. 1, 1815, Doc. No. 3879, Correspondence (original: TC, BL).
24. Talbot to Jane Talbot, Nov. 27, 1814, Doc. No. 628, Correspondence (original: private collection).
25. Talbot to Lady Elisabeth Feilding, Oct. 20, 1814, Doc. No. 620, Correspondence (original: Wiltshire and Swindon History Centre, Chippenham).
26. See H. J. P. Arnold, *William Henry Fox Talbot: Pioneer of Photography and Man of Science* (London: Hutchinson Benham, 1977), 254–66.
27. With this aim in mind the Norfolk botanist Dawson Turner had collaborated with Lewis Weston Dillwyn to produce *The Botanists' Guide through England and Wales*, 2 vols. (London: Phillips and Fardon, 1805), a work familiar to Talbot. See Lewis Weston Dillwyn to Talbot, Nov. 28, 1814, Doc. No. 629, Correspondence (original: TC, BL).
28. Talbot to Lady Elisabeth Feilding, Aug. 13, 1815, Doc. No. 659, Correspondence (original: TC, BL).
29. In typical fashion for second-generation industrialists, Dillwyn's son, John Dillwyn Llewelyn, lived off his inherited wealth and married into landed gentry—in this case, Talbot's cousin Emma Thomasina Talbot of Penrice. Conversely, Emma's brother, the wealthy landowner Christopher Rice Mansel Talbot, encouraged mining as well as agriculture on his estates, invested in an ironworks and railways, and developed the docks at Port Talbot.
30. Anne Secord, "Corresponding Interests: Artisans and Gentlemen in Nineteenth-Century Natural History," *British Journal for the History of Science* 27 (1994): 383–408.
31. Talbot to Lady Elisabeth Feilding, July 19, 1816, Doc. No. 706, Correspondence (original: TC, BL). The *Universal Letter Writer* was just one of many guides to the art of letter writing, most of which supplied model letters for particular situations. For the formal note that accompanied Dawson Turner, *Muscologiae Hibernicae Spicilegium* (London: J. Black, 1804), see Dawson Turner to Talbot, June 21, 1816, Doc. No. 701, Correspondence (original: TC, BL).
32. James Dalton to Talbot, July 29, 1816, Doc. No. 708, Correspondence (original: TC, BL).
33. James Dalton to Talbot, Oct. 2, 1816, Doc. No. 720, Correspondence (original: TC, BL). Carolus Linnaeus had long worked on a natural system, but he never published it. His sexual system of classification was termed artificial because it was not based on characters that grouped plants with the same qualities together.
34. William Jackson Hooker and Thomas Taylor, *Muscologia Britannica: Containing the Mosses of Great Britain and Ireland, Systematically Arranged and Described* (London: Longman, Hurst, Rees, Orme, and Brown, 1818). For Talbot's description of James Edward Smith, *Flora Britannica*, 3 vols. (London: J. Taylor, J. Davis, and J. White, 1800–1804), see Talbot to Jane Talbot, Nov. 27, 1814, Doc. No. 628, Correspondence (original: private collection).
35. William Jackson Hooker to Talbot, Nov. 15, 1816, Doc. No. 732, Correspondence (original: TC, BL); and Hooker to Talbot, Jan. 14, 1817, Doc. No. 741, Correspondence (original: TC, BL).
36. James Dalton to Talbot, Dec. 1, 1816, Doc. No. 730, Correspondence (original: TC, BL).
37. William Jackson Hooker to Talbot, Aug. 20, 1816, Doc. No. 712, Correspondence (original: TC, BL).

38. Hooker was the manager of a Suffolk brewery at this time. In 1841, with Talbot's help, he was appointed director of the Royal Botanic Gardens, Kew. See Arnold, *William Henry Fox Talbot*, 262–63. A number of different specimens in Talbot's moss collection in the British Library are loaded with earth.

39. John Moore, *A Memoir of Mr. Edward Hobson, Author of "Musci Britannici, &c."* (Manchester: Simms & Dinham, and S. Boardman, 1843), title page.

40. William Jackson Hooker to Edward Hobson, June 21, [1817], Botany Department, Manchester Museum.

41. At Bowood House, Wiltshire, Talbot found that the place abounded with rare plants and that the "quantities also of rare Mosses, was quite astonishing." See Talbot to Lady Elisabeth Feilding, Sept. 26, 1815, Doc. No. 667, Correspondence (original: TC, BL). Talbot spent his early years at Penrice Castle, Bowood House, and Melbury House, Dorset; his education took him to Rottingdean, Sussex, Harrow, Leeds, and Cambridge.

42. See botanical specimen album, acc. no. 36419, Add. MS 88942, TC, BL (partially catalogued). In Hooker and Taylor, *Muscologia Britannica*, 120, this moss is described as "Rare in fructification."

43. Jane Talbot to Talbot, June 22, 1815, Doc. No. 650, Correspondence (original: Wiltshire and Swindon History Centre, Chippenham).

44. The albums are in private hands.

45. George Gardner, *Musci Britannici, or Pocket Herbarium of British Mosses* (Glasgow: Allan & Ferguson, 1836); William Jackson Hooker to Talbot, May 16, 1836, Doc. No. 3282, Correspondence (original: TC, BL); and Hooker to Talbot, July 9, 1836, Doc. No. 353, Correspondence (original: TC, BL).

46. Anne Secord, "Pressed into Service: Specimens, Space, and Seeing in Botanical Practice," in *Geographies of Nineteenth-Century Science*, ed. David N. Livingstone and Charles W. J. Withers (Chicago: Univ. of Chicago Press, 2011), 295–97.

47. William Jackson Hooker to Talbot, July 9, 1836, Doc. No. 353, Correspondence (original: TC, BL).

48. William Henry Fox Talbot, *The Pencil of Nature* (London: Longman, Brown, Green and Longmans, 1844), plate X.

49. Secord, "Pressed into Service," 298–306.

50. Gillen D'Arcy Wood, *The Shock of the Real: Romanticism and Visual Culture, 1760–1860* (New York and Basingstoke, UK: Palgrave, 2001), 190.

51. Talbot, "Art of Photogenic Drawing," 199.

52. Carol Armstrong, "Cameraless: From Natural Illustrations and Nature Prints to Manual and Photogenic Drawings and Other Botanographs," in *Ocean Flowers: Impressions from Nature*, ed. Carol Armstrong and Catherine de Zegher (Princeton, NJ, and Oxford: Princeton Univ. Press, 2004), 147–61.

53. William Henry Harvey, *A Manual of the British Algae* (London: John van Voorst, 1841), liv; and Frederick Hanham, *Natural Illustrations of the British Grasses* (Bath: Binns and Goodwin, 1846).

54. William Jackson Hooker to Talbot, June 21, 1839, Doc. No. 3895, Correspondence (original: TC, BL). For the need for botanical illustrations and the expense of producing them, see Anne Secord, "Botany on a Plate: Pleasure and the Power of Pictures in Promoting Early Nineteenth-Century Scientific Knowledge," *Isis* 93 (2002): 34 n.15.

55. William Fox Strangways to Talbot, June 10, 1839, Doc. No. 3890, Correspondence (original: TC, BL).

56. Bruno Latour, "Drawing Things Together," in *Representation in Scientific Practice*, ed. Michael Lynch and Steve Woolgar (Cambridge, MA, and London: MIT Press, 1990), 27–31.

57. Talbot, "Art of Photogenic Drawing," 203–4. See also John Falconer and Louise Hide, *Points of View: Capturing the 19th Century in Photographs* (London: British Library, 2009), 126–27. Photomicrographs are photographic images of magnified objects; microphotographs are small images that require magnification to be seen.

58. Talbot to William Jackson Hooker, Dec. 21, 1847, Doc. No. 6064, Correspondence (original: Royal Botanic Gardens, Kew).

59. William Henry Fox Talbot, "The Traveller's Camera," *Literary Gazette; and Journal of the Belles Lettres, Arts, Sciences, &c.* (hereafter *Literary Gazette*), Nov. 27, 1852, 876.

60. William Jackson Hooker to Talbot, Sept. 11, 1859, Doc. No. 7954, Correspondence (original: TC, BL). William Fox Strangways makes the same criticism of a photographic representation of a fern. See Strangways to Talbot, May 16, 1853, Doc. No. 6790, Correspondence (original: TC, BL).

61. See Anna Atkins, *Photographs of British Algae: Cyanotype Impressions*, 3 vols. (Halstead, UK: privately printed, 1842–59). This work was widely regarded as an example of the use of photography in natural history but was admired by the author's contemporaries more for its aesthetic qualities and for her technical skill with the camera than for its scientific value. See Armstrong, "Cameraless," 110–11.

62. William Henry Fox Talbot, "Calotype (Photogenic) Drawing: To the Editor of the Literary Gazette," *Literary Gazette*, Feb. 13, 1841, 108. For the contentious nature of accusations of de-skilling in this period, see Steve Edwards, *The Making of English Photography: Allegories* (University Park, PA: Pennsylvania State Univ. Press, 2006), 23–44.

63. Robin Kelsey, "Photography, Chance, and *The Pencil of Nature*," in *The Meaning of Photography*, ed. Robin Kelsey and Blake Stimson (Williamstown, MA: Sterling and Francine Clark Art Institute, 2008), 19.

"Merely a Speculation of the Mind"? William Henry Fox Talbot and Mathematics

June Barrow-Green

"Mr Fox Talbot's name is so completely associated with the invention of photography that his mathematical powers are now almost completely unknown except to the few mathematicians who have attended to the subject of elliptic functions."[1] Thus wrote William Henry Fox Talbot's anonymous obituarist for the Royal Astronomical Society, going on to say a few words about Talbot's work in elliptic functions and mentioning in passing that he had written some other mathematical papers at the beginning and end of his career. While this perception still largely pertains today, it obscures the fact that Talbot maintained an interest in mathematics throughout his life.

The story begins properly in the autumn of 1817, when the seventeen-year-old Talbot set out for Trinity College, Cambridge. This would be his academic home for three and a half years, and there his scientific studies would begin in earnest. For Talbot, who had already shown a precocious talent and an interest in mathematics and astronomy, Cambridge—the "Holy City of Mathematics"—was the obvious choice (figs. 27 and 28).[2] Had family tradition or his mother's opinion held sway, Oxford would have been his destination. His late father had been a student at Christ Church, and his mother was worried about the demands of the Cambridge mathematics course: "You seem so mathematically inclined that I ought en bonne mere to send you to Oxford to counteract it that you may not grow into a Rhomboidal shape, walk elliptically, or go off in a tangent. All which evils are imminent if you go to Cambridge."[3]

With a mathematical inclination and a Cambridge education behind him, it is no surprise that Talbot's first scientific publications were mathematical ones. What is surprising, perhaps, is that his last publication, two years before his death, was also on that subject. Although Talbot did not publish prodigiously in mathematics, he sustained an interest in the subject throughout his life, as his publications, mathematical notebooks, and correspondence attest. His mathematical output falls naturally into three phases: the early 1820s, when he published in Joseph-Diez Gergonne's *Annales de mathématiques pures et appliquées*, also known

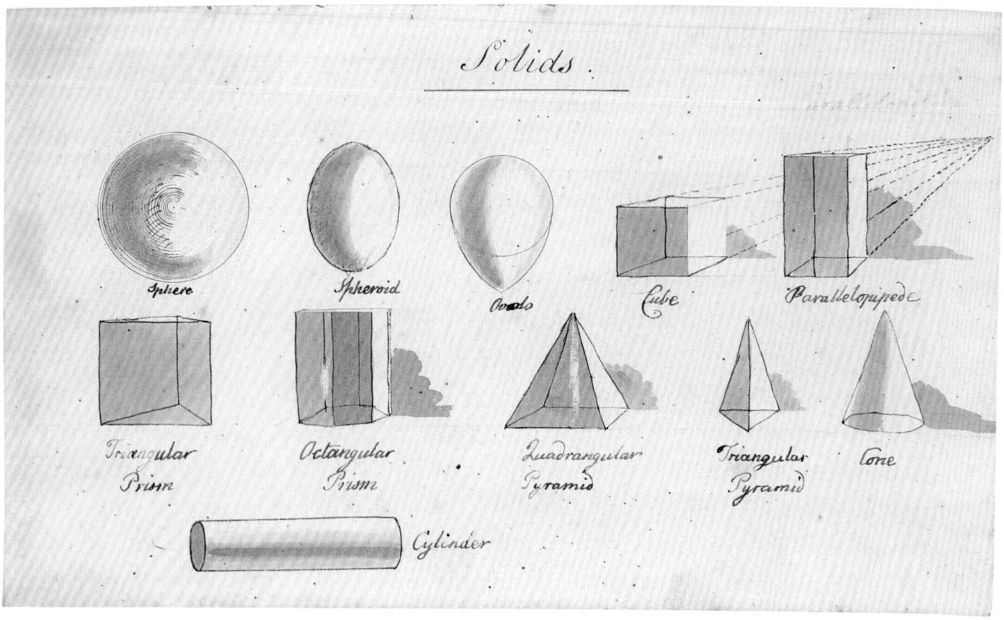

27. William Henry Fox Talbot, "Solids," 1809, in Exercise Book. Add. MS 88942/1/201, Talbot Collection, British Library.

as the *Annales de Gergonne*; the mid-1830s, when he published in the *Philosophical Transactions of the Royal Society of London*; and the 1850s and beyond, when he published in the *Transactions of the Royal Society of Edinburgh.*

THE TRIPOS YEARS: CAMBRIDGE IN 1817–21

At the beginning of the nineteenth century, the normal route to a bachelor of arts degree at Cambridge was through the Senate House Examination, popularly known as the Tripos. The examination was primarily in mathematics but included other subjects, such as logic, philosophy, and theology. It was not until 1824, when a separate Classical Tripos was offered for the first time, that the original examination took on the name Mathematical Tripos. As the century progressed the examination took on an ever-increasing significance. There was a shift from oral to written examinations, with success in the final examination being paramount, and a concomitant rise in the mathematical "coach."

Mathematics was the core of study not because it was preparation for a career as a mathematician but because it provided a fundamental part of a liberal

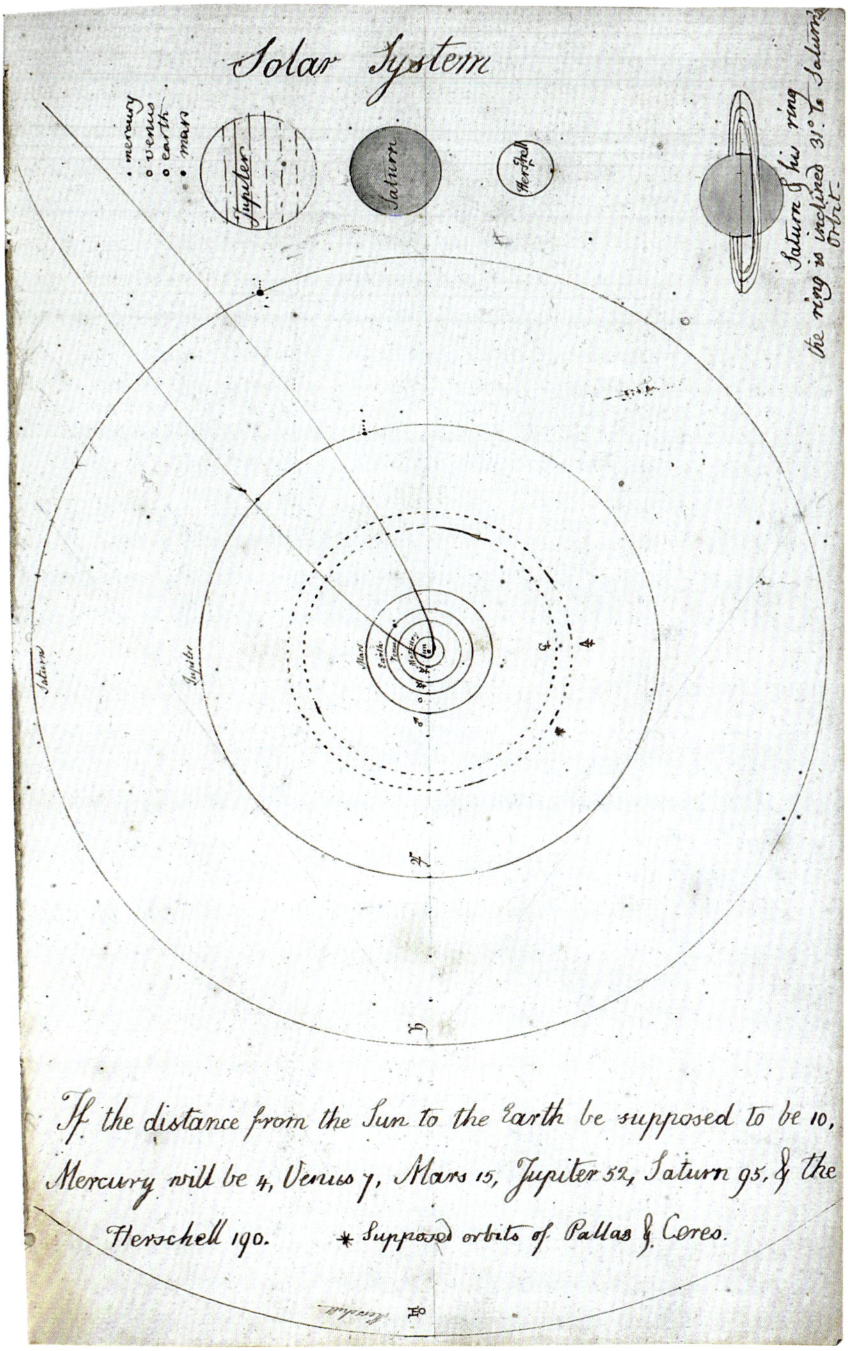

28. William Henry Fox Talbot, "Solar System," 1809, in Exercise Book. Add. MS 88942/1/201, Talbot Collection, British Library.

education. It was no longer just a subject; it became a way of life, the key to open all doors. The reason for studying Euclid's *Elements* was not simply to learn geometry. It was a training of the mind. That being said, knowledge of Euclid provided at least some access to the single most important text a Cambridge mathematics student had to study: Isaac Newton's notoriously difficult *Principia.* Written primarily in the language of geometry, by Cambridge's most famous son, the *Principia* provided the most certain demonstration of man's knowledge of the natural world. It is no wonder that it played a dominating role in the university's undergraduate course. In short, in nineteenth-century Cambridge, mathematics came to be regarded as a form of intellectual, cultural, and social training for men and gentlemen in all walks of life.[4]

In 1821, when Talbot took the Tripos examination, it required eighteen hours of mathematical papers—eleven hours of book work and seven hours of problems—over the course of three days. Students were expected to reproduce lengthy proofs and apply them to increasingly complex problems. When the results were announced the students were ranked in order of merit, with the members of the first class, the wranglers, being headed by the senior wrangler. There was enormous prestige attached to being senior wrangler, and since it was effectively a guarantee for a job in any field, competition for leading positions was extremely fierce. To have any chance of success, speed in the examination was of the essence, and students trained long and hard to hone their memories and develop their examination technique. Summer vacations were taken up with study, and those who could afford it engaged a private tutor to assist them in their quest.[5] It was little wonder that Talbot's mother had qualms about her son's choice of university.

But when Talbot arrived at Cambridge in 1817, the ferocity of the competition was not all he had to contend with: the undergraduate teaching of mathematics was on the cusp of reform. In 1803 Robert Woodhouse, senior wrangler of 1795 and a college fellow, who in 1820 would become the Lucasian Professor of Mathematics, had published the *Principles of Analytical Calculation* in response to a widespread call for Continental methods in the teaching of the calculus. Woodhouse's book, which abandoned Newton's method of fluxions in favor of algebraic analysis based on Taylor series, had little immediate effect. It could be interpreted as an attempt to diminish Newton by replacing Euclidean, geometric methods with algebra; in addition, it was hard to use for teaching since it did not show how the new methods could be applied to actual problems, a key feature of Cambridge education. Nonetheless it did lay the foundations for an eventual change in the curriculum.

The main thrust for reform came from within the student body itself with the formation in 1812 of the Analytical Society, whose leading members were Charles Babbage, John Frederick William Herschel, and George Peacock.[6] Their aim was to revitalize mathematics in Britain, and especially at Cambridge, by introducing the flexible French-style analysis that Woodhouse had introduced to Cambridge a decade earlier. Constrained by the more cumbersome Newtonian methods, British mathematicians had been left far behind their Continental counterparts; it was a situation the Analytical Society was determined to repair. Knowing that success would depend on the production of textbooks, the society proposed to translate Sylvestre Lacroix's introductory treatise on the calculus. Although the society folded in 1814 with only a single volume of memoirs to its name, the Lacroix translation appeared in 1816.[7] Then, in 1817, only months before Talbot's arrival at Cambridge, Peacock, who had been appointed as one of the "moderators" (examiners), put the cat among the pigeons by setting in the examination analytical problems that used Continental notation. Although this unprecedented step angered many of his colleagues, it marked the beginning of change, which gathered increasing momentum. The notation was used again by Peacock in 1819, by William Whewell in the following year, and by Peacock once again in 1821—the year Talbot sat the examination.

Talbot was well aware of, and appreciated, the mathematical reforms that were under way.[8] As a student at Trinity, Talbot would have been taught by Peacock and Whewell, both influential reformers in different ways. Whewell believed in arguing from the particular to the general, while Peacock, in line with the Analytical Society's position, took the opposite view. Although Whewell followed Peacock in using the Continental notation for setting Tripos examination papers in 1820, his endorsement of Peacock was qualified. In a letter to Herschel of March 6, 1817, he explained that Peacock had "stripped analysis of its applications and turned it naked among them."[9] For Whewell utility was paramount: analysis should not be taught and examined only in the abstract; it was when the analysis could be applied to physical problems that it became acceptable.[10] Thus his *Elementary Treatise on Mechanics* of 1819, which became the standard Cambridge text on mechanics, used the Continental notation throughout but remained faithful to geometric diagrams, with systems of pulleys, cogs, and wheels all in place. For Whewell the new methods were a means to an end, and in his follow-up text of 1823, *A Treatise on Dynamics*, he made the end very clear: "The student who feels a proper admiration for the system of the *Principia*, ought to look forward to the complete development of it in Laplace's *Mécanique Celeste*, as the ulterior subject of his labours."[11] It was French mathematics, not British, that carried the prize.

Although Talbot did not leave many traces of his time at Trinity, we can glean a certain amount about what his life was like from one of his exact contemporaries, Solomon Atkinson, who published an account of his life as an undergraduate.[12] Atkinson's background was very different from Talbot's—he described himself as "the son of a Cumbrian peasant" and was largely self-taught.[13] But by dint of hard work and fluency in the examination, he was able to carry off "the single diadem of the Senior Wrangler."[14] His reading list included Euclid's *Elements*, algebra, and plane trigonometry in the first year; elementary differential and integral calculus, statics, dynamics, and some optics and astronomy in the second; and Newton's *Principia*, the higher branches of the differential and integral calculus, and their applications in the third. And he gave freely of his opinion of his teachers: Whewell had a "bold, vigorous and excursive" mind and was the author of the very best elementary treatises in science that Cambridge had ever produced, while Woodhouse and Peacock were acknowledged for their role in introducing the Continental methods to Cambridge.[15]

While Talbot would certainly have read Whewell's *Elementary Treatise* and surely Lacroix, precisely which works by French mathematicians he read at Cambridge is so far unknown. However, among the books that John Wright, an able student just two years ahead of Talbot at Trinity, described reading—admittedly with a certain amount of difficulty—were Joseph Louis Lagrange's *Méchanique analytique* (Analytical Mechanics), first published in 1788, and Pierre-Simon, marquis de Laplace's *Traité de mécanique céleste* (Treatise on Celestial Mechanics), ultimately published in five volumes beginning in 1799. It would therefore seem likely that Talbot at least dipped his toe into these. In addition, a number of works by French authors appear in a list of his books compiled by Giovanni Percij, one of his servants, at the end of December 1823, two years after Talbot had graduated.[16] These include treatises by Charles Dupin and Gaspard Monge (descriptive geometry), Augustin Louis Cauchy (analysis), Joseph Fourier (theory of heat), Lagrange (solving equations), and Alfred Gautier (the three-body problem). Since Talbot made frequent visits to France during his Cambridge period and soon afterward, these could have been purchased on any one of a number of occasions, if not ordered from London.

Talbot and Atkinson must have attended the same lectures and read many of the same texts, but materially their experiences at Cambridge would have been very different. While Atkinson had to take in pupils to help keep his head above the financial parapet, Talbot was able to employ a tutor. During the summer vacations it was common practice for students to make up reading parties and travel with their tutors to somewhere in the country in order to prepare for the academic competition to come. Although beauty spots such as the Lake District and the Norfolk Broads were favored destinations, for the summer of 1819 Talbot had bigger plans: he wanted to go to the Continent. But more important than the destination was the quality of the tutor, as the good ones were snapped up fast. On this occasion Talbot hired John Higman of Trinity, a college fellow and third wrangler

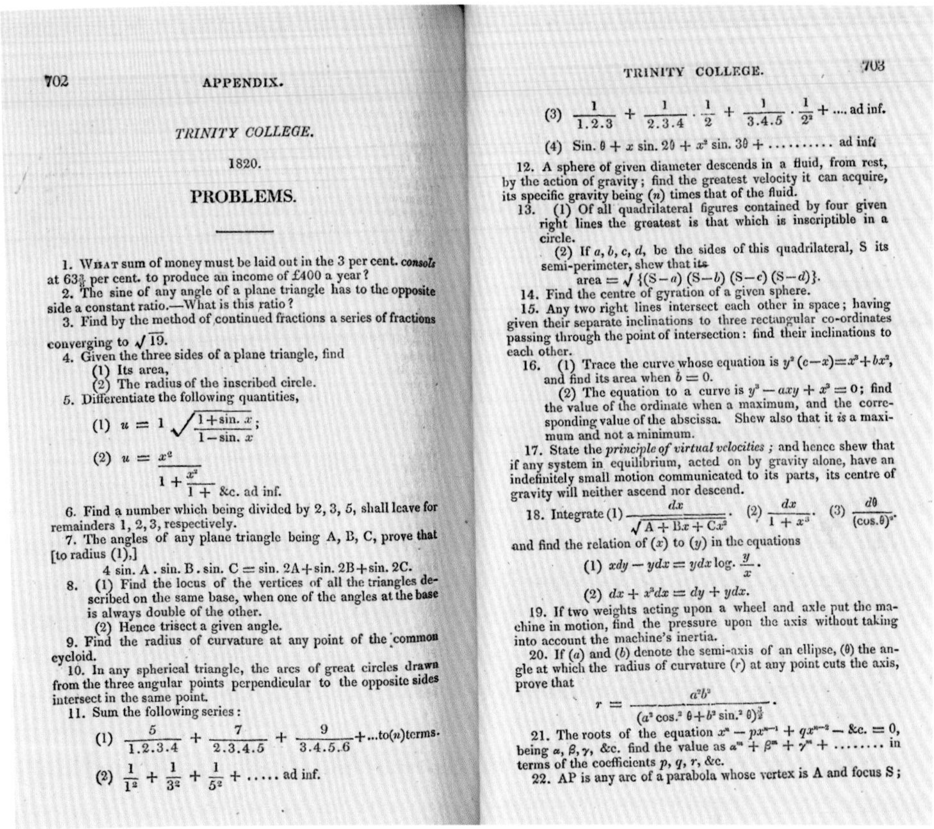

702 APPENDIX.

TRINITY COLLEGE.

1820.

PROBLEMS.

1. WHAT sum of money must be laid out in the 3 per cent. *consols* at $63\frac{3}{4}$ per cent. to produce an income of £400 a year?
2. The sine of any angle of a plane triangle has to the opposite side a constant ratio.—What is this ratio?
3. Find by the method of continued fractions a series of fractions converging to $\sqrt{19}$.
4. Given the three sides of a plane triangle, find
(1) Its area,
(2) The radius of the inscribed circle.
5. Differentiate the following quantities,

(1) $u = 1\sqrt{\frac{1+\sin. x}{1-\sin. x}}$;

(2) $u = \cfrac{x^2}{1+\cfrac{x^2}{1+ \&c. \text{ ad inf.}}}$

6. Find a number which being divided by 2, 3, 5, shall leave for remainders 1, 2, 3, respectively.
7. The angles of any plane triangle being A, B, C, prove that [to radius (1),]

$4 \sin. A . \sin. B . \sin. C = \sin. 2A + \sin. 2B + \sin. 2C.$

8. (1) Find the locus of the vertices of all the triangles described on the same base, when one of the angles at the base is always double of the other.
(2) Hence trisect a given angle.
9. Find the radius of curvature at any point of the common cycloid.
10. In any spherical triangle, the arcs of great circles drawn from the three angular points perpendicular to the opposite sides intersect in the same point.
11. Sum the following series:

(1) $\frac{5}{1.2.3.4} + \frac{7}{2.3.4.5} + \frac{9}{3.4.5.6} + \ldots$ to (n) terms.

(2) $\frac{1}{1^2} + \frac{1}{3^2} + \frac{1}{5^2} + \ldots\ldots$ ad inf.

TRINITY COLLEGE. 703

(3) $\frac{1}{1.2.3} + \frac{1}{2.3.4} \cdot \frac{1}{2} + \frac{1}{3.4.5} \cdot \frac{1}{2^2} + \ldots$ ad inf.

(4) $\text{Sin. } \theta + x \sin. 2\theta + x^2 \sin. 3\theta + \ldots\ldots\ldots\ldots$ ad inf.

12. A sphere of given diameter descends in a fluid, from rest, by the action of gravity; find the greatest velocity it can acquire, its specific gravity being (n) times that of the fluid.
13. (1) Of all quadrilateral figures contained by four given right lines the greatest is that which is inscriptible in a circle.
(2) If a, b, c, d, be the sides of this quadrilateral, S its semi-perimeter, shew that its

area $= \sqrt{\{(S-a)(S-b)(S-c)(S-d)\}}$.

14. Find the centre of gyration of a given sphere.
15. Any two right lines intersect each other in space; having given their separate inclinations to three rectangular co-ordinates passing through the point of intersection: find their inclinations to each other.
16. (1) Trace the curve whose equation is $y^2(c-x) = x^3 + bx^2$, and find its area when $b = 0$.
(2) The equation to a curve is $y^3 - axy + x^3 = 0$; find the value of the ordinate when a maximum, and the corresponding value of the abscissa. Shew also that it *is* a maximum and not a minimum.
17. State the *principle of virtual velocities*; and hence shew that if any system in equilibrium, acted on by gravity alone, have an indefinitely small motion communicated to its parts, its centre of gravity will neither ascend nor descend.
18. Integrate (1) $\frac{dx}{\sqrt{A + Bx + Cx^2}}$. (2) $\frac{dx}{1 + x^3}$. (3) $\frac{d\theta}{(\cos.\theta)^3}$.

and find the relation of (x) to (y) in the equations

(1) $xdy - ydx = ydx \log. \frac{y}{x}$.

(2) $dx + x^2dx = dy + ydx$.

19. If two weights acting upon a wheel and axle put the machine in motion, find the pressure upon the axis without taking into account the machine's inertia.
20. If (a) and (b) denote the semi-axis of an ellipse, (θ) the angle at which the radius of curvature (r) at any point cuts the axis, prove that

$$r = \frac{a^2b^2}{(a^2 \cos.^2 \theta + b^2 \sin.^2 \theta)^{\frac{3}{2}}}.$$

21. The roots of the equation $x^n - px^{n-1} + qx^{n-2} - \&c. = 0$, being α, β, γ, &c. find the value as $\alpha^m + \beta^m + \gamma^m + \ldots\ldots\ldots$ in terms of the coefficients p, q, r, &c.
22. AP is any arc of a parabola whose vertex is A and focus S;

29. John Martin Frederick Wright, *Solutions of the Cambridge Problems, from 1800–1820* (London: Black, Young and Young, 1825), 702–3. Yale University Library.

in 1816, who afterward became a fellow of the Royal Society and college tutor.[17] Unluckily for Talbot the idea of foreign travel did not appeal to his friends, who thought it appeared "outré and inimical to the study of the Mathematics," and the party ended up in Ilfracombe.[18] The following year, feeling sufficiently advanced with his study, he decided it would be better to stay in Cambridge and "tutorize" himself (fig. 29).[19] Atkinson, too, remained in Cambridge, but in his case it was not out of choice. That he was able to triumph nevertheless shows it could be done. But whether Talbot, who ended up as only twelfth wrangler, made the right choice in tutoring himself is not so clear. Nevertheless, he declared that he was "very well pleased" with the place he had got, given that he did not like the examination.[20]

After Talbot came down from Cambridge, all was quiet on the mathematical front for a good eighteen months; then he began to publish.

ANNALES DE MATHÉMATIQUES PURES ET APPLIQUÉES

Talbot's first publications appear as a number of short notes in the *Annales de mathématiques pures et appliquées* (Annals of Pure and Applied Mathematics).[21] Founded in 1810, it was the first journal devoted entirely to mathematics. Although relatively short-lived—it ceased publication in 1831 when its founder-editor, Joseph-Diez Gergonne, became rector of the University of Montpellier in France—it carried articles by several leading mathematicians of the day. The journal was particularly strong on geometry but also included articles on a wide range of both pure and applied mathematics. In addition, it published problems—often requiring a proof of a theorem—to which readers were invited to submit solutions, and it was this aspect of the journal that particularly appealed to Talbot, even though no British mathematician had previously contributed to it. Problems were published anonymously, but it is evident that Talbot proposed problems as well as solving them.[22] Since the journal was available in Cambridge—Wright mentions using it for practice before the Tripos examination—Talbot had no doubt become familiar with it while he was still a student.[23] This might explain why he favored it over British publications such as *Leybourne's Mathematical Repository* or the *Ladies' Diary*, combined with the fact that he already knew Gergonne before he submitted his first contribution.

The two men had met while Talbot was traveling in southern France in the spring of 1822. That October he wrote to Gergonne from Florence, providing a complete and elegant solution to a problem they had already discussed together—finding a curve that simultaneously solved the problem of duplicating the cube and trisecting the angle—and that the editor had since posed in his journal.[24] Gergonne published an extract from the letter as soon as he received it—Talbot's first publication.[25] In his reply Gergonne somewhat presciently posited that there was something in Talbot's method (which involved curve envelopes) that had the potential to be useful in the study of optics.[26] Talbot quickly followed this note with five more, each prompted by a question or article in the journal, dealing with problems of geometry and one, the longest, on the summation of infinite trigonometric series.[27] As Christian Gerini has observed, it is remarkable that most of Talbot's contributions to Gergonne's *Annales* were made not only when he was very young but while he was traveling abroad.[28] That Talbot was able to work in this way attests both to his academic ability and to his mathematical confidence, an attitude cultured at least in part by his Cambridge training. The mobility of pen and paper also helped.

Of all the notes he published in Gergonne's *Annales*, the last made the greatest contribution to Talbot's mathematical legacy.[29] Though effectively only a single page, it considered the curve now known as Talbot's curve, showing that it could be rectified—in other words, that its arc length could be calculated—using a particular elliptic integral. Elliptic integrals, as the name suggests, arose in connection with calculating the arc length of an ellipse and are the simplest integrals that, in general, cannot be evaluated using elementary functions. The significance of Talbot's work was that he had found one of the exceptions—an elliptic integral that could be evaluated easily. References to Talbot's paper and/or curve turn up in the work of various mathematicians in Britain and abroad, showing both the circulation of the *Annales* and appreciation of his contribution. According to Arthur Cayley, one of the leading British mathematicians of the period, the curve was first ascribed to Talbot by the Italian mathematician Barnaba Tortolini in 1855.[30]

After this initial flurry of publications Talbot's relationship with Gergonne seems to have fizzled out. Some two years later Gergonne wrote to Talbot to solicit new manuscripts from him. Having taken care to disparage two rival journals that had recently appeared—August Leopold Crelle's *Journal für die reine und angewandte Mathematik* (Journal for Pure and Applied Mathematics) and Adolphe Quetelet's *Correspondance mathématique et physique* (Mathematical and Physical Correspondence)—Gergonne almost pleaded with him:

> I shall take great care, Sir, over everything you have done me the honour of sending me. Your theorem on similar ellipses was not known to me. Your formula for the parameter is of a most remarkable elegance, and may be usefully employed in astronomy. I hold to the principle that any formula and any method which cannot be explained to a passer-by in the street is, by dint of this very fact, a dull formula or method which does not deserve to see the light of day.[31]

It was to no avail. Talbot published nothing further in Gergonne's *Annales*. But it was not because he had been seduced to publish elsewhere. In fact, more than ten years would elapse before he began publishing in mathematics again. Talbot had not given up mathematics—as his notebooks show—but he had given up on Gergonne. From Talbot's point of view Gergonne been rather overplaying his hand as an editor, "altering" the articles considerably—a habit Talbot "found so perplexing and in all respects inconvenient" that he discontinued the correspondence.[32] Nor was he entirely convinced that Gergonne always acted in Talbot's best interests. On one occasion he became suspicious when a theorem he had sent to the editor later appeared under the name of one of Gergonne's colleagues.[33]

Talbot's publications in Gergonne's *Annales* give us a very partial picture of the kind of mathematics that interested him. Since they were, after all, mostly responses to other people's questions, they do not tell us much about his own specific mathematical interests, apart from the fact that he was motivated by other people's questions and enjoyed demonstrating his own mathematical facility. A glance at his mathematical notebooks for 1822–30, however, broadens the picture considerably.[34] A particular feature is the extent to which mathematics and astronomy are intertwined. One minute Talbot is calculating the path of a comet, and the next he is using analytic geometry to describe the motion of the planets (fig. 30).[35]

Although it is possible that Talbot's experience with Gergonne had reduced his appetite for publishing mathematics, the fact is that he was already pursuing other interests with vigor, notably astronomy and optics. In 1824 he met John Herschel in Munich, a chance meeting that "established a friendship and a scientific collaboration crucial to Talbot's later success, and probably influenced Talbot's turn towards research into light and optical phenomena."[36] He spent several weeks working at the Paris Observatory in 1825, when—if a list of books he owned at that time is anything to go by—botany, Egyptology, classics, and music were also taking up his time.[37] When he returned to publishing mathematics, in 1835, he chose the Royal Society as his venue.

THE ROYAL SOCIETY

By the time Talbot came to publish with the Royal Society in 1835, he had already been a fellow for more than four years.[38] His candidacy had been submitted on December 10, 1830, with the signatures of seven proposers, only three of whom—William Whewell, Michael Faraday, and George Peacock—were men of science. Talbot had, of course, known Whewell and Peacock since his undergraduate days in Cambridge. His acquaintance with Faraday was more recent, but the two had been carrying out experiments together since at least 1827.[39]

The two papers Talbot published in the *Philosophical Transactions of the Royal Society of London* (*PTRS*) were on elliptic integrals, the second being a direct continuation of the first.[40] These papers were of a different caliber altogether from his notes in Gergonne's *Annales*, and they won him the society's Royal Medal in 1838. They were substantial research papers, inherently theoretical but strengthened by calculations. Some indication of how Talbot developed this work can be gleaned from his notebooks and from the abstracts of two unpublished papers that he read at the Royal Society in 1834, which refer to specific results that also appear in the published papers.[41] There was a delay of well over two years between the reading of the second paper, when he announced his intention "to lay before the Society, on a future occasion, an account of the principles on which this branch of analysis

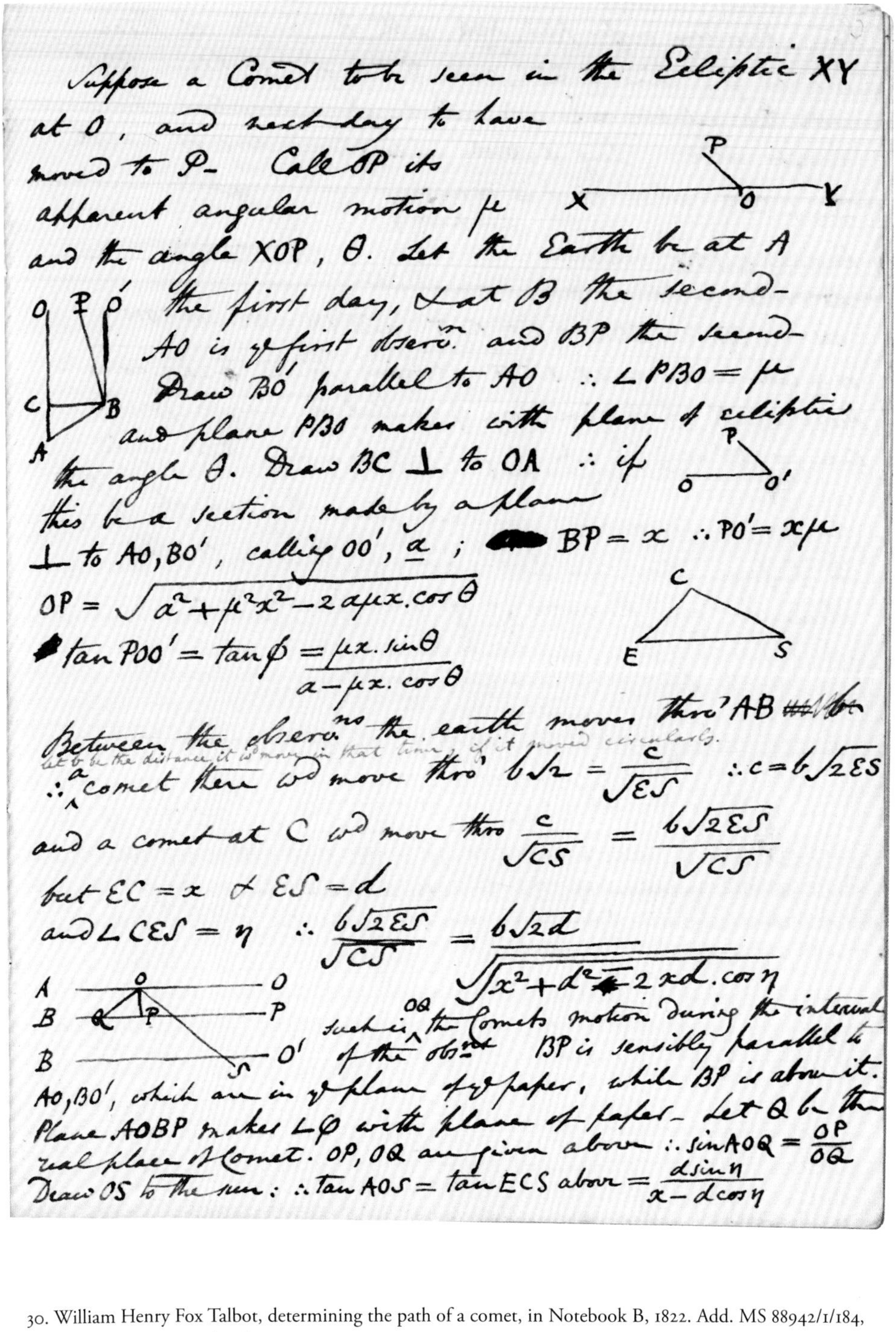

Suppose a Comet to be seen in the Ecliptic XY at O, and next day to have moved to P. Call OP its apparent angular motion μ and the angle XOP, θ. Let the Earth be at A the first day, & at B the second. AO is ye first observ.n and BP the second. Draw BO′ parallel to AO $\therefore \angle PBO' = \mu$ and plane PBO′ makes with plane of ecliptic the angle θ. Draw BC ⊥ to OA $\therefore$ if this be a section made by a plane ⊥ to AO, BO′, calling OO′, a; BP $= x$ $\therefore$ PO′ $= x\mu$

$$OP = \sqrt{a^2 + \mu^2 x^2 - 2a\mu x.\cos\theta}$$

$$\tan POO' = \tan\phi = \frac{\mu x.\sin\theta}{a - \mu x.\cos\theta}$$

Between the observ.ns the earth moves thro′ AB ~~…~~
let b be the distance it w.d move in that time, if it moved circularly.
$\therefore$ a comet there w.d move thro′ $b\sqrt{2} = \frac{c}{\sqrt{ES}}$ $\therefore c = b\sqrt{2ES}$

and a comet at C w.d move thro $\frac{c}{\sqrt{CS}} = \frac{b\sqrt{2ES}}{\sqrt{CS}}$

but EC $= x$ & ES $= d$

and $\angle$ CES $= \eta$ $\therefore \frac{b\sqrt{2ES}}{\sqrt{CS}} = \frac{b\sqrt{2d}}{\sqrt{x^2 + d^2 - 2xd.\cos\eta}}$

such is OQ the Comets motion during the interval of the obs.ns BP is sensibly parallel to AO, BO′, which are in y.e plane of y.e paper, while BP is above it. Plane AOBP makes $\angle\phi$ with plane of paper. Let Q be the real place of Comet. OP, OQ are given above $\therefore \sin AOQ = \frac{OP}{OQ}$

Draw OS to the sun: $\therefore \tan AOS = \tan ECS$ above $= \frac{d\sin\eta}{x - d\cos\eta}$

30. William Henry Fox Talbot, determining the path of a comet, in Notebook B, 1822. Add. MS 88942/1/184, Talbot Collection, British Library.

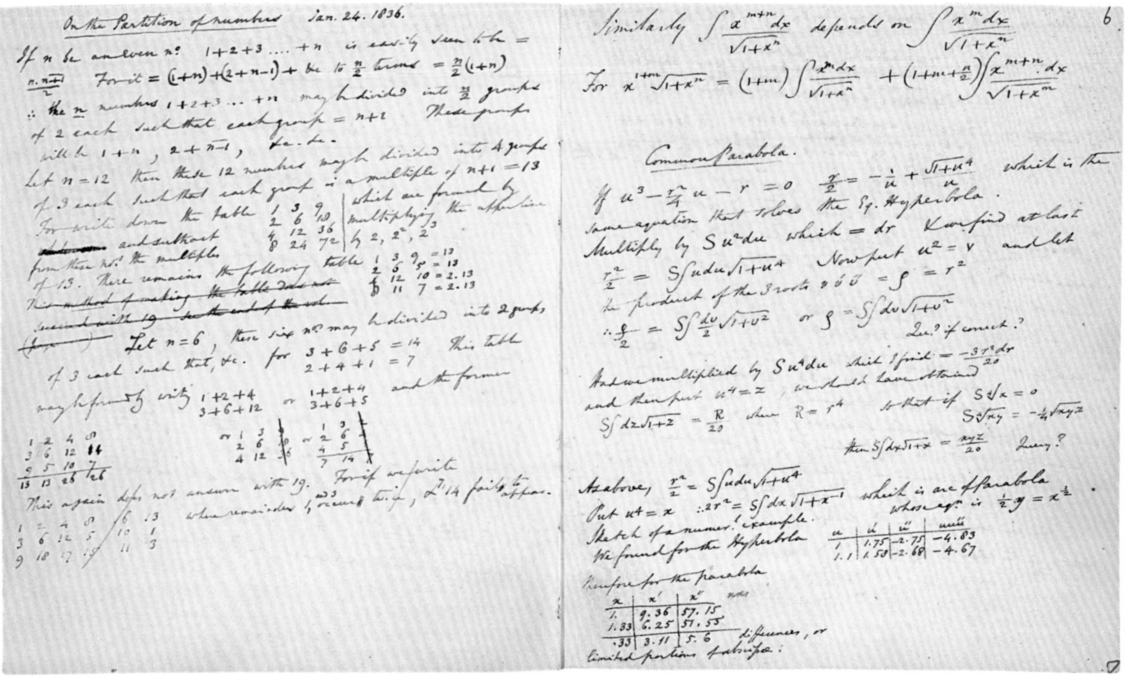

31. William Henry Fox Talbot, the partition of numbers and elliptic integrals, in Notebook N, 1835–1836. Add. MS 88942/1/194, British Library.

is founded," and its publication, partly due to his wife's illness after the birth of their first child in April 1835.[42] In any case, elliptic integrals were not his only academic interest during this period: from 1833 to 1836 he published a number of other papers, mostly on optics and not of a mathematical nature. He also worked on mathematical topics such as the partition of numbers, but he did not publish this work (fig. 31). For reasons that will be explained below, there was no priority issue at stake, which may also have ameliorated any urgency Talbot might otherwise have felt.

Why did Talbot choose the *PTRS* as the venue for these publications? As a generalist science journal published by a society with a then rather meager reputation in pure mathematics, it is not at first sight the most obvious choice for someone who had previously published mathematics in a journal wholly devoted to the subject (even if that journal, no longer in existence, had not in the end lived up to his expectations). It is unlikely that Talbot made his choice in the hope of interesting a large number of fellows in his work. In 1820 Samuel Goodenough, bishop of Carlisle, amateur botanist, and sometime vice president of the society, had remarked that "algebraical pursuits . . . cannot but be in *the reading* very uninteresting, not to say tiresome."[43] Likewise, Herschel wrote in 1830, "In mathematics we have long since drawn the rein, and given over a hopeless race." To reiterate the point, Charles Babbage reproduced Herschel's judgment in his celebrated polemic against the Royal Society later that year.[44]

Certainly there were no comparable British journals, but there were other Continental possibilities.[45] In particular, Crelle's *Journal*, founded in Berlin in 1826, had quickly gained a reputation as the leading mathematics journal. Talbot was familiar with it and, despite Gergonne's disparagement, considered it to be of great value.[46] Moreover, we know that Crelle would have published Talbot's *PTRS* papers, since in November 1837 he asked for Talbot's permission to reprint them, claiming that the *PTRS* was of little use to anybody but mathematicians living in a town with a public library. Talbot acceded to the request, but despite Crelle's assurance that they would be reproduced in volume 19, they never actually appeared.[47] Talbot did admit that that he thought Crelle's journal "too little known" in Britain, but it is unlikely that that would have prevented him from publishing in it.[48] Rather it would seem that Talbot never looked at an alternative to the *PTRS* and that his decision was prompted by the connections he had been cultivating within the Royal Society since his election as a fellow. The society provided not only the ideal confluence of social status and scientific credentials but, more pragmatically, a venue in which to test out his ideas with the two papers he read there in 1834 and ultimately withdrew.[49]

Talbot had first become interested in elliptic integrals at Cambridge, as can be glimpsed in his contributions to Gergonne's *Annales* as well as the mathematical notebooks D to M written on various dates between 1825 and 1835 (fig. 32).[50] As Talbot himself described, his work originated in the publications of the eighteenth-century Italian nobleman Count Giulio Carlo de' Toschi di Fagnano, who in 1718 had found a formula for doubling or bisecting the arc of a lemniscate (a horizontal figure-eight-shaped curve) that involved an elliptic integral.[51] Fagnano's results, which also included theorems on the ellipse and hyperbola, caught the attention of Leonhard Euler, who in 1757, among a welter of papers on the topic, published his famous addition theorem for two elliptic integrals. Hitherto, interest in elliptic integrals had been confined to particular integrals arising in rectification and mechanical problems, but Euler's intervention established them as objects of independent study. Many other mathematicians made contributions, and the whole was gathered together in an organized theory over a forty-year period by the French mathematician Adrien-Marie Legendre, culminating in his three-volume *Traité des fonctions elliptiques* (Treatise on Elliptic Functions) of 1825–28.[52]

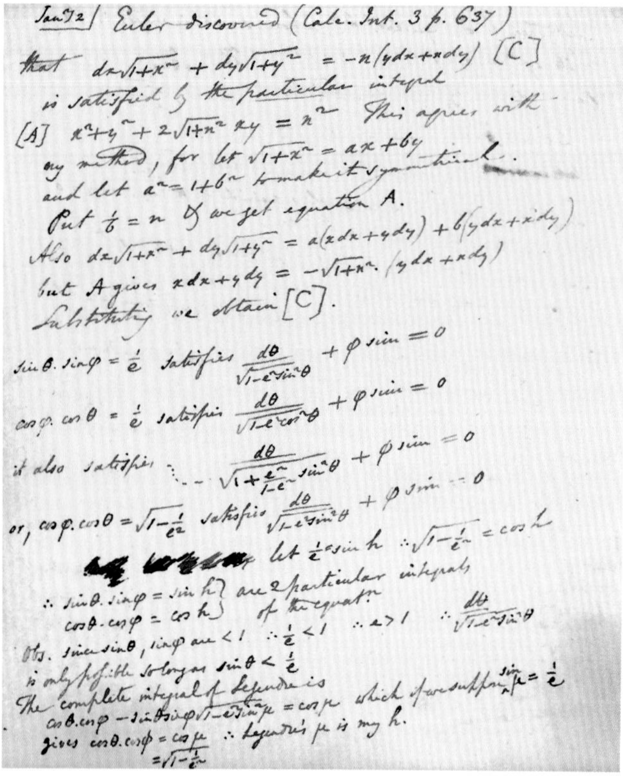

Jan 72] Euler discovered (Calc. Int. 3 p. 637)
that $dx\sqrt{1+x^2} + dy\sqrt{1+y^2} = -n(ydx+xdy)$ [C]
is satisfied by the particular integral
[A] $x^2+y^2+2\sqrt{1+n^2}\,xy = n^2$ This agrees with
my method, for let $\sqrt{1+x^2} = ax+by$
and let $a^2=1+b^2$ to make it symmetrical.
Put $\frac{1}{b} = n$ & we get equation A.
Also $dx\sqrt{1+x^2}+dy\sqrt{1+y^2} = a(xdx+ydy)+b(ydx+xdy)$
but A gives $xdx+ydy = -\sqrt{1+n^2}.(ydx+xdy)$
Substituting we obtain [C].

32. William Henry Fox Talbot, notes on elliptic integrals, in Notebook K, 1833–1834. Add. MS 88942/1/192, Talbot Collection, British Library.

While studying the process by which Fagnano had rectified the hyperbola, Talbot realized that it might "be possible to combine *three* integrals" where Fagnano had combined only two.[53] Indeed, in 1825 Talbot had discovered a new property of the equilateral hyperbola (figs. 33 and 34) that gave the sum of three arcs of the curve, and which he verified by calculating numerical examples.[54] As it turned out, the method that led him to the result could be used for generating many other new results—so many, in fact, that Talbot became stuck trying to classify them or reduce them to a coherent theory. The delay was to prove costly. In 1828, while he was still sorting out his results, the Norwegian mathematician Niels Henrik Abel published an equivalent theorem derived by a different method.[55] Any claim Talbot might have had for priority was immediately lost. Undaunted, he recognized that his method could be applied to a wider class of mathematical problems than Abel's and so continued his work.

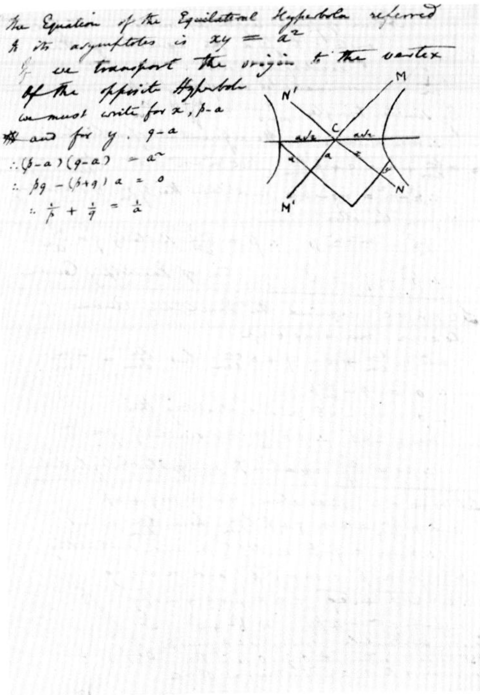

The Equation of the Equilateral Hyperbola referred to its asymptotes is $xy = a^2$

If we transport the origin to the vertex of the opposite Hyperbola we must write for x, $p-a$ and for y, $q-a$

$\therefore (p-a)(q-a) = a^2$

$\therefore pq - (p+q)a = 0$

$\therefore \frac{1}{p} + \frac{1}{q} = \frac{1}{a}$

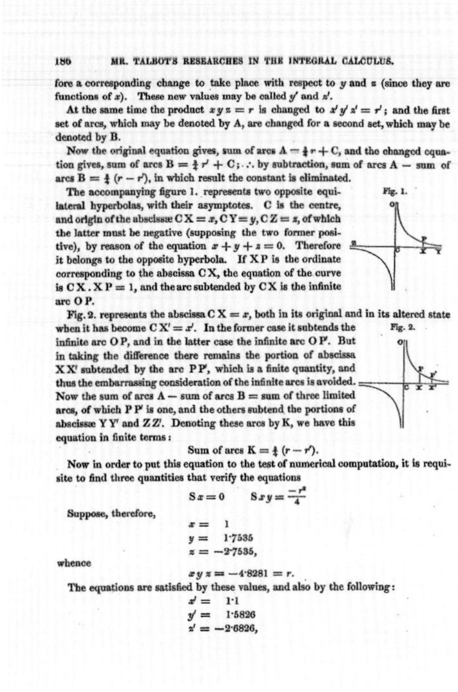

186 MR. TALBOT'S RESEARCHES IN THE INTEGRAL CALCULUS.

fore a corresponding change to take place with respect to y and z (since they are functions of x). These new values may be called y' and z'.

At the same time the product $xyz = r$ is changed to $x'y'z' = r'$; and the first set of arcs, which may be denoted by A, are changed for a second set, which may be denoted by B.

Now the original equation gives, sum of arcs A $= \frac{3}{4}r + C$, and the changed equation gives, sum of arcs B $= \frac{3}{4}r' + C$; $\therefore$ by subtraction, sum of arcs A − sum of arcs B $= \frac{3}{4}(r - r')$, in which result the constant is eliminated.

The accompanying figure 1. represents two opposite equilateral hyperbolas, with their asymptotes. C is the centre, and origin of the abscissæ CX $= x$, CY $= y$, CZ $= z$, of which the latter must be negative (supposing the two former positive), by reason of the equation $x + y + z = 0$. Therefore it belongs to the opposite hyperbola. If XP is the ordinate corresponding to the abscissa CX, the equation of the curve is CX . XP $= 1$, and the arc subtended by CX is the infinite arc OP.

Fig. 1.

Fig. 2. represents the abscissa CX $= x$, both in its original and in its altered state when it has become CX' $= x'$. In the former case it subtends the infinite arc OP, and in the latter case the infinite arc OP'. But in taking the difference there remains the portion of abscissa XX' subtended by the arc PP', which is a finite quantity, and thus the embarrassing consideration of the infinite arcs is avoided. Now the sum of arcs A − sum of arcs B = sum of three limited arcs, of which PP' is one, and the others subtend the portions of abscissæ YY' and ZZ'. Denoting these arcs by K, we have this equation in finite terms:

Fig. 2.

$$\text{Sum of arcs K} = \tfrac{3}{4}(r - r').$$

Now in order to put this equation to the test of numerical computation, it is requisite to find three quantities that verify the equations

$$Sx = 0 \qquad Sxy = \frac{-r^2}{4}$$

Suppose, therefore,

$$x = 1$$
$$y = 1{\cdot}7535$$
$$z = -2{\cdot}7535,$$

whence

$$xyz = -4{\cdot}8281 = r.$$

The equations are satisfied by these values, and also by the following:

$$x' = 1{\cdot}1$$
$$y' = 1{\cdot}5826$$
$$z' = -2{\cdot}6826,$$

33. William Henry Fox Talbot, equilateral hyperbola, in Notebook B, 1822–1824. Add. MS 88942/1/184, Talbot Collection, British Library.

34. William Henry Fox Talbot, equilateral hyperbola, in "Researches in the Integral Calculus: Part I," *Philosophical Transactions of the Royal Society of London* 126 (1836): 186.

Six years elapsed before Talbot had anything to present to the Royal Society. After the couple of false starts alluded to above, he finally submitted his work in March 1836, incorporating part of the paper he had read to the society in 1834.[56] As he had earlier confided to Whewell, he was confident of the results.[57] Restating his original theorem concerning the sum of three arcs of the equilateral hyperbola, he also presented a number of other theorems relating to the sums of the arcs of the conic sections. Peacock, who had been asked to referee the paper, told Talbot that he had read it "with great interest and pleasure," while lamenting that Talbot had "suppressed his discovery for so long."[58] Although Peacock was not a specialist in the subject—he admitted to Talbot that he would have to read Legendre on Abel before he could write his report—he clearly saw the benefit the paper would have brought to the supporters of French analysis in Cambridge had Talbot been able to complete the work in the early 1820s. Talbot's second paper, which was

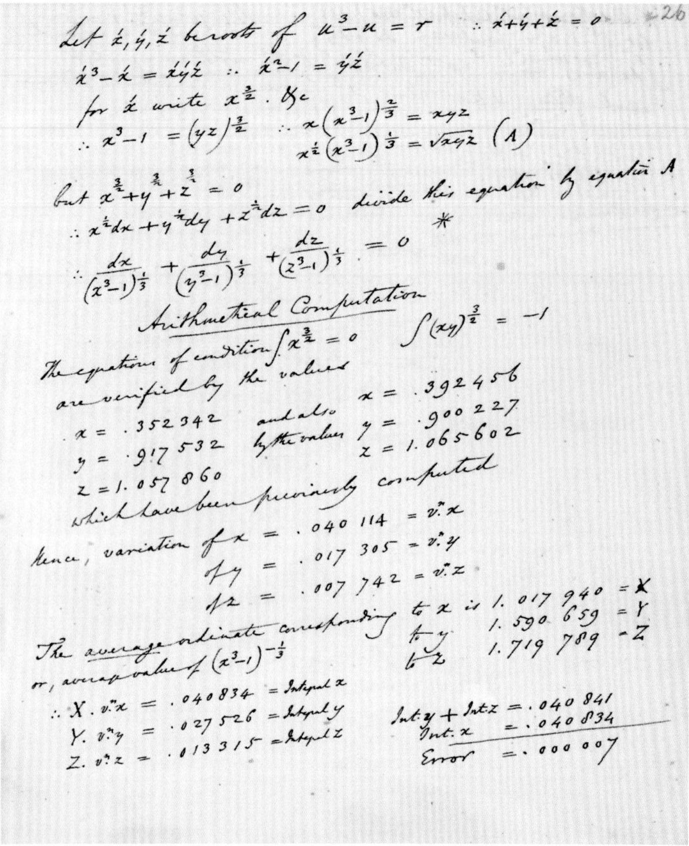

35. William Henry Fox Talbot, part of an example of an arithmetical calculation of a sum of three integrals, in Notebook M, 1834–1835. Add. MS 88942/1/193, Talbot Collection, British Library. A revised form of this calculation also appears in William Henry Fox Talbot, "Researches in the Integral Calculus: Part I," *Philosophical Transactions of the Royal Society of London* (1836): 213–14.

submitted later the same year and a full draft of which still exists, was devoted to examples of the general theorem.[59] Peacock, who was again a referee, considered many of these to be "of great interest and novelty."[60]

However, the fact that Abel had earlier achieved a similar result caused some consternation to those at the Royal Society who had to deal with Talbot's papers, not least because it was quite a struggle for them to come to grips fully with Abel's work, but also because Legendre had only recently put elliptic integrals on a systematic footing. In fact, Talbot had to engage in a lengthy discussion with John Lubbock, the society's vice president, in the pursuance of getting them published.[61]

Although both papers are essentially theoretical, a rather striking feature is the way Talbot, in complete contrast to Abel, used numerical examples to verify his processes of analysis (fig. 35). Indeed, Talbot stressed the value of numerical examples, both for checking solutions and for avoiding errors, when he presented a short account of the first paper at the annual meeting of the British Association for the Advancement of Science in Bristol in 1836.[62] While the inclusion of such examples can be seen as a reflection of his Cambridge training, and in particular his exposure to Whewell's teachings, it can also be perceived as a need to show the potential usefulness of his research.

THE ROYAL MEDAL

As mentioned above, these papers earned Talbot the society's Royal Medal in 1838.[63] The Royal Medal had been founded in 1825, with two medals to be awarded annually "for scientific work of exceptional merit."[64] The only medal to have been awarded for mathematical accomplishment thus far had been presented to James Ivory in the inaugural year of 1826. Thus, on the face of it, one might expect Talbot's papers to contain rather remarkable mathematics. But although they are sound, they contain no new results—essentially they describe a different approach (applied to particular cases) to the general theory already worked out by Abel—and much of their merit lies in the clarity of exposition. Why, then, were Talbot's papers singled out for a medal? In 1846 a rather similar question was asked by Robert Leslie Ellis.[65] If Talbot's papers had appeared in the previous century, he considered, they would have enlarged the boundaries of mathematical science. He added,

> But it cannot be denied that they fall far short of what had been effected at the time they were published, nor does it appear that they contain anything of importance not known before. I have assuredly no wish to speak disparagingly of Mr Talbot; his mathematical writings bear manifest traces of the ability he has shown in so many branches of science. But as in this country they seem to have been thought, and by men not apparently unqualified to judge, to contain great additions to our knowledge, I cannot avoid inquiring whether this be true.[66]

In his subsequent discussion of Talbot's papers, Ellis provided convincing evidence to show that the answer to his inquiry was essentially no: the papers did not "contain great additions to our knowledge."[67] Furthermore, Ellis's comments seem to imply that the award of the medal conferred a status on Talbot's papers that they did not deserve. Thus one is led to ask by what criteria was he awarded the medal? While, like Ellis, we do not wish to disparage Talbot's achievement, it does put a different complexion on the matter to know that only two people were eligible for the medal he won. The regulations in force at the time stipulated that one of the two Royal Medals for 1838 would be given "to the most important unpublished paper in" mathematics, "communicated to the Royal Society for insertion in the Philosophical Transactions, after the present date (November 19, 1835,) and prior to the month of June 1838."[68] As it turned out, the only strictly mathematical papers published during the relevant period were those by Talbot and two by the young and talented Irish mathematician Robert Murphy.[69] A fellow of Caius College, Cambridge, Murphy was, according to the Irish mathematician William Rowan Hamilton, "a mathematician of very high genius."[70] Murphy's papers, one on solutions to equations and the other on operational calculus, are certainly not without interest, and given that Talbot had ceded priority to Abel, there was probably little to choose between them. However, other issues were at stake. Talbot was not only a fellow of the Royal Society; he was a gentleman scholar. Murphy, on the other hand, the son of a shoemaker, was a gambler with a drinking problem whose debts had forced him to leave Cambridge in 1836.[71] There was never any contest.

EDINBURGH PAPERS

For a period of about ten years, from the mid-1850s, Talbot and his family spent their winters in Edinburgh, where he had many mathematical and scientific friends. In 1858 he was elected one of just twenty honorary fellows of the Royal Society of Edinburgh, joining the likes of Whewell, Herschel, Hamilton, and the astronomer royal, George Biddell Airy. So when Talbot began publishing in mathematics again, his papers all appeared in the *Transactions of the Royal Society of Edinburgh* (*TRSE*).

This period also marks a new direction in Talbot's mathematical interests. In the early 1840s he had left elliptic integrals behind and turned his hand to problems in number theory, although his first paper on the subject did not appear until 1857.[72] Like many before him, Talbot was initially attracted to one of its most celebrated problems, Fermat's Last Theorem, producing a short note on a particular case.[73] He followed this with a more general paper presenting a variety of results, including a proof that showed how Fermat's Little Theorem can be deduced from Wilson's Theorem.[74] As in his Royal Society papers, he made rich use of numerical examples in order to verify the theory, although that is where the similarities between the papers end. The questions he was now tackling were at a distinctly lower level than those he had addressed in the 1830s. The fact that he still entertained the hope of recovering Fermat's "lost" proof of the Last Theorem, when some of his contemporaries had good reason for being convinced otherwise, also indicates the current state of his knowledge.[75]

Geometry was also on Talbot's agenda. In 1863 he submitted a paper to the Royal Society of Edinburgh, entitled "The Theory of Parallels," on Euclid's fifth postulate.[76] The question of whether the fifth postulate was in fact a theorem, and so needed to be proved, had greatly vexed mathematicians over the centuries, but in fact the matter had effectively been resolved in the 1830s with the discovery of non-Euclidean geometry. Talbot's paper was already in production—the woodcuts for the diagrams had been prepared—when it was pointed out that he had been anticipated by Legendre.[77] Talbot withdrew the paper, expecting—wrongly—that he could repair it, and now it can be found only in his notebooks.[78]

He had more success with his paper "On Fagnani's [*sic*] Theorem."[79] Although this related to his earlier papers on elliptic integrals inasmuch as it originated in Fagnano's work on the rectification of the ellipse, it was in fact a geometric study of confocal conics. A supplementary note to this paper also included a result that was provided by his son, Charles Henry Talbot, a junior optime in the Mathematical Tripos of 1864.[80]

The subject of Talbot's last geometry paper was Malfatti's Problem: given a triangle, inscribe three circles such that each circle is tangent to the other two and to two sides of the triangle. Like his paper on the parallel postulate, it was flawed by Talbot's incomplete knowledge of earlier developments, although in this case that did not prevent its publication.[81] The problem is named for the Italian mathematician Gianfrancesco Malfatti, who first published it in 1803.[82] The ensuing decades saw many papers on the problem, with, in particular, a geometric (rather than algebraic) proof of the construction being sought. Talbot's paper was also in this vein, aiming to provide an entirely Euclidean-style geometric proof. However, as the American geologist Marcus Baker pointed out in 1877, Talbot was ignorant of certain key publications on the problem, and the truth was that his paper contained nothing new.[83]

Talbot then changed tack once more. His last two papers, published several years apart, concerned solving equations whose solutions are known to have integer form.[84] Although Talbot focused on cubic and quintic equations—that is, equations of the form $ax^3 + bx^2 + cx + d = 0$ and $ax^5 + bx^4 + cx^3 + dx^2 + ex + f = 0$, respectively, where the coefficients a, b, c, and so on are just numbers—his ultimate goal was to develop a method for finding solutions of equations of any degree (not only the third and the fifth). While it is conceivable that such equations could arise within a practical context, Talbot's concern, in common with his other mathematical work, was not with finding a practical solution: "The importance of such a problem is purely theoretical; . . . it is not intended to be used in practice, but is merely a speculation of the mind."[85]

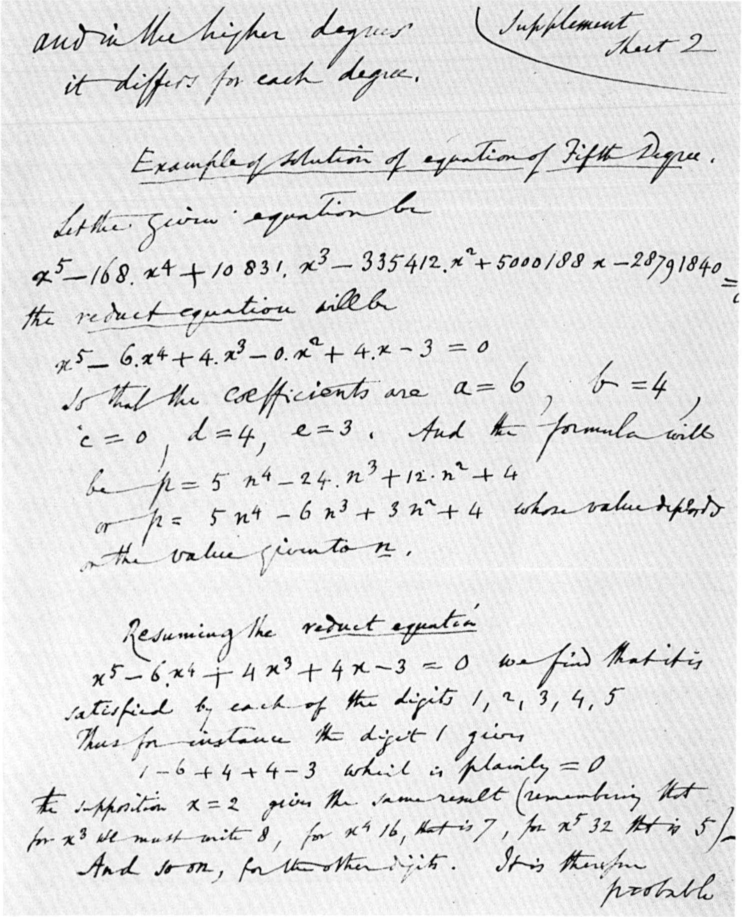
and in the higher degrees
it differs for each degree.
(Supplement
Sheet 2

Example of Solution of equation of Fifth Degree.

Let the given equation be

$x^5 - 168.x^4 + 10831.x^3 - 335412.x^2 + 5000188\,x - 28791840 = 0$

the reduct equation will be

$x^5 - 6.x^4 + 4.x^3 - 0.x^2 + 4.x - 3 = 0$

So that the coefficients are $a = 6$, $b = 4$, $c = 0$, $d = 4$, $e = 3$. And the formula will be $p = 5\,n^4 - 24.n^3 + 12.n^2 + 4$
or $p = 5\,n^4 - 6\,n^3 + 3\,n^2 + 4$ whose value depends on the value given to n.

Resuming the reduct equation
$x^5 - 6.x^4 + 4x^3 + 4x - 3 = 0$ we find that it is satisfied by each of the digits 1, 2, 3, 4, 5
Thus for instance the digit 1 gives
$1 - 6 + 4 + 4 - 3$ which is plainly $= 0$
the supposition $x = 2$ gives the same result (remembering that for x^3 we must write 8, for x^4 16, that is 7, for x^5 32 that is 5)
And so on, for the other digits. It is therefore
probable

36. William Henry Fox Talbot, solution of fifth-degree equation, in Notebook (continuation of "Essay towards a General Solution of Numerical Equations"), Supplement Sheet 2, 1874–1875. Add. MS 88942/1/207, Talbot Collection, British Library. This calculation also appears in William Henry Fox Talbot, "Essay towards a General Solution of Numerical Equations of All Degrees Having Integer Roots," *Transactions of the Royal Society of Edinburgh* 27 (1875): 310–11.

While the topic itself may have had little in common with Talbot's previous mathematical work, his solution was characteristic. Rather than just describe the idiosyncrasies of each case that needed to be considered, he ran riot with numerical examples. Given the proliferation of these often prolix calculations, it is hard not to imagine that Talbot enjoyed doing them (fig. 36).[86] He intended to write a sequel, but it was never completed.

Talbot left notes on almost all his *TRSE* papers. Some represent complete drafts, while others are very much works in progress. The fact that he tackled three quite distinct areas of mathematics—number theory, geometry, and the theory of equations—with no apparent connection between them would seem to indicate that Talbot now had no overall plan for mathematical research, except that his interests, as in other areas, took a decidedly antiquarian turn. Since most of the mathematical papers were self-contained and none of them presented anything remarkable by way of results or methods, it is tempting to see mathematics as a leisure activity for Talbot during this period. His failure to keep properly abreast with the mathematical literature would tend to bear this out. While getting hold of foreign journals cannot always have been easy, he could have turned to his contacts in the London and the Edinburgh Royal Societies for help had he so wished.

As noted in the essay by Anne Secord in this volume, when Talbot left Cambridge he remarked on the advantage in pursuing more than one interest at a time so that he had something to occupy his mind should one of them go wrong, and this idea seems never to have left him. The period when he was contributing to the *TRSE* was exactly when he was also developing his new and intense interest in Assyriology, as discussed in the essay by Eleanor Robson in this volume. Mathematics was the subject with which he was familiar, the subject he had enjoyed all his life. Thus it is not so fanciful to think of it as the subject to which he would now turn for recreation.

CONCLUSION

Talbot's contributions to mathematics are barely remembered by mathematicians. Given that they appeared during a somewhat fallow time for mathematics in Britain, one might expect to find them at least footnoted in histories of the period, but even the recent admirably thorough discussion by Alex D. D. Craik contains no hint of their existence, or indeed of Talbot's.[87] In 1838, however, the Harvard mathematics professor Benjamin Peirce gave a full account of Talbot's Royal Society papers, considering them to be "one of the most fruitful sources for future investigations into the Integral Calculus."[88] Across the English Channel, Joseph Bertrand, professor of analysis at the École Polytechnique in Paris and a lauded teacher, included a clear account of Talbot's work in his treatise of 1870 on the integral calculus.[89] In an article in 1915 on English mathematics, the mathematician George Ballard Mathews drew attention to Talbot—describing him as someone "only vaguely remembered in connection with photography"—and in particular to his Royal Society papers. Although Mathews thought Talbot as a mathematician was "not to be compared with Abel," he was "inclined to believe that the simplest proof of Abel's theorem will ultimately follow the lines that Talbot has indicated." These remarks may now seem rather optimistic, but that was not the point. Rather, Mathews's purpose was to present Talbot as an example of

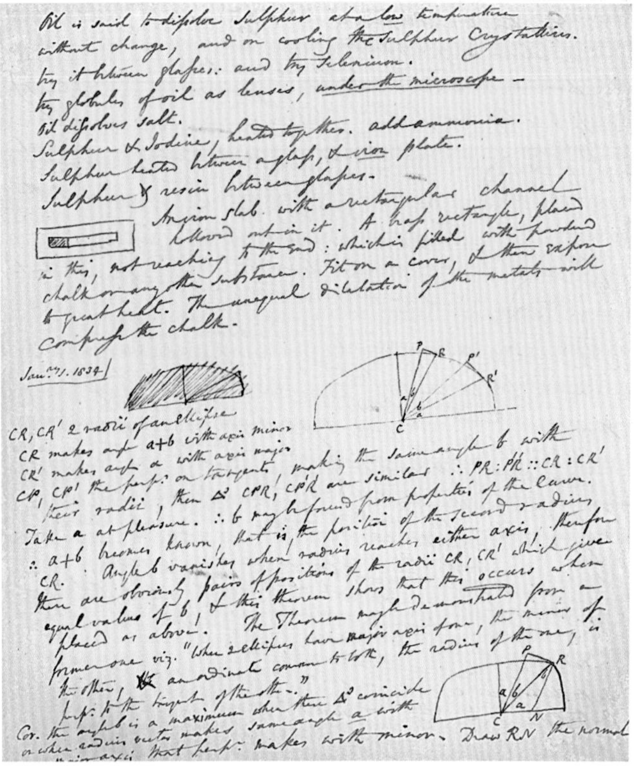
Oil is said to dissolve Sulphur at a low temperature
without change, and on cooling the Sulphur Crystallizes.
try it between glasses: and try Selenium.
try globules of oil as lenses, under the microscope –
Oil dissolves salt.
Sulphur & Iodine, heated together. add ammonia.
Sulphur heated between a glass, & iron plate.
Sulphur & resin between glasses.
An iron slab with a rectangular channel
hollowed out in it. A brass rectangle, placed
in this, not reaching to the end: which is filled with powdered
chalk or any other substance. Fit on a cover, & then expose
to great heat. The unequal dilatation of the metals will
compress the chalk.

Jan.^ry 1. 1834

37. William Henry Fox Talbot, notes on chemistry and an ellipse theorem, in Notebook K, 1833–1834. Add. MS 88942/1/192, Talbot Collection, British Library. This page of notes precedes the page depicted in fig. 32.

an English mathematician "of a distinctly national type, who may, perhaps, be more justly appreciated at home than abroad."[90] Mathews's characterization is well formulated. Although Talbot began his mathematical career publishing in a French journal and corresponding with Continental mathematicians, and continued to keep at least partly abreast of mathematical developments abroad, he confined his later work and mathematical contacts to a more local setting.

In a conventional mathematical story, this would be the point at which to make an assessment of the protagonist's mathematics. But Talbot's mathematical story is not a conventional one. What is significant about Talbot and mathematics is not what he achieved in mathematics but rather what mathematics achieved for him. His mathematical story began in Cambridge, where he not only learned to study and problem-solve in a systematic way but also engaged in mathematics as a collective, albeit competitive, activity with fellow students. As an undergraduate at Trinity under the tutelage of Peacock and Whewell, he was exposed to the excitement of new ideas by two of the leading—but different—exponents of the Continental analysis, and he bore witness to Cambridge mathematics beginning to break free from its Newtonian constraints. He learned mathematics that made him keen to engage with the European mathematical community, and he learned mathematics and natural philosophy that provided him with the necessary tools to develop his interests in optics and astronomy.[91]

With publications in Gergonne's *Annales* under his belt and with European recognition—and in the early 1820s there were not many compatriots who could claim as much—Talbot could hold his own in wider scientific circles from a young age. By the 1830s, when he came to complete his most advanced papers, his mathematical credentials were firmly established, as indeed was his mathematical methodology. The Cambridge model of theory backed up with worked examples, all carefully and clearly laid out, had never left him (fig. 37). Although in essence his Royal Society papers had nothing new to say—he had let his ideas ferment for too long for them to be original—his careful exposition had enabled their subject, elliptic integrals, to become accessible to a wider audience.

For the next twenty years Talbot did no original research in mathematics, but he certainly retained his interest in the subject. In 1844 he told Herschel that elliptic integrals were a "a great favourite" of his and that he had in fact "advanced the subject a good deal further" than it had appeared in the memoirs, but he could never find time to put the material in order for publication, although he very much wanted to continue it.[92] During this period Talbot also received letters from the mathematician James Joseph Sylvester, who in 1851 sent him his recent essay on canonical forms in the hope of encouraging him to engage in some algebraic investigation.[93] But maintaining a keen interest in mathematics and being in touch with some of the country's leading mathematicians are not the same as doing original research. When Talbot started publishing again, he was on a mathematically more pedestrian track, and his interest in new ideas seems to have dried up. While technically there is nothing wrong with his Edinburgh papers, they come across rather as a diversion from the main event, especially when compared to his earlier ones, his paper on Malfatti's Problem being a case in point. To be sure, his solution to the problem filled a gap that he (mistakenly) thought still needed to be filled, but his choice of a problem that relied on Euclidean geometry for its answer could be seen as somewhat conservative. Still, that is not to deny the contribution of these papers to Talbot's scientific life. Publishing in the *TRSE* made Talbot an active participant in the Royal Society of Edinburgh and in the wider society of mathematicians at large. The type and variety of problems Talbot dealt with meant that he was never in danger of being too absorbed by them, leaving him plenty of free time to continue with his other interests.

1. *Monthly Notices of the Royal Astronomical Society* 38 (1878): 148.
2. William Wallace to George Peacock, ca. 1833, in Maria Panteki, "William Wallace: Letter to George Peacock," *Historia Mathematica* 14 (1987): 122. For Talbot's earliest surviving exercise book—from 1809, when he was only nine years old, and containing geometric and astronomical diagrams—see Mirjam Brusius, "Beyond Photography: An Introduction to William Henry Fox Talbot's Notebooks in the Talbot Collection at the British Library," *Electronic British Library Journal* (2010), http://www.bl.uk/eblj/2010articles/pdf/ebljarticle142010.pdf.
3. Lady Elisabeth Feilding to William Henry Fox Talbot, May 19, 1817, Doc. No. 767, The Correspondence of William Henry Fox Talbot, http://foxtalbot.dmu.ac.uk (accessed Feb. 14, 2012), hereafter "Correspondence" (original: Talbot Collection, hereafter "TC," British Library, hereafter "BL"); see also Talbot to Lady Elisabeth, May 18, 1817, Doc. No. 766, Correspondence (original: TC, BL).
4. See Andrew Warwick, *Masters of Theory: Cambridge and the Rise of Mathematical Physics* (Chicago: Univ. of Chicago Press, 2003).
5. For a discussion of private tutoring and the rise of the mathematical "coach" in the nineteenth century, see Warwick, *Masters of Theory*, esp. chap. 5.
6. Harvey W. Becher, "Woodhouse, Babbage, Peacock and Modern Algebra," *Historia Mathematica* 7, no. 4 (1980): 389–400; and Philip C. Enros, "The Analytical Society (1812–1813): Precursor of the Renewal of Cambridge Mathematics," *Historia Mathematica* 10, no. 1 (1983): 24–37.
7. Sylvestre François Lacroix, *An Elementary Treatise on the Differential and Integral Calculus*, trans. Charles Babbage, John Herschel, and George Peacock, with an appendix and notes (Cambridge: J. Deighton and Sons, 1816).
8. Talbot to Charles Feilding, Jan. 14, 1820, Doc. No. 862, Correspondence (original: TC, BL).
9. Isaac Todhunter, *William Whewell, D.D., Master of Trinity College, Cambridge: An Account of His Writings; With Selections from His Literary and Scientific Correspondence*, 2 vols. (London: Macmillan, 1876), 2:16.
10. See Harvey W. Becher, "William Whewell and Cambridge Mathematics," *Historical Studies in Physical Sciences* 11 (1980): 14–19. On Whewell's philosophy of science and other aspects of his life and work, see Menachem Fisch and Simon Schaffer, eds., *William Whewell: A Composite Portrait* (Oxford: Oxford Univ. Press, 1991).
11. William Whewell, *A Treatise on Dynamics: Containing a Considerable Collection of Mechanical Problems* (Cambridge: J. Deighton and Sons, 1823), v.
12. Solomon Atkinson, "Struggles of a Poor Student through Cambridge," *London Magazine and Review*, April 1, 1825, 491–510; and J. A. Venn, *Alumni Cantabrigiensis*, pt. 2 (1752–1900), 6 vols. (Cambridge: Cambridge Univ. Press, 1940–54).
13. Atkinson, "Struggles of a Poor Student," 492.
14. Atkinson, "Struggles of a Poor Student," 510.
15. Atkinson, "Struggles of a Poor Student," 504.
16. Giovanni Percij to Talbot, Dec. 31, 1823, Doc. No. 1142 (original: TC, BL).
17. Talbot to Lady Elisabeth Feilding, April 30, 1819, Doc. No. 847, Correspondence (original: TC, BL). Higman was held in high esteem by Whewell. See, for example, Todhunter, *William Whewell*, 1:11.
18. Talbot to Lady Elisabeth Feilding, Aug. 4, 1819, Doc. No. 852, Correspondence (original: TC, BL).
19. Talbot to Charles Feilding, Feb. 15, 1820, Doc. No. 864, Correspondence (original: TC, BL).
20. Talbot to Charles Feilding, (undated) 1821, Doc. No. 905, Correspondence (original: TC, BL).
21. *Annales de Gergonne*, http://www.numdam.org/numdam-bin/feuilleter?j=AMPA (accessed Feb. 15, 2012).
22. See the footnote by Gergonne in *Annales de mathématiques pures et appliquées* 14 (1823–24): 23, which reveals that Talbot was the author of two problems (requiring the proofs of theorems concerning the arcs of a lemniscate, or horizontal figure-eight-shaped curve) published in *Annales de mathématiques pures et appliquées* 13 (1822–23): 248.
23. John Martin Frederick Wright, *Alma Mater, or Seven Years at the University of Cambridge*, 2 vols. (London: Black, Young and Young, 1827), 27.

24. Duplicating the cube and trisecting the angle are two of the "three classical problems" of Greek geometry, the third being squaring the circle. The Greeks recognized the impossibility of solving these construction problems using only a straightedge and compass but were unable to provide proofs. The quest for proofs led to many new developments in geometry, including the construction of new curves. The proofs came only later in the nineteenth century.
25. See William Henry Fox Talbot, "Questions Résolues: Rectification de l'énoncé du problème de géométrie proposé à la page 321 du XIIe volume des Annales, et traité à la page 115 du présent volume, et solution complète de ce problème," *Annales de mathématiques pures et appliquées* 13 (1822–23): 242–47. See also Joseph-Diez Gergonne to Talbot, June 6, 1822, Doc. No. 977 (original: TC, BL).
26. An envelope of a family of curves in the plane comprises a curve that is tangent to each member of the family at some point.
27. That is, a series in which each term is a trigonometric function, for example, sinx + cosx + sin2x + cos2x + . . . See William Henry Fox Talbot, "Solution des problèmes d'analise transcendante, proposes à la 247 du XIIIe volume du présent recueil," *Annales de mathématiques pures et appliquées* 14 (1823–24): 88–95, 187–90.
28. Christian Gerini, *W. H. F. Talbot (1800–1877): Le mathématicien photographe* (Pierrefeu du Var, France: Editions Nouvelles Mémoires, 2012).
29. William Henry Fox Talbot, "Questions Résolues: Addition à l'article de la page 207 du présent volume," *Annales de mathématiques pures et appliquées* 14 (1823–24): 380–81.
30. Arthur Cayley, "Sur la surface qui est l'enveloppe des plans conduits par les points d'un ellipsoide perpendiculairement aux rayons menés par le centre," *Annali di Matematica pura ed applicata* 2 (1859): 3.
31. Joseph-Diez Gergonne to Talbot, Dec. 16, 1826, Doc. No. 1512, Correspondence (original: TC, BL).
32. Talbot to John Lubbock, Dec. 1834, Doc. No. 3014, Correspondence (original: Lub 38 T4, Royal Society, London, hereafter RSA).
33. Talbot to John Frederick William Herschel, Oct. 29, 1827, Doc. No. 1607, Correspondence (original: HS 17: 267, RSA).
34. Notebooks A–F, Add. MSS 88942/1/183–88, TC, BL.
35. See Notebook B, Add. MS 88942/1/184, TC, BL.
36. Larry J. Schaaf, "William Henry Fox Talbot (1800–1877)," in *Oxford Dictionary of National Biography*, online ed., Sept. 2010, http://www.oxforddnb.com/view/article/26946 (accessed Nov. 29, 2011).
37. P. Waters to Talbot, Aug. 25, 1825, Doc. No. 1303, Correspondence (original: TC, BL). There is also a list of books contained in a letter of 1823, and although these books are predominantly mathematical, several relate to botany and the classics. Giovanni Percij to Talbot, Dec. 31, 1823, Doc. No. 1142, Correspondence (original: TC, BL).
38. Talbot was clearly a joiner of scientific societies. In the early 1820s he joined the recently formed Royal Astronomical Society and the Cambridge Philosophical Society. He was also a life member of the British Association for the Advancement of Science, and in 1836, when its annual meeting was held in Bristol, he hosted a house party with guests including Babbage and Whewell. See H. J. P. Arnold, *William Henry Fox Talbot: Pioneer of Photography and Man of Science* (London: Hutchinson Benham, 1977), 80. Later he joined the Royal Asiatic Society and was a founding member of the Society of Biblical Archaeology; see the essay by Eleanor Robson in this volume.
39. Talbot probably first met Faraday in 1824, when he joined the embryonic Athenaeum Club, of which Faraday was the first secretary. See Talbot to Lady Elisabeth Feilding, March 5, 1824, Doc. No. 1174, Correspondence (original: TC, BL).
40. William Henry Fox Talbot, "Researches in the Integral Calculus: Part I," *Philosophical Transactions of the Royal Society of London* 126 (1836): 177–215; and Talbot, "Researches in the Integral Calculus: Part II," *Philosophical Transactions of the Royal Society of London* 127 (1837): 1–18.
41. William Henry Fox Talbot, "On a New Property of the Arcs of the Equilateral Hyperbola," *Abstracts of the Papers Printed in the Philosophical Transactions of the Royal Society of London* 3 (1834): 258; and Talbot, "On the Arcs of Certain Parabolic Curves," *Abstracts of the Papers Printed in the Philosophical Transactions of the Royal Society of London* 3 (1834): 287–88.

42. Talbot, "Integral Calculus: Part I." For Talbot's explanation of the delay in publication, see Talbot to John William Lubbock, Aug. 30, 1835, Doc. No. 3123, Correspondence (original: LUB 38 T8, RSA).
43. Quoted in Warwick, *Masters of Theory*, 36–37.
44. Charles Babbage, *Reflections on the Decline of Science in England: And Some of Its Causes* (London: B. Fellowes, 1830), viii.
45. No British journal was devoted solely to higher-level mathematics until 1865, when the newly founded London Mathematical Society began publishing its *Proceedings*. The *Transactions of the Cambridge Philosophical Society* (1821–) published mathematical articles only rarely.
46. Talbot to John William Lubbock, Dec. 19, 1834, Doc. No. 3025, Correspondence (original: LUB 38 T2, RSA).
47. August Leopold Crelle to Talbot, Nov. 24, 1837, Doc. No. 3619 (original: TC, BL). Crelle apologized to Talbot for the delay, citing lack of space in the journal, so perhaps he never did manage to find space for them.
48. Talbot to John William Lubbock, Dec. 19, 1834, Doc. No. 3025, Correspondence (original: Lub 38 T2, RSA).
49. In the context of the 1839 election for president of the Royal Society, he wrote to his mother: "There is a sad dearth of scientific aristocracy; & in a country, the mainsprings of whose wealth and commerce are so intimately connected with the improvements of science!" See Talbot to Lady Elisabeth Feilding, Nov. 17, 1838, Doc. No. 3756, Correspondence (original: TC, BL).
50. Talbot, "Integral Calculus: Part I," 181; Talbot, "Démonstration du théorème de géométrie énoncé à la page 212 du présent volume," *Annales de mathématiques pures et appliquées* 13 (1822): 319–20; Talbot, "Questions Résolues, 380–81; and Notebooks A–K, Add. MSS 88942/1/183–92, TC, BL. For a discussion of Talbot's notebooks, see Brusius, "Beyond Photography."
51. Talbot wrote "Fagnani" in his published works. See the sources cited in notes 40 above (part I), 79 and 80 below.
52. Although Legendre used the term "elliptic functions," he studied what today we call elliptic integrals, the inverse to elliptic functions. For the history of elliptic integrals and elliptic functions, see Adrian Rice, "In Search of the 'Birthday' of Elliptic Functions," *Mathematical Intelligencer* 30, no. 2 (2008): 48–56.
53. Talbot, "Integral Calculus: Part I," 181.
54. William Henry Fox Talbot, "A Brief Account of Some Researches in the Integral Calculus," in *Notices and Abstracts of Communications to the British Association for the Advancement of Science at the Bristol Meeting, August 1836* (London: John Murray, 1837), 1–4.
55. Niels Henrik Abel, "Remarques sur quelques propriétés générales d'une certaine sorte de fonctions transcendantes," *Journal für die reine und angewandte Mathematik* 3 (1828): 313–23.
56. Talbot, "Integral Calculus: Part I"; and Talbot, "On the Arcs of Certain Parabolic Curves."
57. Talbot to William Whewell, Oct. 7, 1835, Doc. No. 5515 (original: Add. MS a 213/95, Trinity College Library, Univ. of Cambridge).
58. George Peacock to Talbot, May 23, 1836, Doc. No. 3286, Correspondence (original: TC, BL).
59. Talbot, "Integral Calculus: Part II." For the draft, see untitled notebook, Add. MS 88942/1/218, TC, BL.
60. "Rev. Geo. Peacock's Report on Mr. Fox Talbot's Paper (Researches in the Integral Calculus) Part II," RR.1.238, RSA.
61. See, for example, Talbot to John William Lubbock, Nov. 25, 1834, Doc. No. 3011, Correspondence (original: LUB 38 T1, RSA).
62. Talbot, "Some Researches in the Integral Calculus," 4.
63. *Abstracts of the Papers Printed in the Philosophical Transactions of the Royal Society of London* 4 (1838): 101–2.
64. For a history of the Royal Medals, 1826–1914, see Roy MacLeod, "Of Medals and Men: A Reward System in Victorian Science," *Notes and Records of the Royal Society of London* 26 (1979): 81–105.
65. Ellis, who had been senior wrangler in 1840, was a fellow of Trinity College, Cambridge.
66. Robert Leslie Ellis, "Report on the Recent Progress of Analysis (Theory of the Comparison of Transcendentals) for the British Association for the Advancement of Science," in *The Mathematical and Other Writings of Robert Leslie Ellis*, ed. W. Walton (Cambridge: Deighton, Bell, 1863), 249–50.

67. Ellis, "Recent Progress of Analysis," 250–52.

68. The original notice, published in the *Philosophical Transactions of the Royal Society of London* 126 (1836): ix, stated "Physics," but an erratum was later inserted, in *Philosophical Transactions of the Royal Society of London* 127 (1837): vi.

69. By "strictly mathematical" is meant papers in pure mathematics, excluding, for example, James Ivory's paper of 1838 on the gravitational attraction of ellipsoids.

70. William Rowan Hamilton, "On the Argument of Abel, Respecting the Impossibility of Expressing a Root of Any General Equation above the Fourth Degree, by Any Finite Combination of Radicals and Rational Functions," *Transactions of the Royal Irish Academy* 18 (1839): 256.

71. Alex D. D. Craik, *Mr Hopkins' Men: Cambridge Reform and British Mathematics in the 19th Century* (London: Springer-Verlag, 2007), 169–70.

72. See Philip Kelland to Talbot, April 2, 1842, Doc. No. 4479, Correspondence (original: TC, BL); and James Joseph Sylvester to Talbot, Oct. 12, 1847, Doc. No. 6375, Correspondence (original: TC, BL).

73. William Henry Fox Talbot, "On Fermat's Theorem," *Transactions of the Royal Society of Edinburgh* 21 (1857): 403–6. Fermat's Last Theorem states that the equation $x^n + y^n = z^n$ has no non-zero integer solutions for $n > 2$. In about 1630 Pierre de Fermat famously recorded that he had a proof of the theorem but that his margin was not big enough to contain it. Talbot dealt with the case when one of x, y, z is a prime number. The theorem was eventually proved by Andrew Wiles in 1994.

74. William Henry Fox Talbot, "On the Theory of Numbers," *Transactions of the Royal Society of Edinburgh* 23 (1862): 45–52.

75. Talbot, "On Fermat's Theorem," 403.

76. Euclid's fifth postulate states: If a straight line falling on two straight lines makes the interior angles on the same side less than two right angles, then the straight lines, if produced indefinitely, meet on the side containing these angles. It is often called the parallel postulate, even though it makes no specific mention of parallel lines; this is because of the way it is used in the *Elements*.

77. Talbot to John Hutton Balfour, April 9, 1863, and May 17, 1863, Doc. Nos. 8682 and 317, Correspondence (original: Balfour Correspondence, v. XI T11, T22, Royal Botanic Garden Edinburgh). In 1864 Philip Kelland published a paper that bore the same title as Talbot's but contained a rather speculative discussion on the nature of non-Euclidean geometry.

78. Notebook, "Some Additional Remarks [on the Theory of Parallel Lines]," Add. MS 88942/1/204, TC, BL; Notebook, "On the Theory of Parallel Lines," Add. MS 88942/1/205, TC, BL.

79. William Henry Fox Talbot, "On Fagnani's Theorem," *Transactions of the Royal Society of Edinburgh* 23 (1863): 285–98.

80. William Henry Fox Talbot, "Note on Confocal Conic Sections," *Transactions of the Royal Society of Edinburgh* 24 (1865): 53–57.

81. William Henry Fox Talbot, "Researches on Malfatti's Problem," *Transactions of the Royal Society of Edinburgh* 24 (1865): 127–38.

82. Malfatti originally published a three-dimensional form of the problem. He then assumed, incorrectly as it turned out (although this was proved only in 1994), that this problem could be reduced to the two-dimensional planar one that now bears his name. See Marco Andreatta, András Bezdek, and Jan P. Boroński, "The Problem of Malfatti: Two Centuries of Debate," *Mathematical Intelligencer* 33 (2010): 72–76.

83. Marcus Baker, "The History of Malfatti's Problem," *Bulletin of the Washington Philosophical Society* 20 (1881): 113–23. I am grateful to Jemma Lorenat of Simon Fraser University for drawing my attention to Baker's critique of Talbot's paper.

84. William Henry Fox Talbot, "Some Mathematical Researches on Cubic Equations," *Transactions of the Royal Society of Edinburgh* 24 (1867): 573–90; and Talbot, "Essay towards a General Solution of Numerical Equations of All Degrees Having Integer Roots," *Transactions of the Royal Society of Edinburgh* 27 (1875): 303–12.

85. Talbot, "Researches on Cubic Equations," 574.

86. See Notebook, "Some Mathematical Researches," Add. MS 88942/1/206, TC, BL; and untitled notebook, Add. MS 88942/1/209, TC, BL; they contain drafts and notes for these two papers.
87. Craik, *Mr Hopkins' Men.*
88. Benjamin Peirce, "An Account of Mr Talbot's 'Researches in the Integral Calculus' with a More General Solution of the Principal Problem," *Mathematical Miscellany* 1 (1838): 407.
89. Joseph Bertrand, *Traité de calcul différential et de calcul intégral*, vol. 2 (Paris: Gauthier-Villars, 1870), 382–88.
90. George Ballard Mathews, "English Mathematics," *Nature* 95 (1915): 219.
91. Although Talbot's papers in optics and astronomy are not explicitly mathematical in nature, they do contain the occasional mathematical diversion, such as, for example, a use of the integral calculus to calculate the average brightness of a luminous zone. See William Henry Fox Talbot, "Experiments on Light," *London and Edinburgh Philosophical Magazine and Journal of Science* 5 (1834): 332.
92. Talbot to John Frederick William Herschel, April 8, 1844, Doc. No. 5058, Correspondence (original: HS 17:317, RSA).
93. James Joseph Sylvester to Talbot, July 8, 1851, Doc. No. 6438, Correspondence (original: TC, BL).

In the Shadow of Scott: Talbot and the Wizard of the North

Graham Smith

In March 1813, during his second year at Harrow, William Henry Fox Talbot wrote a remarkable letter to his mother, Lady Elisabeth Feilding (fig. 38).[1] He was barely thirteen at the time, and his letter offers a somewhat breathless account of what he had been doing, interspersed with requests for a phosphorus box, matches, mercury, dates, and French plums. Six lines in the letter concern books, and they record that Talbot had been reading Walter Scott's *The Lady of the Lake* and Shakespeare's *King John*, *Richard II*, *Henry IV*, and *Henry V*. Talbot was evidently a precocious child, but his letter also shows a typical public schoolboy's preoccupation with things to eat—dates, plums, and pancakes—and mentions an injury of the sort that is commonplace among young boys. The letter is worth quoting in full, for it brings Talbot to life while also introducing the principal topic addressed in this paper: his admiration for Scott's writings from his school days until 1845, when he issued what he termed, in a letter to the London publisher John Murray, a "photographic work called Sun Pictures in Scotland."[2]

> My dear Mamma,
>
> I want my blue phosphorus box extremely, so pray send it. I think you had better send me with it 100 of the matches, which is only 8[d] I believe, and ask [Friedrich] Accum whether the stuff they are dipped into, is sulphuric acid or not. — I forget where the key of the black trunk was put — Has M[r] Lemons ticket No 3185 I think, come up a blank or a prize? I have been making some nitrate of quicksilver. It forms pretty yellow crystals, by heating which in the candle, they turn to a bright red powder, which when it is taken away turns yellow-orange colour. Pray send me a *little* good quicksilver, which can be got at Accum's & I will try & make vermillion out of it for you. — I have also been dissolving a halfpenny in nitric acid it makes lovely little crystals, blue, of this size [illustration], of this sort of shape but smaller, [illustration]. I have got a small steel bow — I have just finished the Lady of the Lake. I read all the notes. I have also read these plays of Shakespeare.

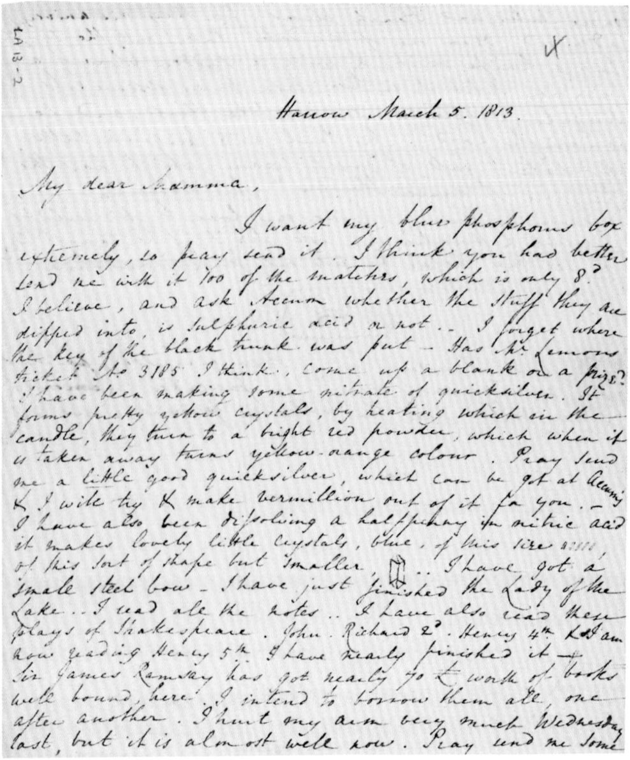

LA13-2

X

Harrow March 5. 1813.

My dear Mamma,

I want my blue phosphorus box extremely, so pray send it. I think you had better send me with it 100 of the matches, which is only 8d I believe, and ask Accum whether the stuff they are dipped into, is Sulphuric acid or not.. I forget where the key of the black trunk was put – Has Mr Lemon's ticket No 3185 I think, come up a blank or a prize? I have been making some nitrate of quicksilver. It forms pretty yellow crystals, by heating which in the candle, they turn to a bright red powder, which when it is taken away turns yellow-orange colour. Pray send me a little good quicksilver, which can be got at Accum's & I will try & make vermillion out of it for you. – I have also been dissolving a halfpenny in nitric acid it makes lovely little crystals, blue, of this size of this sort of shape but smaller .. I have got a small steel bow – I have just finished the Lady of the Lake.. I read all the notes.. I have also read these plays of Shakespeare. John. Richard 2d. Henry 4th & I am now reading Henry 5th I have nearly finished it – Sir James Ramsay has got nearly 70 £ worth of books well bound, here. I intend to borrow them all, one after another. I hurt my arm very much Wednesday last, but it is almost well now. Pray send me some

38. William Henry Fox Talbot, first page of letter to his mother, Lady Elisabeth Feilding, from Harrow School, March 5, 1813. Letter 27704, Add. MS 88942, Talbot Collection, British Library (partially catalogued).

John. Richard 2d. Henry 4th & I am now reading Henry 5th. I have nearly finished it — Sir James Ramsay has got nearly 70 £ worth of books well bound, here. I intend to borrow them all, one after another. I hurt my arm very much Wednesday last, but it is almost well now. Pray send me some more dates and French plums, for they are universally admired. Open one of the dates, take out the stone lay a knife on it, & with a sudden blow of a hammer split it in two, then pray observe its singular substance, so very like flint, and so amazingly hard. My Paul and Virginia calls the date "full of sweet cream, which has the perfume of orange flowers." My dates had no cream in them — Give my love to Mr Feilding Caroline & Horatia; & ask Mr F. if he cannot come & see me, before he sets out for Rutlandshire —

Your Affectionate Son

W H F Talbot

P.S. — Last Tuesday being Shrove Tuesday we had our anniversary dinner on pancakes — Lemons & sugar [rolled] over the tables in promiscuous confusion.

THE LADY OF THE LAKE

Talbot's reference to *The Lady of the Lake* is noteworthy, for the poem had been published less than three years before, when it became Scott's first great success and led in 1813 to the prince regent's offering him the position of poet laureate.[3] Readers were transfixed by the poem's narrative and enchanted by Scott's descriptions of the wild beauty of The Trossachs and Loch Katrine, with the result that hordes of sightseers traveled there to experience the magic for themselves. In 1811 James Hogg, "the Ettrick Shepherd," poked fun at this phenomenon, writing:

> Mr Scott has superseded the possibility of ever more pleasing, by a second description of the Trossacks, but in so doing he has certainly added to the pleasure arising from a view of them. Whoever goes to survey the Trossacks, let him have the 11th, 12th, and 13th divisions of the first canto of the *Lady of the Lake* in his heart; a little Highland whisky in his head; and then he shall see the most wonderful scene that nature ever produced.[4]

Talbot's juxtaposition of Scott and Shakespeare was also prescient, for in 1871, the centenary of Scott's birth, the *Illustrated London News* would pronounce him "the greatest original author of the nineteenth century," equal in stature to Homer and Shakespeare.[5]

Talbot may have memorized sections of *The Lady of the Lake*, for this was customary in public schools at the time.[6] It was certainly the case that he remembered the poem some thirteen years later, for, writing from the Adriatic coast of Italy after having spent several weeks exploring the Ionian Islands, he quoted eight lines from the first canto of *The Lady of the Lake* to evoke the beauties of the island of Corfu.[7] "It is a lovely wilderness," he observed in a letter to his mother, "quite unlike any other place I have seen, and one thinks as one rambles thro' it, what a Paradise it *might* be." Next he transcribed the lines from the fifteenth stanza in which James Fitz-James looks out over Loch Katrine:

> From the steep promontory gazed
> The stranger raptured and amazed
> And, What a scene were here! He cried,
> For princely pomp or churchman's pride!
> On this bold brow, a lordly tower;
> In that soft vale, a lady's bower;
> On yonder meadow, far away,
> The turrets of a Cloister grey.

Talbot closed this quotation with a fragment from the first lines of the sixteenth stanza: "Blithe were it *then* to wander here! / But *now*! . . . Walter Scott."[8]

It is difficult today to comprehend the extent of Scott's fame in the first decades of the nineteenth century, but readers of Jane Austen's *Sense and Sensibility*, published in 1811, will remember how Elinor Dashwood teases her sister Marianne after John Willoughby calls at their cottage: "You know what he thinks of Cowper and Scott; you are certain of his estimating their beauties as he ought, and you have received every assurance of his admiring Pope no more than is proper."[9] Three years later, in *Mansfield Park*, Fanny Price quotes from Scott's narrative poem *The Lay of the Last Minstrel* when she and Edmund Bertram stand in the barren chapel at Sotherton Court.[10] And in *Persuasion*, published in 1818, Anne Elliott and Captain Benwick discuss the respective merits of *The Lady of the Lake* and another epic poem by Scott, *Marmion*.[11] These episodes show that familiarity with Scott was a sign of accomplishment among young women during this period and constituted a form of currency in social exchanges between the sexes. In short, Scott's poetry was woven into the fabric of late Georgian and Regency society.[12]

THE WAVERLEY NOVELS

The publication in 1814 of *Waverley*, a historical novel set in Scotland during the Jacobite Rebellion of 1745, inaugurated a new phase in Scott's career. Thereafter his fame was largely constructed upon his novels: *Guy Mannering*, *The Antiquary*, *Rob Roy*, *The Heart of Midlothian*, *The Bride of Lammermoor*, *Ivanhoe*, *The Abbot*, *The Fortunes of Nigel*, *Quentin Durward*, *Redgauntlet*, and *Woodstock*. As Thomas Carlyle put it in "Sir Walter Scott," an essay inspired in 1838 by John Gibson Lockhart's *Memoirs of the Life of Sir Walter Scott, Bart.*, the "'Waverley series,' swift-following one on the other apparently without end, was the universal reading; looked for like an annual harvest, by all ranks, in all European countries."[13]

Carlyle questioned whether Scott had been "a great man" but acknowledged that "there can be no question with any one that he was a most noted and even notable man." He added:

> In this generation there was no literary man with such a popularity in any country; there have only been a few with such, taking in all generations and all countries. Nay, it is farther to be admitted that Sir Walter Scott's popularity was of a select sort rather; not a popularity of the populace. His admirers were at one time almost all the intelligent of civilised countries; and to the last included, and do still include, a great portion of that sort.[14]

Later in the same essay Carlyle observed, "In a word, the Waverley Novels circulated and reigned triumphant; to the general imagination the 'Author of *Waverley*' was like some living mythological personage, and ranked among the chief wonders of the world."[15]

Talbot and his family, like most members of their class, admired the Waverley Novels and read them as they appeared. In August 1819, for instance, in a letter to Lady Elisabeth, written when he was in his second year at Cambridge, Talbot volunteered that he had just read *The Bride of Lammermoor* and liked it very much.[16] He wrote similarly from Cambridge on December 26, asking: "Have you read *Ivanhoe*? I have no time to read it now & have merely gazed upon the outside. By the report of all who read it, it is very good."[17] References to Scott's prose writings continue during the 1820s, appearing repeatedly in correspondence with his mother. In 1828, for instance, he wrote, "I liked Walter Scott's 2^d series of the Canongate '*the fair maid of Perth*' so much that I procured the 1st series. There are two very good stories in it 'the Highland widow' and 'the two Drovers' — The other story 'the Surgeon's daughter' I do not like, being violently improbable and altogether like the production of some common Novelist."[18] Other titles mentioned in this correspondence are *Peveril of the Peak*, *Quentin Durward*, *Woodstock*, *Anne of Geierstein*, and *Waverley*.[19]

Scott's novels occupied an important place in the social lives of Talbot's half sisters Caroline and Horatia. "We had *tableaux* again last Tuesday at Lady Dudley Stuart's," reported Horatia Feilding in a letter of December 1830. "They made me act *Rowena* in the last scene in *Ivanhoe* & Lady Dudley was Rebecca in a beautiful oriental dress."[20] Two years later, in a letter of December 26, 1832, Caroline Edgcumbe informed Talbot that the Marquess and Marchioness of Salisbury planned "to have tableaux next month [at Hatfield House], & pressed us very much to stay for them." "They wanted L^d V[alletort] to represent the Earl of Leicester [from *Kenilworth*], & me Flora McIvor [from *Waverley*]," she wrote, "but I do not think either would suit particularly well — they are all to be selected from Walter Scott's novels."[21] About two months earlier Lady Elisabeth had referred to a mysterious "Box" that the queen consort, Adelaide, had sent to Caroline: "It was *Tableaux* of all the Waverley Novels & ending with a Pageant including the Tomb & all the Characters. It was really beautiful."[22]

One of the Waverley Novels even played a role in Talbot's courtship of Constance Mundy. We learn once more from *Sense and Sensibility* that a couple's tastes in literature, drawing, and music were taken to provide an indication of their future compatibility as husband and wife. Marianne Dashwood is dogmatic on this point, insisting near the beginning of the novel that she "could not be happy with a man whose taste did not in every point coincide with my own. He must enter into all my feelings; the same books, the same music must charm us both."[23] Whereas Marianne, who is only sixteen and possesses the certainty of youth, expects any suitor to mirror her own taste, Constance, who was twenty, was clearly amenable to learning from Talbot, accommodating herself to his preferences. Three letters that she wrote to Talbot in the weeks before their marriage in December 1832 show her eagerness to please him by reading *The Abbot*:

> I have not yet begun to read the Abbot as I preferred waiting till nearer the time when I imagined it possible that you might take it up also — besides, my thoughts were so busily engaged in looking back upon the events of the last few days & enjoying over again all that I then enjoyed, that I had no inclination to turn them immediately upon any other subject. — Last night I contented myself with reading [Baron Friedrich de La Motte-Fouqué's] Undine, but this evening I think of indulging myself with the Abbot.[24]

> I have found some very hard Scotch words & other things in the Abbot which you would have explained to me in a moment, you would have told me all about the Legend of the Witch of Berkeley & many other things of the same kind that I never heard of before. — Independently of these difficulties I like the Abbot extremely.[25]

> I meant to have sent you word that I had quite finished the Abbot but I am still a *little* way from the end. — I have been excessively interested in it, particularly from the moment of entering Lochleven Castle but I am sorry to confess that I do not think Queen Mary is represented in quite so amiable a light as I have always fancied her.[26]

These extracts show that Constance was eager to align herself with Talbot's taste in literature and that she was disposed to defer to him.

LEGENDARY TALES, IN VERSE AND PROSE

Talbot's first book, *Legendary Tales, in Verse and Prose,* may have owed something to Scott, for it was written in part in Edinburgh during October and November 1829.[27] In a letter sent from Edinburgh on November 8, Talbot excused himself from joining his mother at Hopetoun House, near South Queensferry, to the west of the capital, on the grounds that he was "employed in writing some Tales in prose

& verse."[28] In another letter, written several weeks later, he informed his mother that he had "settled with Murray to publish my legendary tales."[29] Lady Elisabeth was evidently ambivalent about her son's occupying himself in this way, for she asked in her reply to see "the Legendary Tales before they are given up in publication."[30] In February she wrote frankly, "knowing as I do *your sterling* abilities *I do not like these tales & poems* to be taken in the world for the Standard of *what you can do.*" She concluded, "I know you have too much *grandeur d'âme* to be annoyed at being put in possession of the truth or least of what I think so."[31] Perhaps because of his mother's reservations, Talbot asked Murray to return his manuscript.[32] Nevertheless, James Ridgway of Piccadilly published *Legendary Tales* that summer, and the book earned a mildly positive notice in the *Edinburgh Literary Journal.* "Mr H. Fox Talbot, an assumed name we suppose," suggested the reviewer, "need not blush to reveal his real name."[33] In light of what had gone before, Lady Elisabeth may not have been amused by the epigraph Talbot chose for his title page: *Dulce est desipere . . .* (Sweet it is to be frivolous).[34]

Talbot sent copies of *Legendary Tales* to friends and to members of his family, and they responded by sending standard compliments. More informative, however, were observations made in a letter to Lady Elisabeth by Amélina Petit de Billier ("Amandier"), a close friend and longtime companion to the Feilding and Talbot families: "I have read all Mr Talbot's *Legendary Tales* and they have greatly entertained me. I don't know if I am wrong or not, but I prefer those in verse. Some of them put me in mind of the originality of Walter Scott."[35] The last observation is percipient, for it draws attention to another connection between Talbot and Scott, one that goes beyond the former's enjoyment of the latter's writings and encompasses instead their shared interest in romanticized pasts—a passion that Scott expressed in *The Minstrelsy of the Scottish Border* and Talbot in *Legendary Tales.*[36]

Vered Maimon has recently drawn attention to Talbot's intellectual and temperamental affinity with romantic forms of history popularized by Scott's poems and novels while also aligning the two men's imaginative conceptions of history with creative approaches to historical writing advocated more or less contemporaneously by Carlyle and by Thomas Babington Macaulay.[37] Maimon's observations embed Talbot's admiration for Scott in a larger intellectual context and amplify the cultural significance of *Sun Pictures in Scotland,* Talbot's most extended tribute to Scott.

TALBOT IN SCOTLAND

Between September 26 and October 2, 1844, Talbot was in York attending the annual meeting of the British Association for the Advancement of Science. After this meeting he traveled north to Newcastle and then on to the Borders region of Scotland, visiting Berwick, Kelso, Dryburgh, Melrose, Jedburgh, and Selkirk (fig. 39) before continuing to Edinburgh, where he spent the nights of October 10 and 11 at Rampling's Waterloo Hotel (fig. 40).[38] On October 12 he set off once more, traveling by train to Glasgow and then spending some days in the west of Scotland. On October 18, having returned to Edinburgh the day before, he journeyed west to Stirling and on to Callander, which he used as his base for an expedition to Loch Katrine. He was back in Edinburgh on October 22 and began his homeward journey the following day, spending the nights of October 23 and 24 at Galashiels. On October 24 he visited Melrose Abbey and Abbotsford, where he signed the visitors' book.[39] From there he went to Kelso and made a second pilgrimage to Dryburgh Abbey, Scott's burial place, before setting out for Carlisle, Newcastle, and York, which he reached on October 28 after a circuit lasting just under four weeks.

Bad weather on the journey north prevented Talbot from photographing in the Borders, as he reported to Constance in a letter from Jedburgh: "Extremely dark weather both yesterday and today have prevented me from making any views of the ruined Abbey here, and I proceed this evening to Selkirk."[40] The next day Constance acknowledged a letter Talbot had sent from Berwick:

> I now want to hear most of your health & whether you have followed my advice about eating. — I must condole with you on the gale yesterday accompanied by almost incessant rain — and today has been scarcely any better *here* — but as there were thunder storms today which are partial I hope they may not have visited Scotland — Every evening my thoughts are with you hovering over Melrose Edinburgh &c — for I am reading in my Scottish History of the events in the beginning of George 1st's reign, when the standard of James 8th was raised — & the chief operations of the insurrectionary forces under the Earl of Mar were carried on in *your* environs.[41]

She added, "*Do* tell me whether your journey has begun to do you good," and concluded. "I am so anxious it should independently of *the Talbotypes*."[42] She wrote again three days later: "I received your letter from Jedburgh yesterday it was 3 days on the road which I think rather long in these days of steam conveyance — I was afraid the dark weather would be a sad hindrance to you — but as the environs of

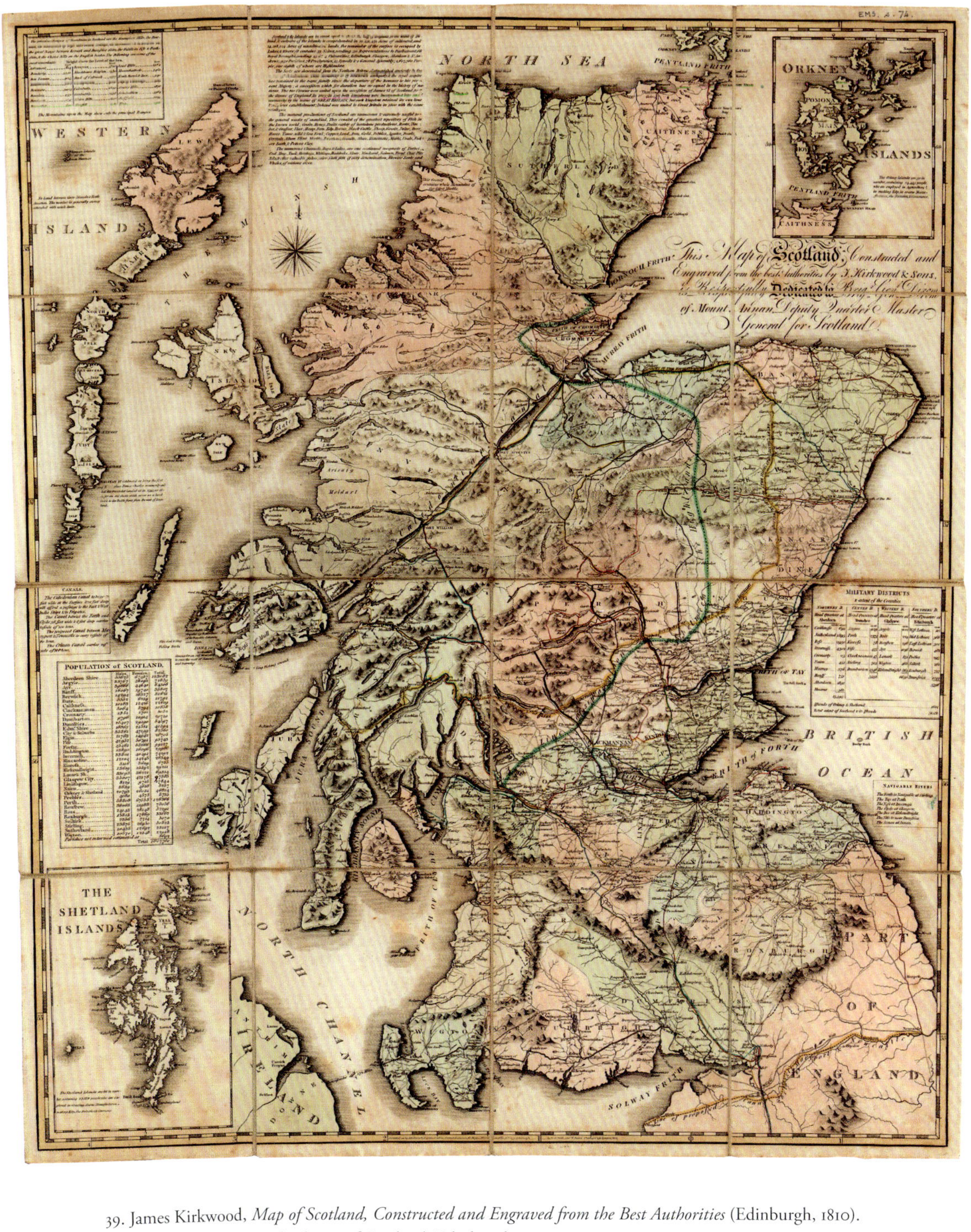

39. James Kirkwood, *Map of Scotland, Constructed and Engraved from the Best Authorities* (Edinburgh, 1810). Map Collection, National Library of Scotland, Edinburgh.

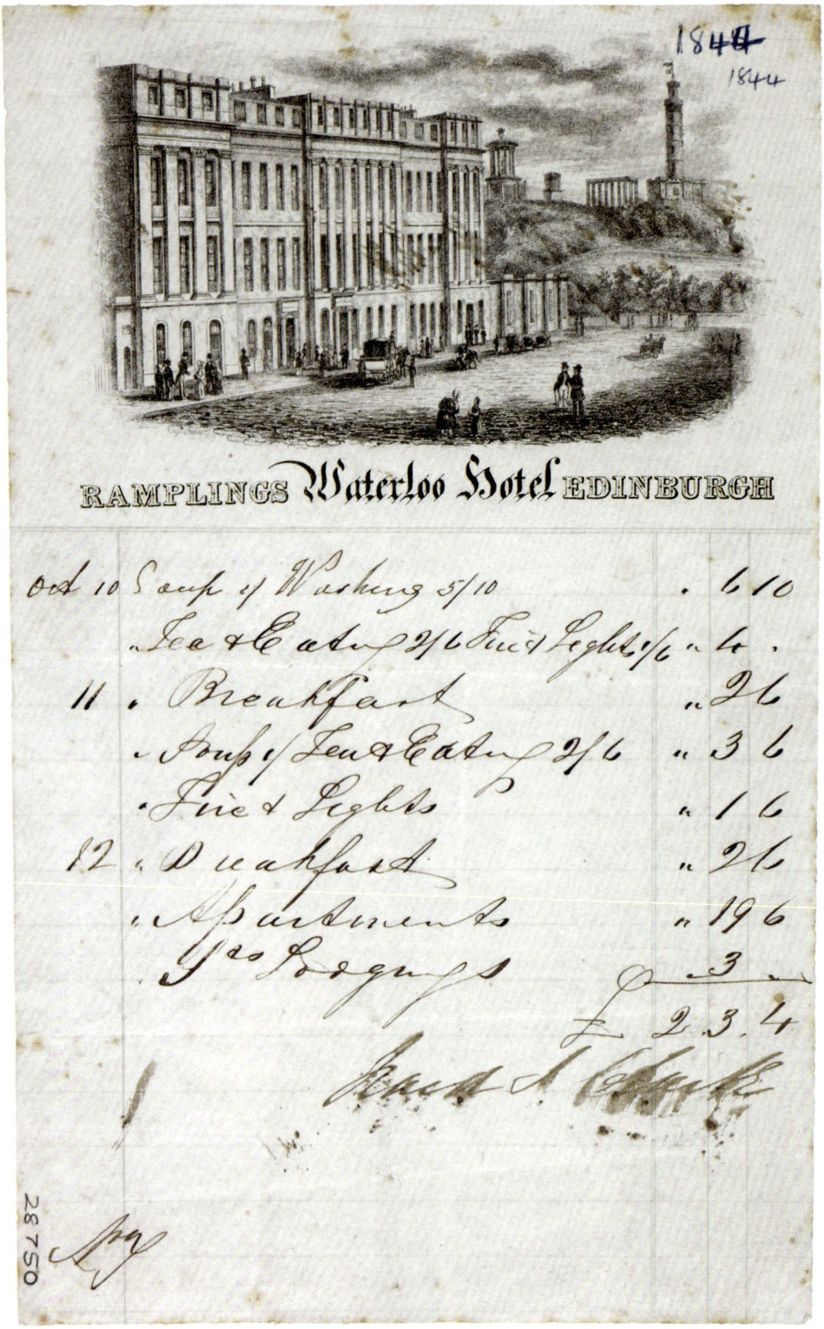
1844
1844

RAMPLINGS Waterloo Hotel EDINBURGH

Oct 10	Soap & Washing 5/10	.	6 10
	Tea & Coffee 2/6 Fire & Lights 1/6	..	4 .
11	Breakfast	..	2 6
	Soup / Tea & Coffee 2/6	..	3 6
	Fire & Lights	..	1 6
12	Breakfast	..	2 6
	Apartments	..	19 6
	Servants Lodgings		3 .
		£	2.3.4

28750

40. Receipt from Rampling's Waterloo Hotel, Edinburgh, Oct. 12, 1844. Talbot Collection, British Library (partially catalogued).

Jedburgh are so pretty as you describe them, I dare say you filled up your time well."[43] In a letter sent to Edinburgh three days after that, Constance commiserated with Talbot once more over "the present bad weather . . . because I know you must be, all this while, so extremely contrarié."[44]

On October 20 Constance expressed her dismay over not having received a letter from her husband in a while. "I am greatly troubled by your long silence — I have heard nothing since Jedburgh — a week yesterday! — Since which I have written three times."[45] Nevertheless, Constance accompanied Talbot in spirit by reading Scottish history while he traveled in Scotland. "Last night," she wrote, "I was at Inverness with Prince Charles Edward — & fought the battle of Culloden — I took leave of him at his death & concluded my book by rejoicing over the improved state of the Highlands." She concluded, "*Pray* do write a *little* oftener to your affectionate Constance."[46] The next day she wrote: "Your letter from Edinburgh just received has recovered my spirits — for I was quite troubled at your previous silence — (ever since Jedburgh) — and only yesterday I wrote again to Edinburgh to entreat you to be more communicative. . . . I am so very glad your weather has been finer than ours — Has the improvement which we now experience come from you to us?"[47] On October 26 she wrote once more: "I did receive yesterday your charming letter from the banks of Loch Achray — In it you draw the Sunny side of the picture — but your letter from Edinburgh today discloses the sad reverse . . . and reveals the cold discomforts of the Inns — I quite pity you for lying awake shivering with cold! — what must those wretched places be in really severe weather?"[48] Three days later she responded to a cheering letter from Kelso, writing: "We are so glad that you have been successful at Abbotsford & Dryburgh Abbey — and we think the latter part of your tour has made amends for disappointments at the beginning."[49]

Reading this correspondence in conjunction with Talbot's itinerary shows that he had relatively few opportunities to make photographs during his tour of Scotland, and that those he did make may be collected in three groups: the Edinburgh plates, which must have been taken on October 11, between his arrival in Edinburgh and his departure for Glasgow; the views of Loch Katrine, which must have been taken on October 20 or 21, during his second expedition away from Edinburgh; and the photographs of Melrose Abbey and Abbotsford, which were made on October 24, the day after Talbot left Edinburgh to begin his homeward journey, as well as the view of Scott's tomb at Dryburgh Abbey, which was taken on October 26.

Talbot evidently made those photographs with a view to collecting them in *Sun Pictures in Scotland*, his second photographically illustrated book, finished copies of which were distributed in July 1845.[50] Unlike *The Pencil of Nature* (of which the first four fascicles, out of six, appeared between June 1844 and July

1845), *Sun Pictures* is a unitary work, for its twenty-three plates are all connected with the life and writings of Sir Walter Scott. It is also Talbot's monument to Scott, comprising an anthology of "sites of memory" in which the writer's memory "crystallizes and secretes itself."[51] The remainder of this study considers how Scott's memory and Talbot's are embedded in the latter's photographs depicting Loch Katrine.

THE TROSSACHS AND LOCH KATRINE

Although the Ettrick Shepherd gave the impression that Scott invented The Trossachs, Tom Furniss has shown that the region's reputation for wild, picturesque landscape and "aesthetic geology" had already been established by the Reverend James Robertson's "Parish Report on Callander," which was published in 1794 in Sir John Sinclair's monumental *Statistical Account of Scotland*.[52] The *Statistical Account* did not circulate widely, but Robertson must be credited with situating The Trossachs at the heart of Romantic Scotland, for it appears that he issued a separate, brief account of the "Romantic Prospects" of his parish. Dorothy Wordsworth mentioned this guidebook in her recollections of the tour of Scotland that she made in 1803 with her brother William: "After having taken up at least six closely-printed pages with the Trossachs, [the minister of Callander] concludes thus: 'In a word, the Trossachs beggar all description'—a conclusion in which everybody who has been there will agree with him."[53]

By the time Talbot made his pilgrimage, Loch Katrine had been admired by a great many literary travelers and ordinary tourists.[54] Of those the most notable in the present context was surely J. M. W. Turner, who visited the loch in 1831 and made approximately thirty pencil sketches, some of which he would later develop into watercolors.[55] Representing The Trossachs, Benvenue, and Ben A'an as seen from Loch Achray, and Ellen's Isle and Benvenue seen across Loch Katrine, those watercolors (figs. 41 and 42) served in turn as models for the frontispiece and title-page vignette of the eighth volume of Robert Cadell's edition of *The Poetical Works of Sir Walter Scott*, the one containing *The Lady of the Lake*.[56]

We know that Talbot had read *The Lady of the Lake* when he was a boy, but it is also quite likely that he carried a copy with him to read at Callander or even on the shores of Loch Katrine. If he did not know Robertson's report in the *Statistical Account*, he probably knew the brief version of it or some other popular guidebook based upon it. It is also likely that he knew the engravings after Turner; indeed, he may have owned Cadell's edition of *The Poetical Works*. Primed in this way by reading and looking, he may have found that The Trossachs and Loch Katrine

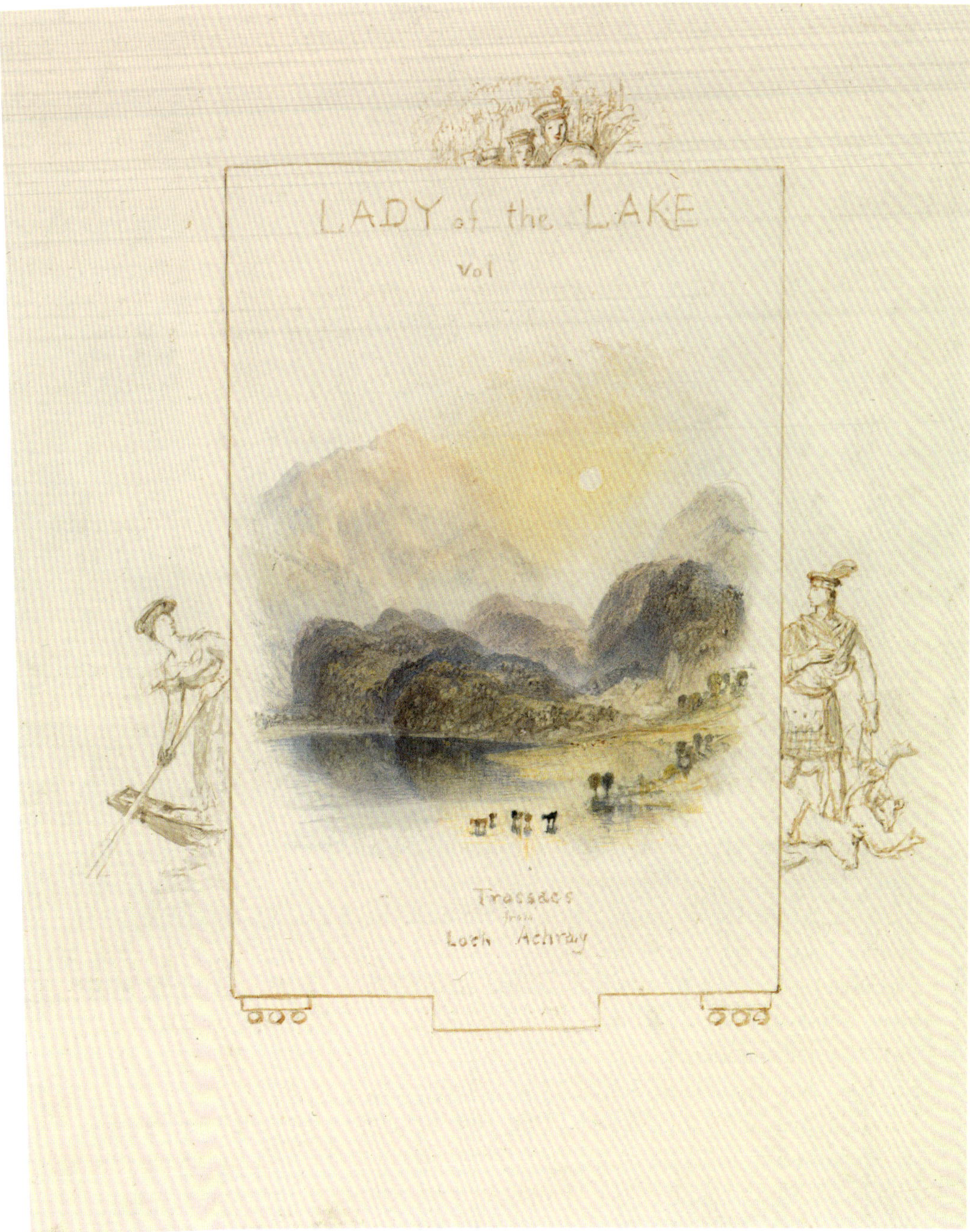

41. Joseph Mallord William Turner, *Design for an Illustration for Walter Scott's "Lady of the Lake," Loch Achray*, ca. 1832, watercolor, graphite, and pen and brown ink on wove paper, 11¼ x 8½ in. (28.6 x 21.6 cm). Yale Center for British Art, Paul Mellon Collection.

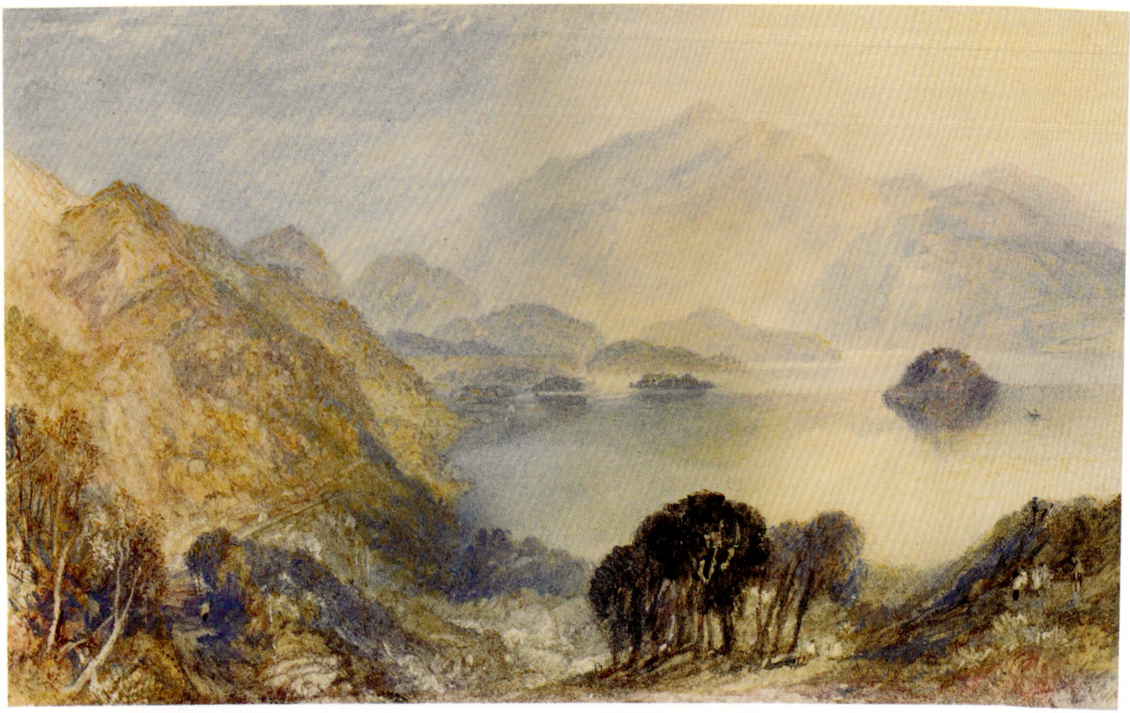

42. Joseph Mallord William Turner, *Loch Katrine*, ca. 1832, watercolor on paper support, 3¾ x 5⅞ in. (9.6 x 14.9 cm). British Museum.

appeared before him like mirages. In short his preparedness may have altered in some respects the actuality of what he was seeing. Certainly he had experienced the loch through what John Hale aptly termed the "gauze of association."[57] In essence Talbot's preexisting knowledge of the site comprised a palimpsest of words and images into which he had to assimilate his personal responses to the actual view. Many nineteenth-century visitors to Loch Katrine would have drawn on similar visual and literary associations, but for Talbot this conventional knowledge had been enhanced by involuntary memories of reading Scott's poem when he was a young boy and by recollections of the tour he had made of the Ionian Islands, when the "lovely wilderness" of Corfu reminded him of Scott's pictures of Loch Katrine.[58] Those memories had complicated and enriched Talbot's perception of Scott's landscape, placing it under what Marcel Proust once termed the "magnifying glass of memory."[59]

Talbot exposed seven negatives at Loch Katrine, and despite—or perhaps because of—the realistic character of photography, he succeeded in creating pic-

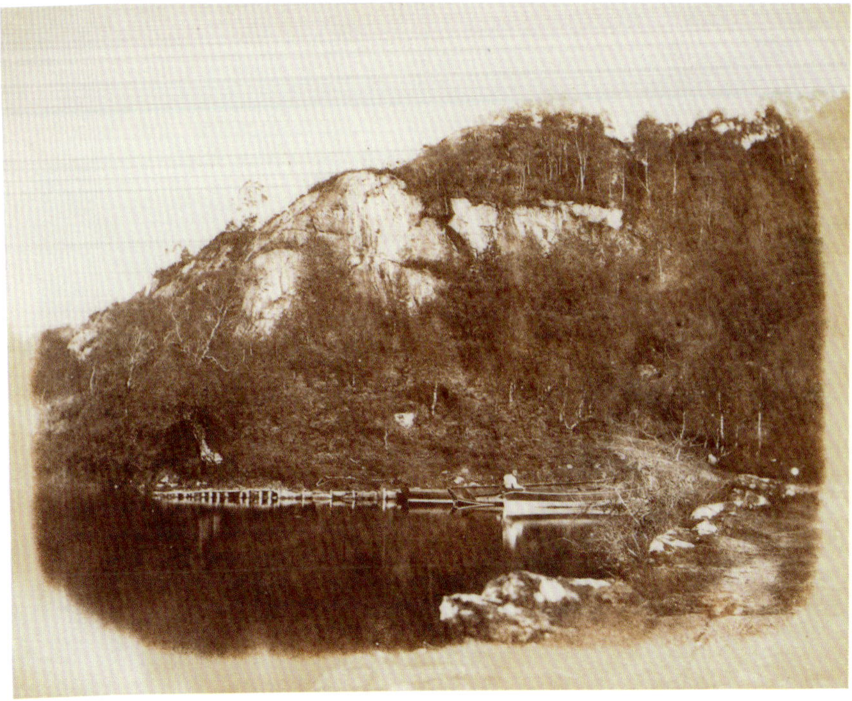

43. William Henry Fox Talbot, *Loch Katrine*, 1844–45, salted paper print from a calotype negative, 6¾ x 8⅛ in. (17.1 x 20.6 cm). Harry Ransom Center, University of Texas at Austin.

tures that capture the primordial nature of the site while also evoking the Elysian character of certain aspects of the landscape. Two large views show the east end of Loch Katrine and the pier, where travelers would hire boats to take them the length of the loch, heading west along the south shore and returning by the north side. The first of these shows boats moored at the pier, over which looms a massive outcrop covered in large part by trees (fig. 43). The second is a limpid, sooty view, in which the mirror-like surface of the loch accepts reflections of the somber hillside on the left and the "time-withered" trees on the right (fig. 44).[60] By exploiting particular characteristics of his paper process, Talbot succeeded in making a poetic, evocative image that is more than a topographical transcription.[61] This is also the sort of image that, in Talbot's words, may "awaken a train of thoughts and feelings, and picturesque imaginings."[62] The view is also very similar to a sketch by Turner of Loch Katrine (1831, Tate), showing once more Talbot's assimilation of the notion of the picturesque view.[63] A third photograph, looking across Loch Katrine to Benvenue (fig. 45), evokes lines in the first canto of *The Lady of the Lake*:

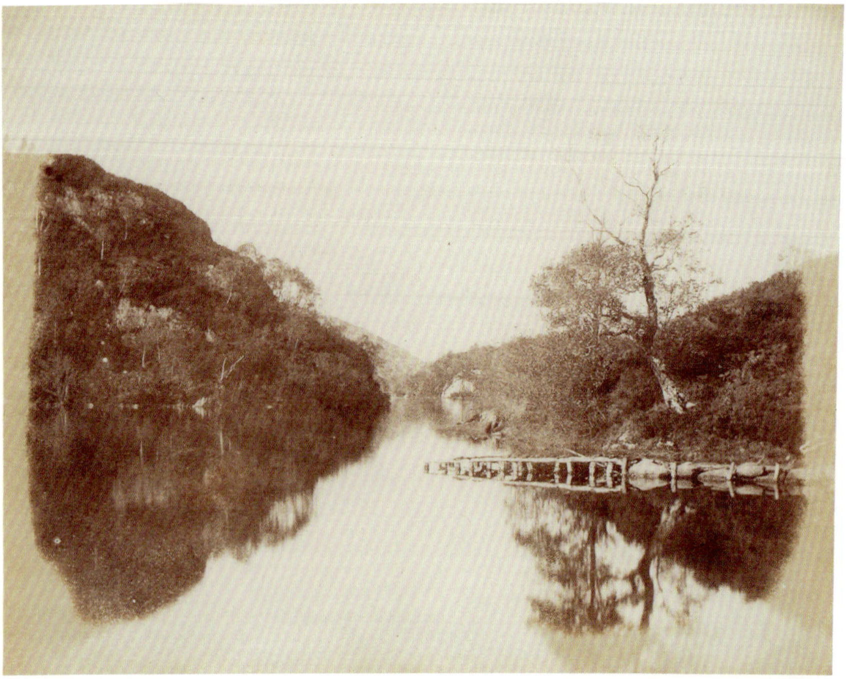

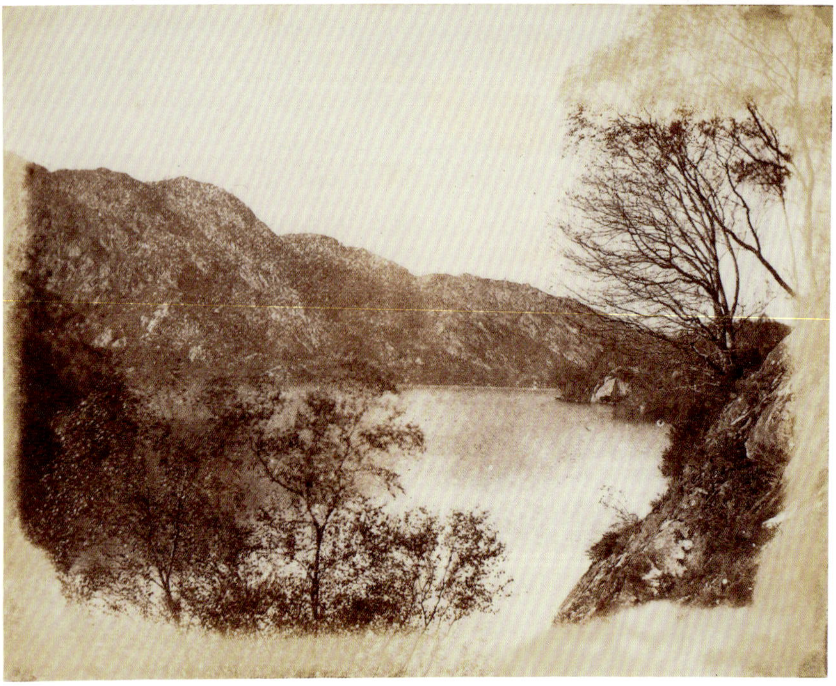

44. William Henry Fox Talbot, *Loch Katrine*, 1844–45, salted paper print from a calotype negative, 6¾ x 8¼ in. (17 x 21 cm). Harry Ransom Center, University of Texas at Austin.

45. William Henry Fox Talbot, *Loch Katrine*, 1844–45, salted paper print from a calotype negative, 6⅝ x 8⅛ in. (16.7 x 20.5 cm). Harry Ransom Center, University of Texas at Austin.

High on the south, huge Benvenue
Down to the lake in masses threw
Crags, knolls, and mounds, confusedly hurled,
The fragments of an earlier world;
A wildering forest feathered o'er
His ruined sides and summit hoar.

Finally, this passage recalls Robertson's description of the approach to The Trossachs, where there is "such an assemblage of wildness and of rude grandeur" that it "seems as if a whole mountain had been torn to pieces, and frittered down by a convulsion of the earth."[64]

CONCLUSION

Talbot's second plate in *Sun Pictures of Scotland* is the photograph of the Scott Monument in Edinburgh, "as it appeared when nearly finished, in October 1844" (fig. 46). The view is to the southeast, for the arches of North Bridge, the viaduct connecting the Old and New Towns, can be made out in the background in especially well-preserved prints.[65] This photograph is anomalous in some respects, for it records Scott's cenotaph and so is not connected with his life and writings in the way that the other images are. It is true, however, that the completed monument does accommodate a large number of small statues representing characters from Scott's writings. In this respect its decoration has something in common with the didactic sculptural programs of the great French Gothic cathedrals. The positioning of the plate in the album is also puzzling, for one might have expected it to appear either first, before that showing Heriot's Hospital, or after that representing Scott's tomb at Dryburgh Abbey.

If the view of the Scott Monument would have been a fitting codicil to *Sun Pictures*, it is also an appropriate image with which to conclude the present study, for not only is it a memorial to Scott but in a sense it is also a tribute to Talbot himself. As Roland Barthes argued in *Camera Lucida*, the photograph was originally a "certificate of presence," constituting a new form of monument.[66] In this sense the photograph authenticates the existence and appearance of the Scott Monument, showing it as it was on October 11, 1844, while also ratifying Talbot's actual experience of it and marking his long-standing engagement with Scott and his writings.

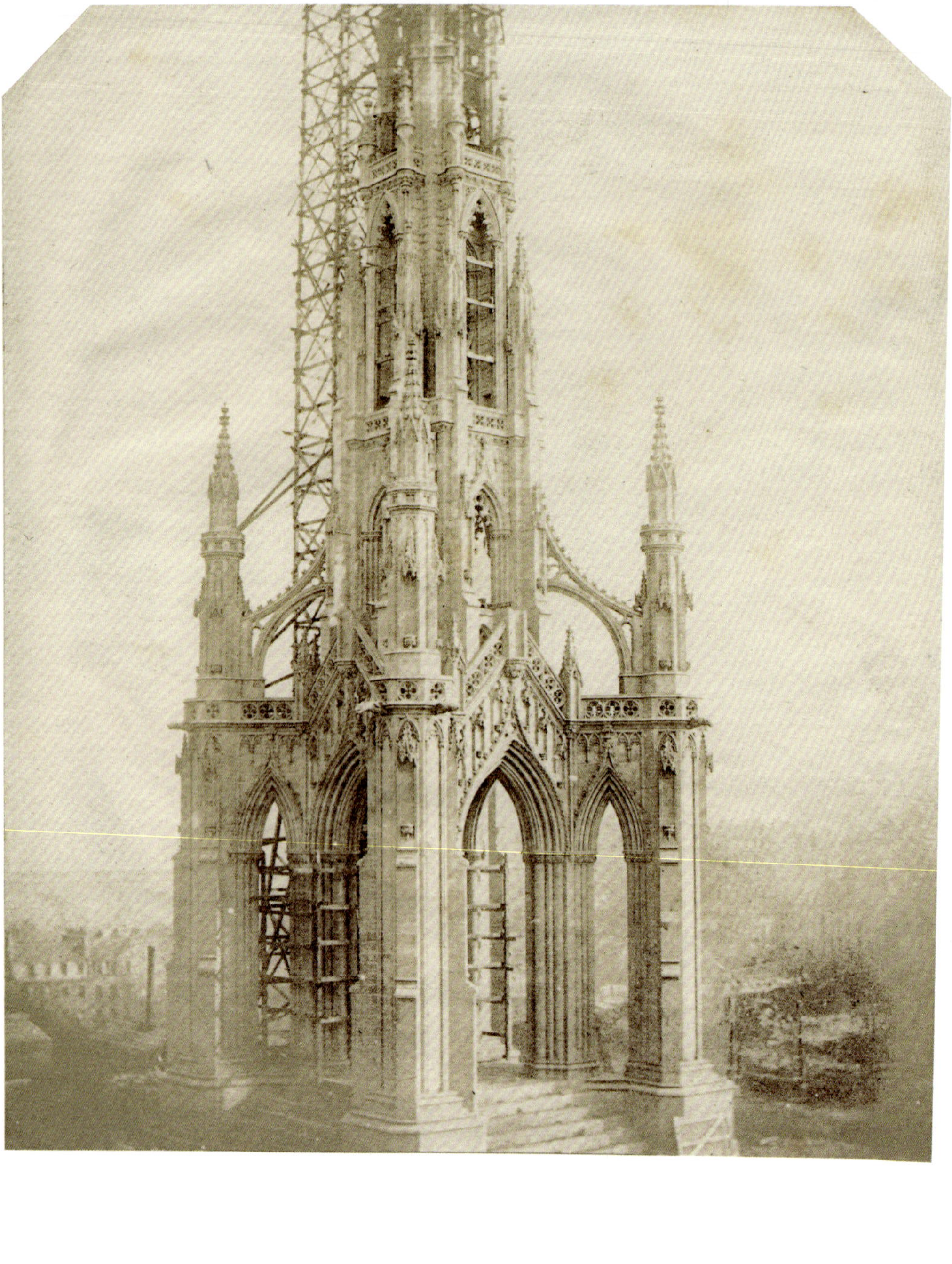

46. William Henry Fox Talbot, *Sir Walter Scott's Monument, Edinburgh; as It Appeared When Nearly Finished, in October 1844*, 1844–45, salted paper print from a calotype negative, 7⅝ x 6⅛ in. (9.6 x 14.9 cm). Harry Ransom Center, University of Texas at Austin.

I am grateful to Mirjam Brusius, Katrina Dean, and Chitra Ramalingam for inviting me to participate in the conference held at Cambridge in 2010 and for the opportunity to publish this much-revised version of my paper here; they are also due warm thanks for their hard work and efficiency as editors of this volume. I am grateful to participants in the conference for constructive observations and to Katherine Peveraro, Larry J. Schaaf, and Sara Stevenson for particular information.

The present study overlaps from time to time with my "H. Fox Talbot's 'Scotch Views' for *Sun Pictures in Scotland*," in *The Photobook from Talbot to Ruscha*, ed. Patrizia Di Bello, Colette Wilson, and Shamoon Zamir (London: I. B. Taurus, 2012), 17–34. I believe the two essays complement each other positively, and I hope readers will find this to be the case.

1. William Henry Fox Talbot to Lady Elisabeth Feilding, March 5, 1813, Doc. No. 589, The Correspondence of William Henry Fox Talbot, http://foxtalbot.dmu.ac.uk (accessed Jan. 31, 2012), hereafter "Correspondence" (original: Talbot Collection, hereafter "TC," British Library, hereafter "BL").
2. Talbot to John Murray, May 16, 1846, Doc. No. 9887, Correspondence (original: National Library of Scotland, Edinburgh): "I request your acceptance of a copy of my photographic work called Sun Pictures in Scotland. This has not been published but the copies were distributed by private subscription."
3. Walter Scott, *The Lady of the Lake: A Poem* (Edinburgh: John Ballantyne, 1810).
4. Quoted in Elizabeth A. Bohles and Ian Duncan, eds., *Travel Writing, 1700–1830: An Anthology* (Oxford: Oxford Univ. Press, 2005), 179–80.
5. Quoted in Catherine Gordon, "The Illustration of Sir Walter Scott: Nineteenth-Century Enthusiasm and Adaptation," *Journal of the Warburg and Courtauld Institutes* 34 (1971): 297.
6. In John Galsworthy's *The Silver Spoon* (Harmondsworth, UK: Penguin Books, 1970), 200, Soames Forsyte recalls winning a prize at his first school for reciting two hundred lines of *The Lady of the Lake*. Since Soames was born in 1855, this episode must have taken place about 1870. It was still usual in the first half of the twentieth century for pupils to memorize passages from *The Lady of the Lake*. See, for instance, Alfred M. Hitchcock, ed., *Scott's Lady of the Lake (English Readings for Schools)* (New York: Henry Holt, 1911), 169: "Memorize the fourteenth stanza or some other that you like better."
7. Talbot to Lady Elisabeth Feilding, May 7, 1826, Doc. No. 1430, Correspondence (original: TC, BL). Corfu and the other Ionian Islands were governed by the British from 1815 until 1864. Talbot "botanized" intensively during the time he spent on the islands.
8. *The Lady of the Lake*, 1.16.1–2. Variations between Talbot's text and published versions of the poem suggest that he may have been quoting from memory.
9. Jane Austen, *Sense and Sensibility*, vol. 1 of *The Oxford Illustrated Jane Austen*, ed. R. W. Chapman (Oxford and New York: Oxford Univ. Press, 1988), 47.
10. Jane Austen, *Mansfield Park*, vol. 3 of *The Oxford Illustrated Jane Austen*, ed. R. W. Chapman (Oxford and New York: Oxford Univ. Press, 1988), 85–86. Austen returned to *The Lay of the Last Minstrel* near the end of the second volume of her novel, when she cast Fanny in the role of the Lady of Branxholm Hall (280–81).
11. Jane Austen, *Persuasion*, in *Northanger Abbey and Persuasion*, vol. 5 of *The Oxford Illustrated Jane Austen*, ed. R. W. Chapman (Oxford and New York: Oxford Univ. Press, 1988), 100–101.
12. Ian Duncan has observed that "from the end of the Napoleonic Wars to the publishing crash of 1825–26 . . . Scott's novels saturated the British fiction market, outselling all other novels combined." Ian Duncan, *Scott's Shadow: The Novel in Romantic Edinburgh* (Princeton, NJ: Princeton Univ. Press, 2007), xi.
13. [Thomas Carlyle], "Sir Walter Scott," *London and Westminster Review* 12 (1838): 293–345. Reprinted in Thomas Carlyle, *Critical and Miscellaneous Essays*, vol. 5 (London: Chapman and Hall, 1859), 255.
14. Carlyle, *Essays*, 255.
15. Carlyle, *Essays*, 256.
16. Talbot to Lady Elisabeth Feilding, Aug. 10, 1819, Doc. No. 853, Correspondence (original: TC, BL). In this letter Talbot also confessed that he had not yet read *The Legend of Montrose*.
17. Talbot to Lady Elisabeth Feilding, Dec. 26, 1819, Doc. No. 858, Correspondence (original: TC, BL). *Ivanhoe*, the first of Scott's novels to be set in England, was issued in late December 1819, although 1820 was given on the title page as the year of publication.

18. Talbot to Lady Elisabeth Feilding, Oct. 29, 1828, Doc. No. 1732, Correspondence (original: TC, BL).
19. Lady Elisabeth Feilding to Talbot, Feb. 28, 1823, Doc. No. 1038, Correspondence (original: TC, BL); Lady Elisabeth to Talbot, Dec. 10, 1823, Doc. No. 1135, Correspondence (original: TC, BL); Talbot to Lady Elisabeth, June 29, 1826, Doc. No. 1451, Correspondence (original: TC, BL); Talbot to Lady Elisabeth, Sept. 7, 1829, Doc. No. 1854, Correspondence (original: TC, BL); and Constance Talbot to Talbot, June 22, 1838, Doc. No. 339, Correspondence (original: TC, BL).
20. Henrietta Horatia Maria Feilding to Talbot, Dec. 2, 1830, Doc. No. 2094, Correspondence (original: TC, BL).
21. Caroline Edgcumbe, née Feilding, to Talbot, Dec. 26, 1832, Doc. No. 2535, Correspondence (original: TC, BL). For an account of the Hatfield tableaux, which were invented and designed by David Wilkie, see Sara Stevenson and Helen Bennett, *Van Dyck in Check Trousers: Fancy Dress in Art and Life, 1700–1900* (Edinburgh: Scottish National Portrait Gallery and National Museum of Antiquities of Scotland, 1978), 46–48.
22. Lady Elisabeth Feilding to Talbot, Nov. 1, 1832, Doc. No. 2454, Correspondence (original: TC, BL).
23. Austen, *Sense and Sensibility*, 17.
24. Constance Mundy to Talbot, Oct. 30, 1832, Doc. No. 2447, Correspondence (original: TC, BL).
25. Constance Mundy to Talbot, Nov. 5, 1832, Doc. No. 2458, Correspondence (original: TC, BL).
26. Constance Mundy to Talbot, Nov. 7, 1832, Doc. No. 2461, Correspondence (original: TC, BL).
27. William Henry Fox Talbot, *Legendary Tales, in Verse and Prose, Collected by H. Fox Talbot Esq.* (London: James Ridgway, 1830).
28. Talbot to Lady Elisabeth Feilding, Nov. 8, 1829, Doc. No. 1886, Correspondence (original: TC, BL). In a letter written the same day to his half sister Horatia, Talbot termed Edinburgh "this splendid metropolis." Talbot to Horatia Feilding, Nov. 8, 1829, Doc. No. 1887, Correspondence (original: TC, BL).
29. Talbot to Lady Elisabeth Feilding, Dec. 23, 1829, Doc. No. 1918, Correspondence (original: TC, BL).
30. Lady Elisabeth Feilding to Talbot, Dec. 25, 1829, Doc. No. 1920, Correspondence (original: TC, BL).
31. Lady Elisabeth Feilding to Talbot, Feb. 25, 1830, Doc. No. 1960, Correspondence (original: TC, BL).
32. John Murray to Talbot, Thursday [month and day not specified] 1830, Doc. No. 1936, Correspondence (original: TC, BL).
33. *Edinburgh Literary Journal; or, Weekly Register of Criticism and Belles Lettres* 91 (Aug. 7, 1830): 95.
34. Horace, *Odes* 4:12, 28.
35. Lady Elisabeth Feilding to Talbot, Aug. 3, 1830, Doc. No. 2037, Correspondence (original: TC, BL): "J'ai lu tous les Legendary Tales de Mr Talbot & ils m'ont fort amusée. Je ne sais si j'ai tort ou non, Mais je prefère ceux en vers. Quelques uns me rappellent ~~l'orinigi~~ l'originalité de Walter Scott. remerciez le bien pour moi de son aimable envoi."
36. This aspect of Talbot's personality is also apparent in the friendship that he and his family developed with Thomas Moore, the Irish writer and musician. Moore came to reside close to Bowood, a few miles from Lacock, and was a frequent guest at Lacock Abbey, in addition to appearing with great frequency in the Talbot family correspondence.
37. Vered Maimon, "Displaced 'Origins': William Henry Fox Talbot's *The Pencil of Nature*," *History of Photography* 32, no. 4 (2008): 314–25. For the overlapping "of fictional narrative and historical fact, of invention and representation" in Scott's historical novels, see Ann Rigney, "Hybridity: The Case of Sir Walter Scott," in *Imperfect Histories: The Elusive Past and the Legacy of Romantic Historicism* (Ithaca, NY: Cornell Univ. Press, 2001).
38. This account of Talbot's travels in Scotland is drawn from Katherine Peveraro, "Views of the North: Visions of Scotland from Talbot, Turner and Scott" (MLitt thesis, Univ. of St. Andrews, 2004), 23–28. Peveraro's sources include a diary (Talbot, diary, acc. no. 24550, Add. MS 88942, TC, BL [partially catalogued]) and an account book kept by Talbot's valet Nicolaas Henneman (Account book, "Nicholls expenses from Sept 25th til Nov 2," Add. MS 88942/1/284, 1844, TC, BL).
39. Reproduced in Larry J. Schaaf, *Sun Pictures Catalogue Seventeen: William Henry Fox Talbot: Selections from a Private Collection* (New York: Hans P. Kraus, Jr., 2007), 36.
40. Talbot to Constance Talbot, Oct. 9, 1844, Doc. No. 5096, Correspondence (original: TC, BL).
41. Constance Talbot to Talbot, Oct. 10, 1844, Doc. No. 5098, Correspondence (original: TC, BL). "James 8th" refers to James Edward Stuart, "the Old Pretender," son of the exiled James II of England (who was king

of Scotland as James VII). The "Earl of Mar" is John Churchill, Earl and later first Duke of Marlborough.

42. Constance Talbot to Talbot, Oct. 10, 1844, Doc. No. 5098, Correspondence (original: TC, BL).
43. Constance Talbot to Talbot, Oct. 13, 1844, Doc. No. 5102, Correspondence (original: TC, BL).
44. Constance Talbot to Talbot, Oct 16, 1844, Doc. No. 5103, Correspondence (original: TC, BL).
45. Constance Talbot to Talbot, Oct. 20, 1844, Doc. No. 5106, Correspondence (original: TC, BL).
46. Constance Talbot to Talbot, Oct. 20, 1844, Doc. No. 5106, Correspondence (original: TC, BL). "Prince Charles Edward" refers to Charles Edward Stuart, "the Young Pretender," also called Bonnie Prince Charlie, the son of James Edward Stuart and grandson of James II.
47. Constance Talbot to Talbot, Oct. 21, 1844, Doc. No. 5107, Correspondence (original: TC, BL).
48. Constance Talbot to Talbot, Oct. 26, 1844, Doc. No. 5110, Correspondence (original: TC, BL).
49. Constance Talbot to Talbot, Oct. 29, 1844, Doc. No. 5111, Correspondence (original: TC, BL).
50. William Henry Fox Talbot, *Sun Pictures in Scotland* (London, 1845). For a more detailed account of the production and composition of the volume, see Smith, "H. Fox Talbot's 'Scotch Views,'" 19–20.
51. Pierre Nora, "Between Memory and History: *Les Lieux de Mémoire*," *Representations* 26 (Spring 1989): 7.
52. Tom Furniss, "'Plumb-Pudding Stone' and the Romantic Sublime: The Landscape and Geology of the Trossachs in *The Statistical Account of Scotland* (1791–9)," in *Romantic Localities: Europe Writes Place*, ed. Christophe Bode and Jacqueline Labbe (London: Pickering and Chatto, 2010), 51–66.
53. Quoted in Furniss, "Plumb-Pudding Stone," 58.
54. See, most recently, Nicola J. Watson, "Readers of Romantic Locality: Tourists, Loch Katrine and The Lady of the Lake," in Bode and Labbe, *Romantic Localities*, 67–79.
55. The drawings are contained in Turner's notebook and can be viewed online at the Tate Gallery Turner site, http://www.tate.org.uk/research/prints-and-drawings-rooms/turner-bequest.
56. Walter Scott, *The Poetical Works of Sir Walter Scott*, vol. 8, ed. J. G. Lockhart (Edinburgh: Cadell, 1834).
57. John R. Hale, ed., *The Italian Journal of Samuel Rogers* (London: Faber and Faber, 1956), 95.
58. Gillen D'Arcy Wood has suggested that the act of photographing Loch Katrine may have reminded Talbot of the occasion of his own inspiration for photography, when, in October 1833, he attempted to make sketches at Lake Como using a camera lucida. Gillen D'Arcy Wood, *The Shock of the Real: Romanticism and Visual Culture, 1760–1860* (New York and Basingstoke, UK: Palgrave, 2001), 187. Talbot's emotions at Loch Katrine may also have produced what Roland Barthes termed a counter-memory, recalling for him his own reactions to the scenery of Corfu. See Roland Barthes, *Camera Lucida: Reflections on Photography*, trans. Richard Howard (New York: Hill and Wang, 1981), 91.
59. Marcel Proust, *The Fugitive*, trans. Peter Collier, vol. 5 of *In Search of Lost Time* (London: Penguin Books, 2003), 639.
60. Commenting on *The Open Door*, the sixth plate in *The Pencil of Nature*, Talbot observed: "A painter's eye will often be arrested where ordinary people see nothing remarkable. A casual gleam of sunshine, or a shadow thrown across his path, a time-withered oak or a moss-covered stone may awaken a train of thoughts and feelings, and picturesque imaginings."
61. Gillen D'Arcy Wood criticized the "unromantic clarity" of Talbot's photographs of Loch Katrine, concluding incorrectly that these "denuded landscape images dissolve the Romantic aura of Scott mythology, giving us the Highlands as the 'real.'" D'Arcy Wood, *Shock of the Real*, 191–92.
62. See note 60 above.
63. See Martin Kemp, "Talbot and the Picturesque View: Henry, Caroline and Constance," *History of Photography* 21, no. 4 (2007): 270–82. Eric Shanes suggests that "Broken trees fulfill an associative function, to further the mood of lonely desolation and asceticism." Eric Shanes, "Landscapes of Meaning," in *Turner's Human Landscape* (London: Heinemann, 1990), 151.
64. Quoted in Furniss, "Plumb-Pudding Stone," 56.
65. Katherine Peveraro has proposed that the photograph was taken from an upper window of the Royal Hotel, at 53 Princes Street, and this is supported by the angle and direction of Talbot's view. Peveraro, "Views of the North," 8–9.
66. Barthes, *Camera Lucida*, 87, 93.

INVENTION AND DISCOVERY

Nature as Drawing Mistress

Herta Wolf

On Sunday, January 6, 1839, the *Gazette de France* in Paris published the first article ever dedicated to a photographic process. Published the evening prior to the presentation of the daguerreotype at the weekly meeting of the Académie des Sciences, the article functioned as a kind of press release for the remarks by François Arago, the academy's permanent secretary for the mathematical sciences, on the basic idea of Louis-Jacques-Mandé Daguerre's discovery. Even before his comments could be published in the *Comptes rendus hebdomadaires des séances de l'Académie des Sciences*—as usual, a week later, hence on the following Monday, January 14—the text that Hippolyte Gaucheraud had written on the "French Invention" was translated into English and published on January 12 in the *Literary Gazette* (a weekly journal of literature, the fine arts, and science).[1] On February 2 the *Literary Gazette* dedicated more space to coverage of the photographic process, in a section entitled "The New Art" and comprising contributions by different authors. This time it published a letter to the editor, William Jerdan, from William Henry Fox Talbot on his own photographic process, photogenic drawings; another account of the French invention; and separate reports on the most recent meetings of the Royal Institution and Royal Society regarding the new imaging technologies.[2] This material discussed not only photogenic drawings and daguerreotypes—though there was nothing yet about the differences between them—but also the characteristics of the medium and the consequences of these new automatic recording techniques for visual politics and culture. "No human hand has hitherto traced such lines as these drawings displayed; and what man may hereafter do, now that dame Nature has become his drawing mistress, it is impossible to predict," states the report on the meeting of the Royal Institution.[3] The present essay will discuss the metaphor of Nature as a drawing mistress and its implications for the medium and the history of science, more specifically with its designation and hence qualification of Nature as a drawing mistress.

It was not unusual for the *Literary Gazette* to paraphrase lectures at the popular Royal Institution. Indeed, as early as 1827 it had begun to summarize both the lectures and the meetings of learned societies and to report on "advanced information about scientific discoveries."[4] Two years later the *Athenaeum* took up this practice as well, which is why both journals provided space for developments in this "new art" previously related in the above-mentioned *Comptes rendus hebdomadaires des séances de l'Académie des Sciences* and in the *Transactions of the Royal Society*. Whereas such organs of scientific societies addressed "academics" in a wide variety of disciplines, their target audience was an interested bourgeois public—"a wide middle-class audience," as Maurice Crosland has called it, which could not so easily be divided into "practitioners" and "dilettantes."[5] Nevertheless, as will be demonstrated below, they too—or rather, they especially—had specialized knowledge that would ultimately become the fundamental prerequisite for the rapid spread of photographic imaging technologies.

Discussions of Talbot's photography generally draw on the paper he submitted to the Royal Society, "Some Account of the Art of Photogenic Drawing,"[6] and especially on his introduction to his first book of photographs, *The Pencil of Nature*, published beginning in 1844.[7] This historiography often misses two points: first, that Talbot developed two different photographic processes—photogenic drawings and, from the 1840s on, the calotype; second, that he was able to "invent" photography on paper only with recourse to the findings and work of others. But no process requiring a knowledge of such varied fields could have been invented by a single scientist or craftsman. Just the length of time required to gather and assimilate the knowledge on which the new visual recording techniques were based would suggest that they had multiple authors.

A close reading of the articles published in the first months of 1839 makes this perfectly apparent.[8] With their diverse viewpoints they provide a better sense of the cultural, technical, and scientific background than does Talbot's frequently quoted accompanying text to *The Pencil of Nature*—an account that must be regarded as having been written in hindsight. That his first photographs were the result of a lengthy series of experiments based on the available knowledge, which had to be (re)combined in order to reach a successful result, is clear enough from his own notebooks (fig. 47).[9] His reliance on experimental practices and existing knowledge is equally obvious from the "photogenic drawings" he produced as variants and "replicas" of picture subjects (see below). Most important, all these early documents confirm that, from the perspective of the history of science, photography as a form of knowledge and of experimentation, of representation and recording, should be associated with the *Leitwissenschaften* (core sciences) of physics and chemistry that shaped the first third of the nineteenth century.[10]

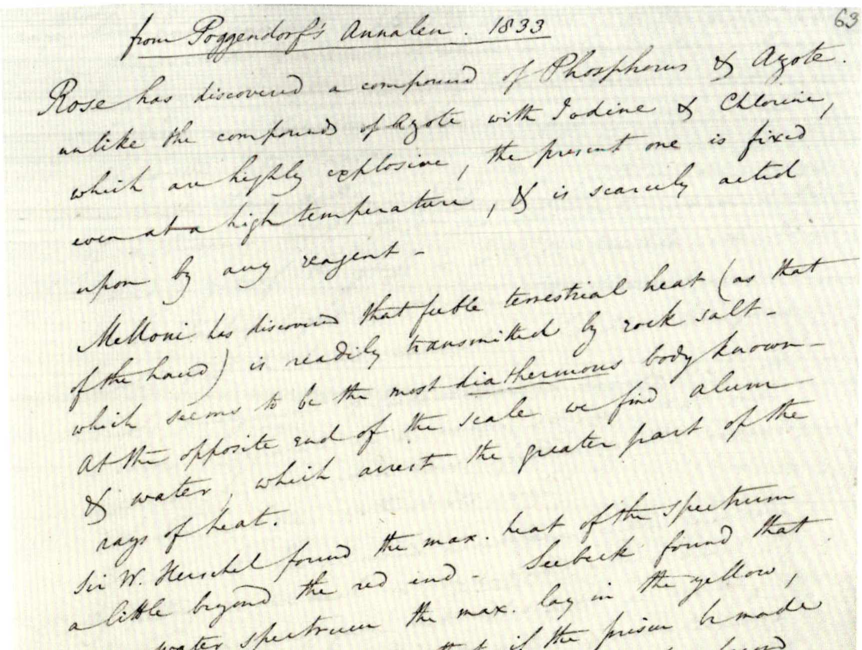
from Poggendorf's Annalen. 1833 63

Rose has discovered a compound of Phosphorus & Azote. unlike the compounds of Azote with Iodine & Chlorine, which are highly explosive, the present one is fixed even at a high temperature, & is scarcely acted upon by any reagent -

Melloni has discovered that feeble terrestrial heat (as that of the hand) is readily transmitted by rock salt - which seems to be the most diathermous body known - At the opposite end of the scale we find alum & water, which arrest the greater part of the rays of heat.

Sir W. Herschel found the max. heat of the spectrum a little beyond the red end - Seebeck found that in a water spectrum the max. lay in the yellow, now Melloni has discovered that if the prism be made of rock salt the max. of heat lies as far beyond the red as the red is from the violet, or is distant from the red end by the whole length of the spectrum. He finds that smoky topaz or quartz semiopaque transmits much more radiant heat than alum 20 times thinner & quite transparent. Also that this topaz scarcely absorbs any more heat when it is thicker, whereas with light the opacity increases rapidly. He finds that of 1000

47. William Henry Fox Talbot, Notebook L, "London, 8th January 1834." Loan Collection, British Library (uncatalogued). Here Talbot is referring to different scientists and their experiments on light mentioned in articles of the German journal *Annalen der Physik* (see notes 31, 38, 41 below). The passage reads: "Sir W. Herschel found the max. heat of the spectrum a little beyond the red and Seebeck found that in a water spectrum the max. lay in the yellow, now Melloni has discovered that if the prism be made of rock salt the max of heat lies as far beyond the red as the red is from the violet, or is distant from the red end by the whole lengths of the spectrum. He finds that smoky topaz or quartz semiopaque transmits much more radiant heat than alum twenty times thinner & quite transparent. Also that this topaz securely absorbs any more heat when it is thicker, whereas with light this opacity increases rapidly. He finds that of 1000 rays of heat which have traversed a plate of citric acid a far greater number are transmitted by a plate of white or coloured glass, than of 1000 rays emanating directly from a lamp — *But green glass forms an exception*, transmitting much fewer, than without the citric acid."

READINGS OF THE METAPHOR

The terms that appear in the title of this essay, "Nature" and "drawing mistress," suggest that early photography was not merely based on these core sciences; it was immediately recognized as a medium that would become part of the history of image-making techniques. The trope implies that the photographic image is not only a "natural" phenomenon but also "similar to nature." This in turn implies the potential to be a simulacrum, which is why the connection it makes between nature and drawing, or drawing technique, is of particular interest in any discussion of photography's potential as a scientific—that is, precise—way of recording reality. Moreover, this closeness to nature makes explicit that photographic processes are anchored in culture. Unlike all previous representational techniques it was based not just on the aforementioned scientific findings but also on the new practices for generating scientific knowledge such as trial and error and experimentation.[11]

The "drawing mistress" metaphor also invites reflection on the role of drawing and drawing instruction in the last third of the eighteenth century. Because the term "mistress" implies the didactic effort to impart a particular skill, it serves better than many others to illuminate the early history of photography from new perspectives.[12] One cannot think of a teacher without a pupil, and this conceptual coupling calls to mind other pairs of terms encountered in the historiography on early photography: "creator" and "imitator," for example, "inventor" and "plagiarist," or "Frenchman" and "Englishman." Emphasis on such polarities in some previous accounts of the early history of photographic processes only obscures the medium's indebtedness to a number of specialized skills and fields of knowledge: for example, glassmaking, lens grinding, achromatic optics, and—the focus of the present essay—chemistry.

The designation of Nature as drawing mistress is associated with Talbot, but he was not responsible for it. The phrase appears in one of the pieces that followed his letter to the editor of the *Literary Gazette* in which he introduced his photogenic drawings. The author of the report on the meeting of the Royal Institution on January 25, 1839, was moved to ask this anxious question after studying the examples of Talbot's photogenic drawings displayed in its library and wondering about the consequences of this new automatic drawing process.

His metaphor applies equally well both to Talbot's technique and to that of Daguerre.[13] Above all it is of interest in that it can be interpreted as a particularly English conceit. Nowhere in Europe had drawing been accorded such educational, social, and especially rhetorical importance, at the latest since the last third of the eighteenth century, as in England. This did not—or, more precisely, did not yet—refer to the "polite art" (photography) taught to young ladies, especially those of the gentry.[14] Rather, it still referred to drawing as an "applied art," to which great significance had been attached in England since the late seventeenth century—not least for economic reasons. After all, if it was to keep up with the (pre)industrial developments of other European nations, English society needed skills in depiction, as the architect Christopher Wren pointed out in a letter to the treasurer of Christ's Hospital, London, in 1694: "Our Natives want not Genius, but education in that which is the ffoundation of all Mechanick Arts, a practise in designing or drawing, to which everybody in Italy, Ffrance and the Low Countreys pretends to more or less."[15] All the drawing instructions published after 1700 were meant to ensure that those whose professions required them to draw were capable of reproducing especially the effects of light and shadow, so that they might render nature with some accuracy. Therefore, Paul Sandby, after being engaged as chief drawing master to the Royal Military Academy, Woolwich, in 1768, had to teach his students "putting Perspective in Practice by Copying from Drawings, which qualifies them for Drawing from nature; teaches them the effect of Light and Shade; and makes them acquainted also with Aerial Perspective."[16] And in 1839, in the summary of the paper presented to the Royal Society on January 31, Talbot claimed that photography was able to realize precisely that: "Nay, even a shadow, the emblem of all that is most fleeting in this world, is fettered by the spell of the invention, and remains perfect and permanent long after it has been given back to the sunbeam which produced it."[17]

Talbot himself spoke of acting on the advice of "scientific friends" before displaying some of his "photogenic drawings" at a meeting of the Royal Institution.[18] The review of that meeting suggests that it was Michael Faraday who urged Talbot to show his work and who subsequently referred to "dame Nature" as "drawing mistress" in describing it. From the memoirs of Vernon Heath we learn that, at the close of the meeting, Faraday spoke of the parallel inventions and invited everyone to examine Talbot's pictures in the library.[19] Faraday's notes merely include the comment that "photogenic drawings from Mr. Fox Talbot" were shown (fig. 48).[20] Since 1836 the editor of the *Literary Gazette*, William Jerdan, had been regularly invited to meetings of the Royal Institution in order to report on them. But it is also known that Faraday had himself supplied Jerdan with the minutes of those meetings for some time.[21] Because Faraday did not repeat the phrase presumably coined by him in either his notebooks or his correspondence from 1839, it is impossible to determine which man actually wrote the piece about the Royal Institution meeting. In his letters from that year Faraday referred to photographic methods only three times. One to John Frederick William Herschel asks him about his "varnishes," or chemical "coatings," for photographic processes,[22] and two letters to the French chemist Jean-Baptiste-André Dumas mention

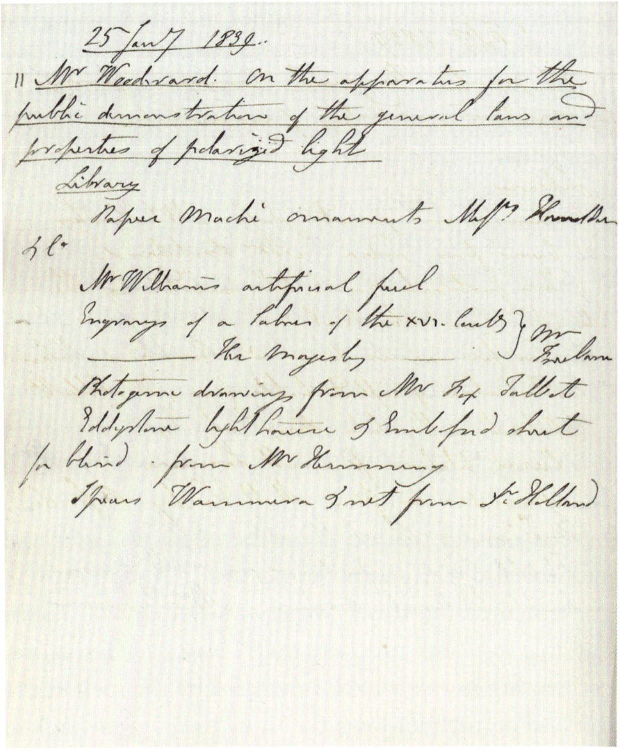

48. Michael Faraday, "Notes on Lectures," Jan. 25, 1839, Notebook RI MS F4 £, 45. Royal Institution, London. That "Photogenic drawings from Mr. Fox Talbot" were shown in the "Library" can be read in Faraday's "notes relating to the Friday Evening Discourse here on 25 January 1839."

a "Daguerre apparatus," the first indicating that Faraday is expecting it, the second assuring him that Faraday had received it.[23] His decision to acquire the French apparatus was not mere curiosity.

NATURA NATURANS, OR "THE PICTURE WHICH MAKES ITSELF"

In Talbot's letter to Jerdan at the *Literary Gazette* he revealed a few of the fundamental assumptions behind his technique, not so much in his explicit descriptions as in the conclusions he drew from them. His often-quoted statement that with his process "it is not the artist who makes the picture, but the picture which makes ITSELF"[24] was intended to emphasize the epistemic rift between the imaging technologies he had developed and the earlier camera obscura, both of which operated based on the same parameters of projection—that is, on natural laws. That assertion is not as interesting from our present point of view as his description of the course of events in this act of self-creation, which allows us to appreciate a

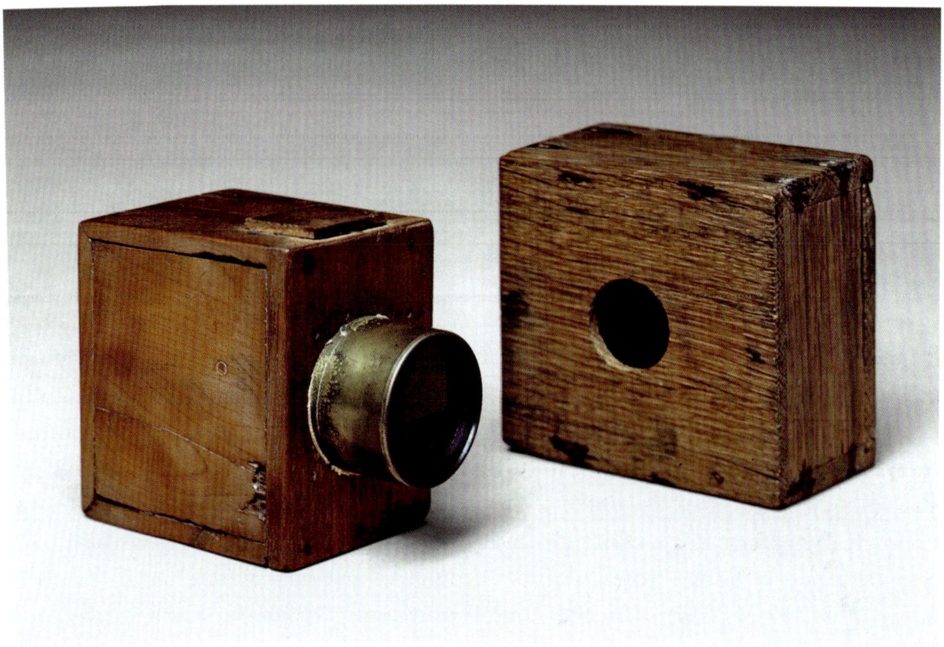

49. First small cameras used by Talbot, ca. 1835–39, National Media Museum, Bradford, in John Ward with Sara Stevenson, *Printed Light: The Scientific Art of William Henry Fox Talbot and David Octavius Hill with Robert Adamson* (Edinburgh: HMSO for the Scottish National Portrait Gallery, 1986), 11, fig. 6.

specific feature of Talbot's photography. When he wrote that the artist need only place the apparatus before the object to be pictured and leave it alone for a time, after which he can return to remove the finished image from the machine,[25] it does not simply explain why his wife, Constance, is said to have referred to the oldest camera ever employed as a "mousetrap" (fig. 49). Rather, he also pointed up the most fundamental problem with his first photographic technique: because a *finished* image could be removed from the camera, we are confronted with a manifest, or physical, picture-making process that—unlike Daguerre's—did not involve any latent picture or subsequent chemical developing. Because of this, the quality of Talbot's pictures was such that Alexander von Humboldt would refer to them dismissively as "tristes chlorures" (mirthless chlorides).[26]

Despite the quality of his pictures, to Talbot one of the most important conditions of his method was the existence of a light-sensitive chemical substance on which pictures can be imprinted. It was only because such chemicals are present in nature that it was possible to realize the desire to replicate natural objects. In his letter to the editor of the *Literary Gazette*, he wrote, "The very foundation of the art therefore consists in this—eminently curious—natural fact, viz., that there exists a substance so sensitive to light as to be capable of receiving even its faint impressions. The whole possibility of the process depends upon this; for if no such substance existed in *rerum natura*, the notion of thus copying objects would be nothing more than a scientific dream."[27] Talbot's reflections amount to a virtual explication of the "drawing mistress" metaphor attributed to Faraday. Because Nature supplies substances that allow views of herself to be imprinted onto two-dimensional picture supports—note the recurrence of the metaphor—she is capable of functioning as the initiator of a realistic pictorial process. She appears here, as in chemical treatises from about 1800, as *Natura naturans*, a generative nature that not only imparts the ability to produce such simulacra but holds in store the materials with which she is capable of producing pictures of herself.

Nature's "sensitive substances" did not simply allow a picture to be generated by way of an imprint on what Talbot described as "sensitive" paper.[28] They were also responsible for the coloring of the resulting pictures. As Talbot explained, even if a picture negative has a white ground, this "white" can be miraculously tinted: "sky-blue" or "yellow," "rose-colour," "brown, of various shades," and "black." Green was the only color a picture could not assume.[29] Here he was not referring to the reproduction of natural colors, to be sure, only of the coloring of the picture in response to different chemicals. That the chemical layer is imperfect—that is, that it reacts to blue light faster, to green much more slowly, and to red hardly at all—is explained in the texts from the beginning of 1839. Yet Humphry Davy, the first professor of chemistry at the Royal Institution, between 1802 and 1812, had already pointed out the varied effects of different light frequencies in his report on his attempts, undertaken with Thomas Wedgwood, to capture pictures in silver nitrate and silver chloride.[30] Above all, in quoting the "exposure times" of light of different colors from the third volume of Jean Senebier's *Mémoires physico-chymiques* (1782), Davy provided a striking example of how widely scientific knowledge was circulated in the early nineteenth century.[31]

"PENCIL OF NATURE"

The most famous of all the early designations of photography appears in the title of one of the "New Art" articles published in the *Literary Gazette* on February 2, 1839: "French Discovery—Pencil of Nature." This comes as a surprise, inasmuch as historians have led us to believe that it was Talbot who coined the phrase. The article, apparently written by the weekly's editor, summarizes and takes to an extreme, "at the risk of repetition," the report on Daguerre's invention that the journal had published three weeks before.[32] It begins by comparing the "new art" to the camera

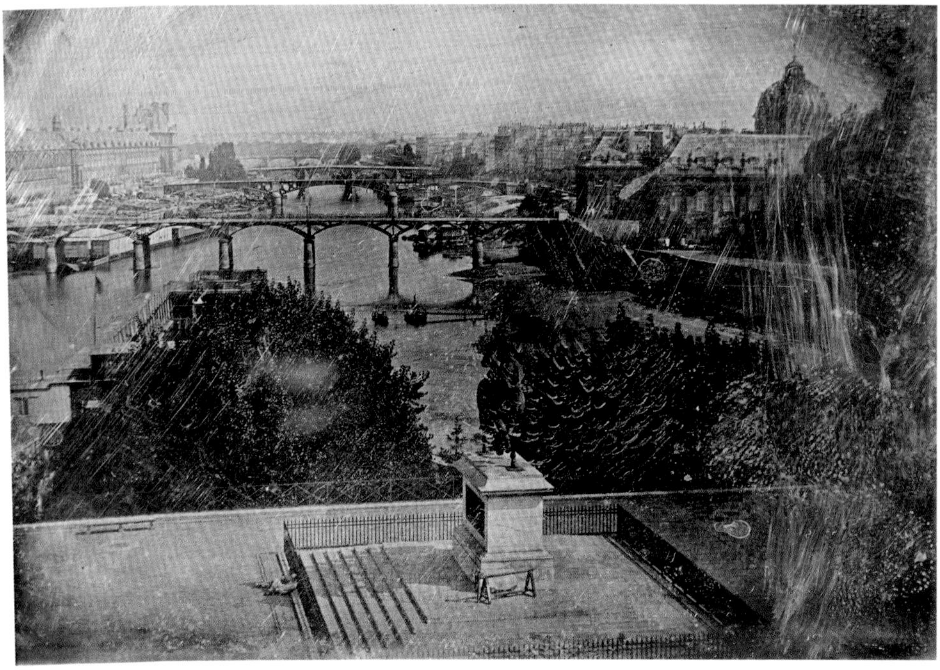

50. Louis-Jacques-Mandé Daguerre, *Vue du Pont-Neuf*, 1839, daguerreotype, 2⅞ x 4 in. (7.3 x 10 cm), Musée des Arts et Métiers, Paris. In Quentin Bajac, *Le daguerréotype français: Un objet photographique* (Paris: Musée d'Orsay, 2003), 155, fig. 36.

obscura, the "splendid and wonderful representations" of which had been universally admired and, though formerly fleeting, could now be "captured."[33] In these camera images everything was reproduced with the greatest perfection, just as in a drawing executed with mathematical precision (figs. 50 and 51). Lenses, mirrors, and cameras were what was required by this "pencil of nature," and it was they that directed and controlled it. Light rays—in fact, a synonym of *pencil*—imprinting themselves on light-sensitive substances by way of optical systems could produce pictures without parallel. "The creations of nature triumph. Far from betraying any defect, the highest magnifier only tends to shew more clearly its vast superiority."[34]

The distinction made in these articles between the English "photogenic drawing" and Daguerre's technique, described as a "pencil of nature," was not based only on the differences in their pictorial quality. British experiments in the years 1839 to 1841 concentrated on the chemistry of the light-sensitive layer (perhaps because it was clear to the proponents of the technique that it was inadequate), but in the propagation of Daguerre's method this was of only secondary importance. Although it would be mistaken to separate chemical experiments from

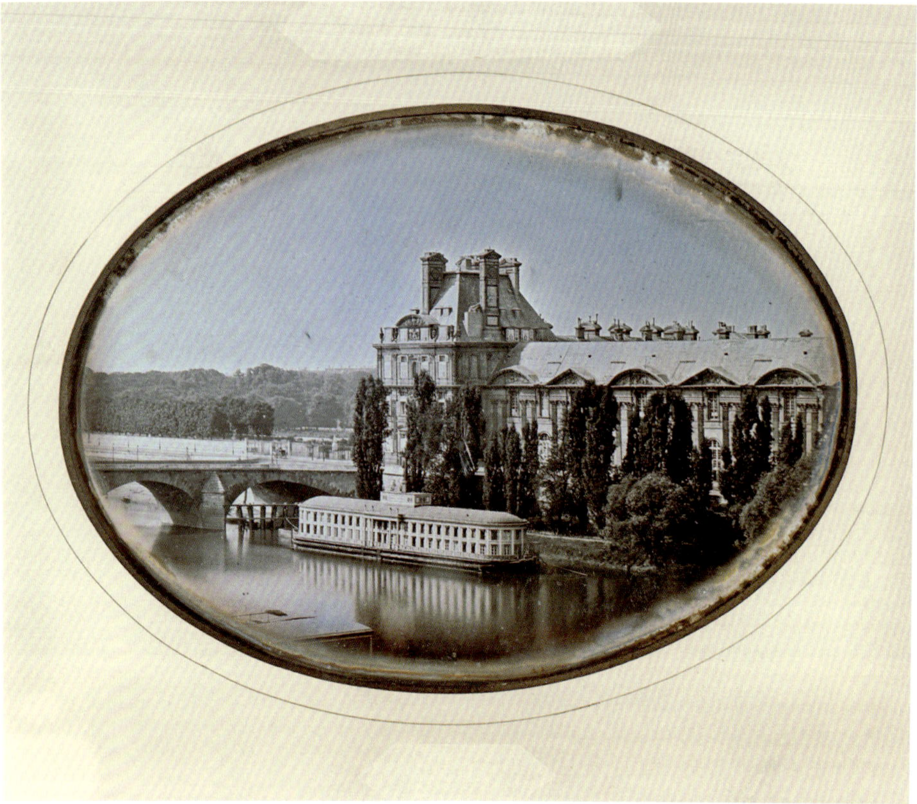

51. Louis-Jacques-Mandé Daguerre, *Le Pavillon de Flore et le Pont-Royal*, 1839, daguerreotype, 6⅜ x 8⅜ in. (16.3 x 21.2 cm), with frame 8⅝ x 10⅝ in. (22 x 27 cm), Musée des Arts et Métiers, Paris. In Quentin Bajac, *Le daguerréotype français: Un objet photographique* (Paris: Musée d'Orsay, 2003), 155, fig. 35.

optical ones, since both were important in the development and improvement of photographic techniques, it is clear that the introductory treatises on Talbot's "photogenic drawings," when lauding their draftsmanly potential, were particularly concerned with the different light-sensitive substances. By contrast, in dealing with photographic techniques as examples of "physique appliquée" (applied physics), French discussions of them—most notably by Arago in all his introductory lectures from 1839—relegated the "new art" to the history of optical devices.[35] With the term "physique appliquée" Arago not only emphasized the role of geometric optics and of optical instruments in the generation of photographic pictures but also referred explicitly to its possible applications—that is, its practical uses. Unlike physics, which is based on theoretical assumptions and mathematical proofs, modern chemistry, as developed since 1800, is a practical and pragmatic science. As Isabelle Stengers put it, "chemistry's specificity . . . derives from the fact that the operations it performs and the questions it puts are determined not by theory but by utility. Chemistry is the hybrid produced by the compromise between practical concern and physical intelligibility."[36]

THE SAMPLE IMAGES OF AN EXPERIMENTAL SCIENCE

In much early nineteenth-century writing the "core science" of chemistry is presented as a discipline that can be practiced—like photography—without expert knowledge. All one needs is the patience and perseverance to make experiments. While a camera is not necessarily required to produce photographs—as evidenced by Herschel's (fig. 52) and Talbot's camera-less photographs—the "shadows of things" cannot depict themselves without light-sensitive substances. What is more, a knowledge of the effect of light on silver chloride and silver nitrate made possible the experimental investigation of light itself. Herschel had already reviewed the history of light in chemistry (or the history of chemistry in the study of light) in the closing pages of his book-length article "Light" in the *Encyclopaedia Metropolitana*.[37]

Like Davy before him, Herschel recounted in the final paragraph of his study the first attempts by Johann Wilhelm Ritter, in 1801 in Jena, to visualize ultraviolet light by means of photosensitive paper.[38] Herschel also wrote of Carl Wilhelm Scheele and his discovery in 1777 "that muriate of silver is much more powerfully blackened in the violet rays than any other part of the spectrum."[39] Moreover, Herschel pointed out that, as "satisfactorily demonstrated" by Thomas Young in 1801 and 1803, and by Arago in 1811 and with Augustin Fresnel in 1816, the chemical rays "conform to the same laws of interference, whether polarized or otherwise, that are obeyed by the luminous rays similarily circumstanced. Thus, a set of fringes formed by the interference of two solar pencils with a common origin, being kept very steadily projected for a long time on one and the same part of a sheet of paper rubbed with muriate of silver, a series of black lines became traced on it." In his summary Herschel explained—as he would again in 1839—that "solar rays . . . possess at least three distinct powers: those of heating, illuminating, and effecting chemical combinations."[40] He did not neglect to mention that his father, William Herschel, had already distinguished between illuminating and heating rays.[41]

It was not only in their writings and notebooks that all those who can be associated with the publication of photography in 1839 referred to the photochemical experiments of their predecessors. From the iconography of the first photographs, it is obvious that some of the English experimenters appropriated the very picture subjects of earlier researchers as well as their chemicals.[42] Although

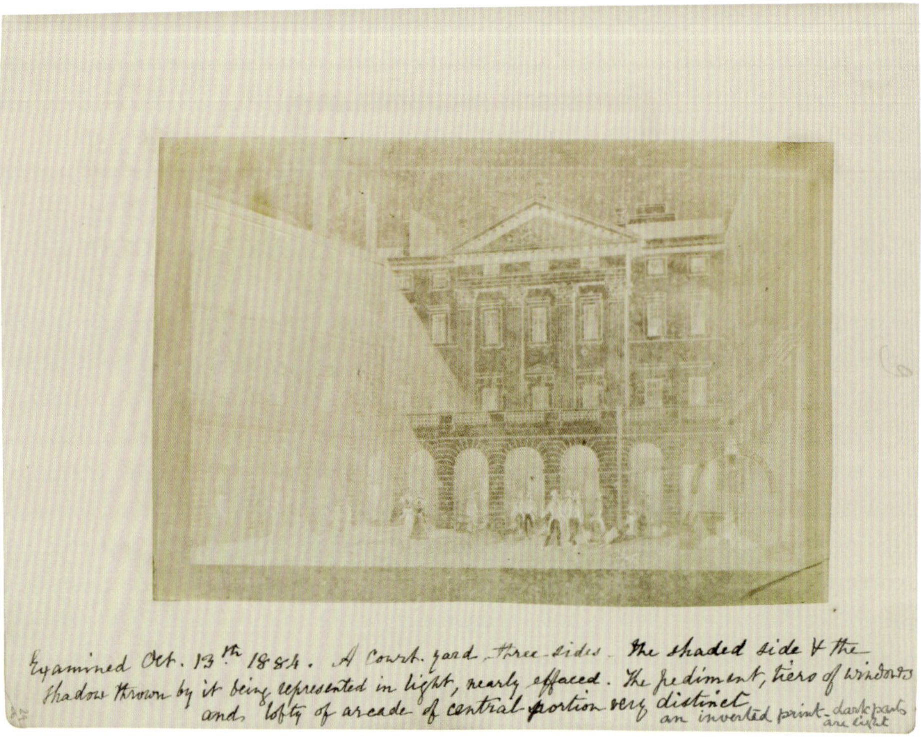

52. John Frederick William Herschel, *Specimen 9*, photograph on paper negative, notebook dimensions: 7 x 8⅝ in. (17.8 x 21.7 cm), in "Specimens of Photography to Accompany Sir J. Herschel's Paper," read at the meeting of the Royal Society on March 14, 1839," notebook, 1. National Media Museum, Bradford.

Talbot denied having been aware of Davy and Wedgwood's photographic experiments when he began his own,[43] they nevertheless served as the patterns for his first photogenic drawings. The subjects of the pictures he displayed in the library of the Royal Institution on January 25 could easily be mistaken for those that Davy and Wedgwood had listed as being suitable for the copying process: "the outlines and shades of paintings on glass may be copied, or profiles of figures procured. . . . The woody fibres of leaves, and the wings of insects may be pretty accurately represented by means of it."[44] Davy continued: "I have found that the images of small objects, produced by means of the solar microscope, may be copied without difficulty."[45] Talbot described the subjects of his sample pictures in his letter to Jerdan at the *Literary Gazette*: "pictures of flowers and leaves [figs. 53 and 54]; a pattern of lace; figures taken from painted glass [fig. 55]; a View of Venice copied from an engraving; some images formed by the Solar Microscope, viz. a slice of wood very

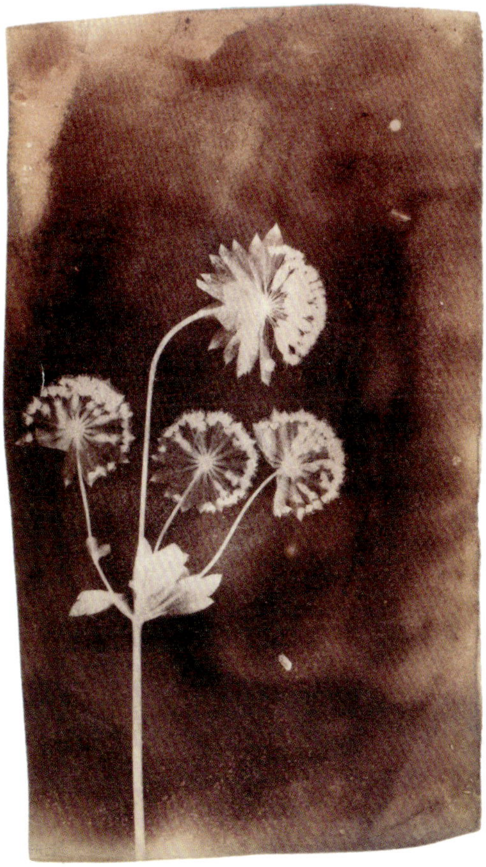

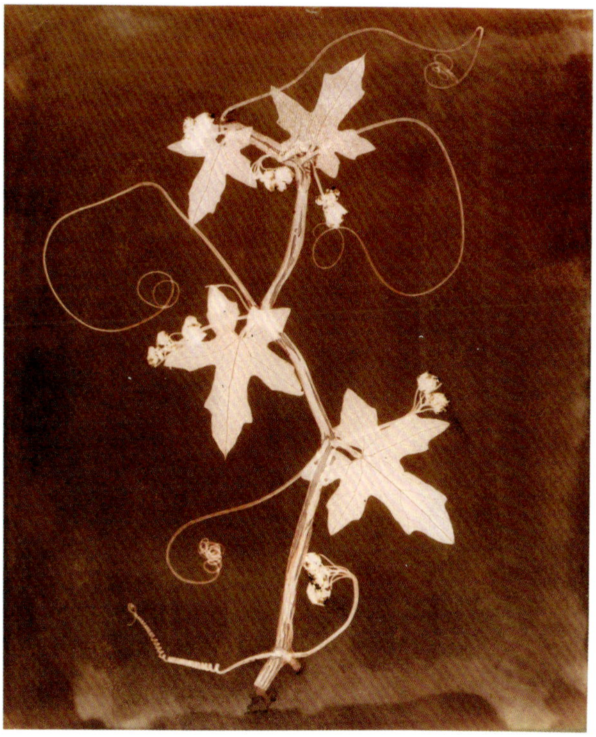

53. William Henry Fox Talbot, *Astrantia Major*, "the melancholy gentleman," Nov. 13, 1838, photogenic drawing, negative, 6⅞ x 3¾ in. (17.3 x 9.5 cm), print irregularly inscribed in ink on verso: "fx," "ab.," and "nov. 13/38." Collection of The Royal Photographic Society, National Media Museum, Bradford.

54. William Henry Fox Talbot, "Botanical Specimen — *Bryonia Dioca*, The English Wild Vine," ca. 1835–36, salted paper print, 8⅞ x 7¼ in. (22.4 x 18.3 cm), National Media Museum, Bradford. The "pictures of flowers and leaves" by Talbot were displayed at the Royal Institution on January 25, 1839; Humphry Davy and Thomas Wedgwood agreed that the "woody fibres of leaves" were very well depicted.

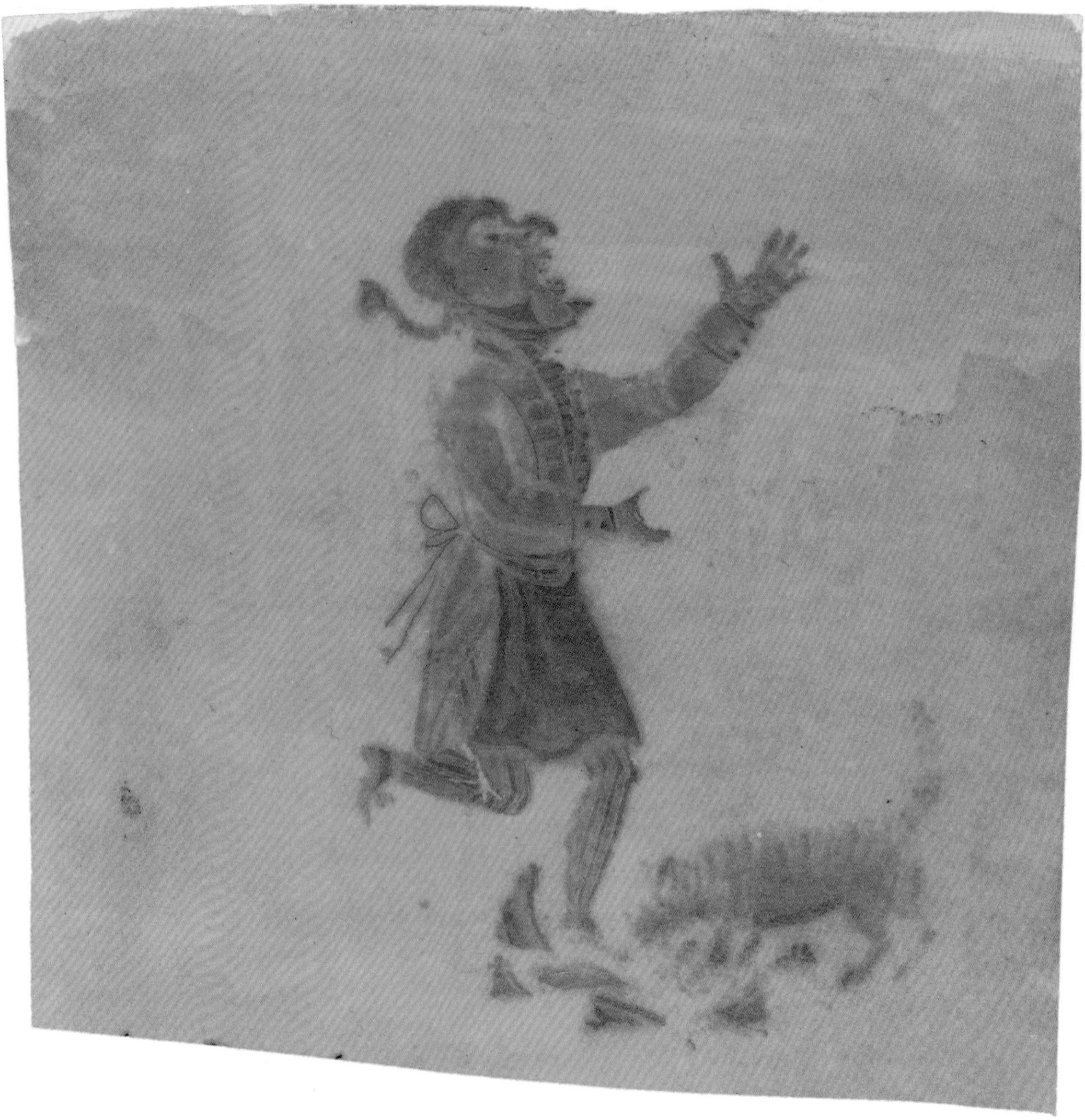

55. William Henry Fox Talbot, Copy of a joker and cat, ca. 1835, photogenic drawing negative, 27⅛ x 27⅝ in. (69 x 70 cm), overall 39 x 27⅝ in. (99 x 70 cm), mounted on black paper. National Media Museum, Bradford. Humphry Davy and Thomas Wedgwood wrote that the "outlines and shades of paintings on glass may be copied." In his letter to William Jerdan at the *Literary Gazette* on January 30, 1839, Talbot stated that he had displayed "figures taken from painted glass" in the library of the Royal Institution.

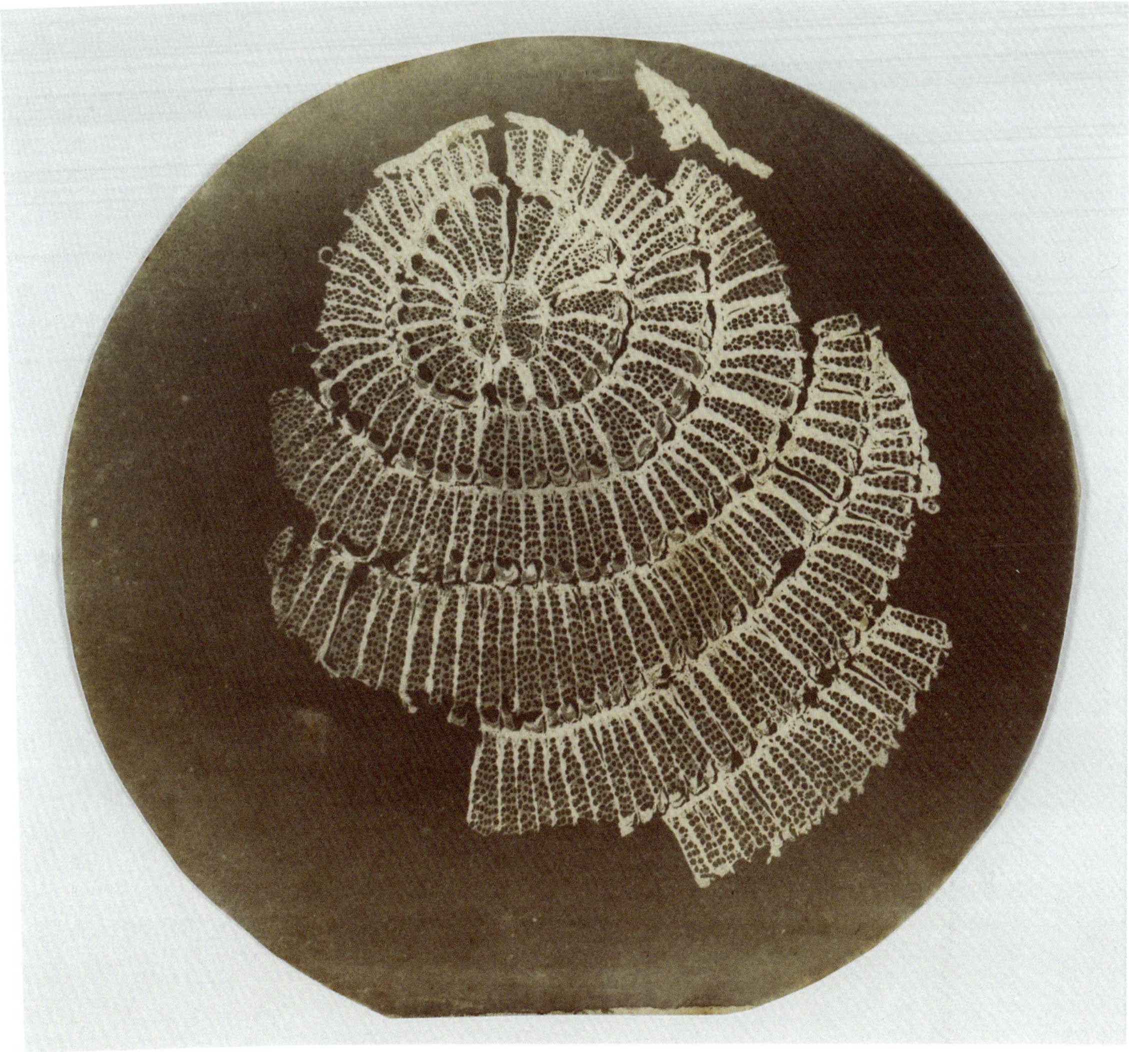

56. William Henry Fox Talbot, photomicrograph of transverse section of stem taken with a solar microscope, ca. 1839, photogenic drawing negative, 42⅞ x 41¾ in. (109 x 106 cm). National Media Museum, Bradford. According to Davy, this image—which Talbot described as "a slice of wood highly magnified"—is one of the "images of small objects, produced by means of the solar microscope, [which] may be copied without difficulty" and that Talbot showed on January 25, 1839.

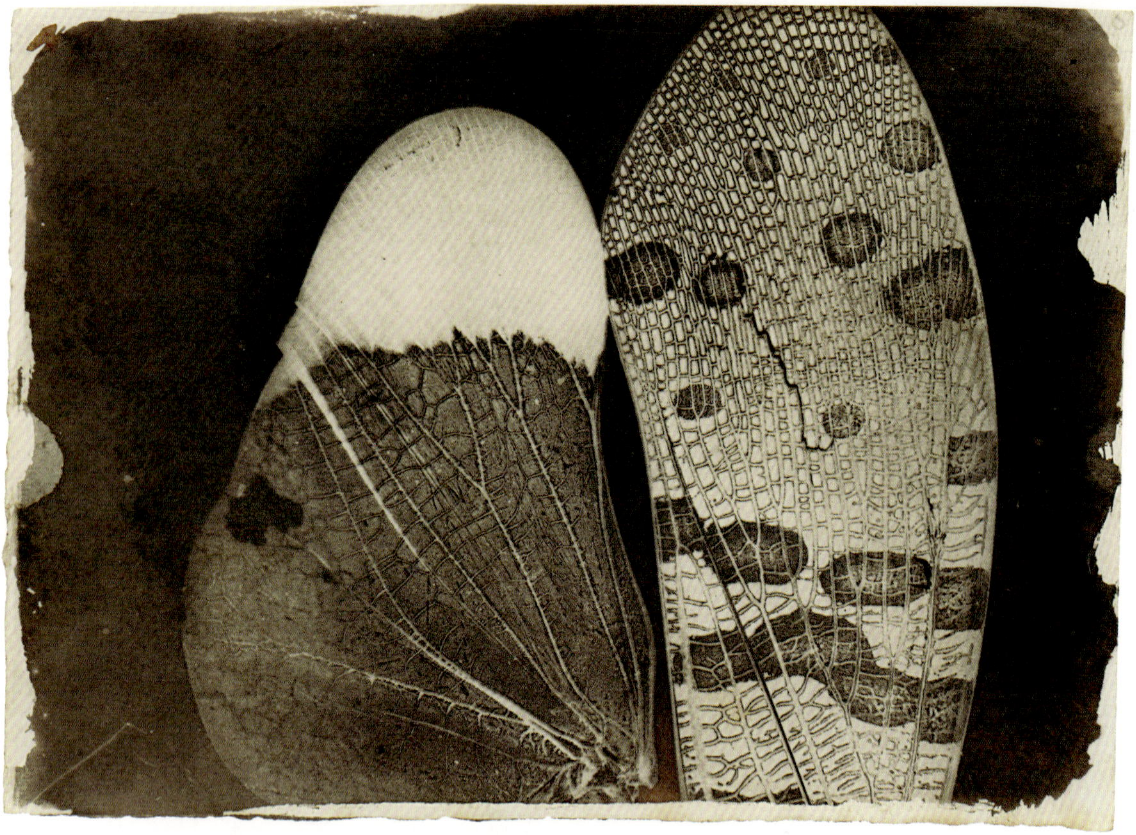

57. William Henry Fox Talbot, insect wing (lantern fly), 1839, photomicrographs, negative, 4½ x 6⅛ in. (11.5 x 15.5 cm). National Media Museum, Bradford. On January 25, 1839, Talbot also displayed a "Microscopic sketch exhibiting the reticulations on the wing of an insect" and, in his letter to William Jerdan, referred to Humphry Davy's and Thomas Wedgwood's "wings of insects," which they had considered to be photogenic.

highly magnified [fig. 56], exhibiting the pores of two kinds, one set much smaller than the other, and more numerous. Another Microscopic sketch exhibiting the reticulations on the wing of an insect [fig. 57]."[46] Whereas the silver nitrate and camera optics available around 1800 were still too insensitive to register "the images formed by means of a camera obscura,"[47] Talbot's "various pictures, representing the architecture of my house in the country; all these made with the Camera Obscura in the summer of 1835"[48] attest to the advances in the applied sciences of optics and chemistry in the meantime as well as to his later introduction of new subject matter (fig. 58).

58. William Henry Fox Talbot, *An Oriel Window at Lacock Abbey*, 1835, photogenic drawing negative, 3¼ x 4¼ in. (8.3 x 10.7 cm) (irregular). The Metropolitan Museum of Art, New York, The Rubel Collection (1997.382.1). Talbot's specimens included various pictures depicting the architecture of Lacock Abbey, which he photographed with the camera obscura in the summer of 1835.

PHOTOGRAPHY AS APPLIED CHEMISTRY

Let us turn once again to early 1839, "The New Art," and the *Literary Gazette*. In another letter to Jerdan, dated April 8 and published on April 13, Talbot qualified his activities leading to the exhibit at the Royal Institution and his account to the Royal Society as an overreaction occasioned by the announcement of the French invention.[49] That same April number of the *Literary Gazette* published another letter to the editor, this one from a certain C[harles] Toogood Downing. His intention in writing it was to provide everyone with the knowledge with which to produce a successful photograph. To that end he not only described the conditions under which photographic images could be produced but also provided precise information about the quantities and proportions of the chemicals required. Downing's instructions were meticulous, perfectly in accordance with the most up-to-date chemical knowledge and employing the symbolism of modern chemistry:

> In the first manner of preparing the paper, it is recommended that it should be alternately washed with solutions of table salt and lunar caustic, in order to form a coating of chloride of silver. The chemical title of table salt is, *as you well know*, chloride of sodium; and as the atomic numbers of these elementary substances are respectively 35.42 and 23.3, according to Berzelius and Turner, the equivalent of the salt is 58.72. The caustic, consisting of (n+50+ag [argent] +o), has, according to the same authorities, the equivalent of 170.15. It must be evident, therefore, that if these two salts are mixed together for the purpose of exact mutual decomposition, their relative proportions should be as 58.72 to 170.15.[50]

Both the pressure Talbot felt and Downing's explanation of the chemistry involved in photography can be related to the contents of a letter that Arago addressed to the French interior minister, Charles-Marie Tanneguy, comte Duchâtel.[51] In it he tried to convince the minister, the official responsible for state pensions, to purchase Daguerre's technique for the French state, arguing that it was impossible to reveal anything about the technique of photography without making it available to all. Its publication would enable anyone to make daguerreotypes, for which reason marketing it by way of subscription was an impossibility. The feverish exertions in England in the first months of 1839 not only show Arago to have been correct in his concern but also point out a problem from the point of view of the history of science. Although Downing's letter appears to support Arago's argument that anyone at all might practice photography, not just scientists such as Herschel and Talbot with experience working with silver compounds, Downing himself had to have received a specific kind of professional training in order to be able to communicate—at a moment's notice, as it were—this kind of photochemical information.[52]

Few details of Downing's life are known, but one can deduce from his writings that in 1836–37 he worked as a resident physician for the East India Company in China.[53] According to its regulations a resident physician was required to demonstrate a knowledge of chemistry as well as surgery and physics, as spelled out in the *London Manual of Medical Chemistry* of 1831.[54] Downing was awarded membership in the Royal College of Surgeons on May 20, 1837, which indicates that—in accordance with its bylaws—he had attended at least one course of lectures on chemistry.[55]

The headline for all materials on photographic processes in the *Literary Gazette* in early 1839 was "The New Art." In 1830, in his *Preliminary Discourse on the Study of Natural Philosophy*, John Frederick William Herschel wrote, "Art is the application of knowledge to a practical end," and added that "if the knowledge be merely accumulated experience, the art is *empirical*; but if it be experience reasoned upon and brought under general principles, it assumes a higher character, and becomes a *scientific art*. In the progress of mankind from barbarism to civilised life, the arts necessarily precede science."[56] In Herschel's terms photography, developed for practical purposes—that is to say, use-related and arrived at empirically—was still in an experimental stage at the beginning of 1839. Yet Herschel himself did much to subject the experiences gained in dealing with photography to general principles. For example, immediately after hearing about Daguerre's discovery from Captain Francis Beaufort on January 22, 1839, he systematically began to investigate the possible ramifications of what he had learned (see fig. 52).[57] Photography became what Herschel considered a "scientific art" in England only after Talbot had learned that much better pictures could be produced by the calotype process, involving a chemical developer, which he patented in 1841. Those pictures were in some respects as good as those of French daguerreotypy—if not superior, in that they could be reproduced.[58]

When analyzing the first texts devoted to the technique of photography from the spring of 1839, we are confronted with a concept of nature that—if one thinks of nature as active and operative—was rooted in the model of chemical causation and chemical experiment. As Herschel explained, chemistry was "the most popular and the most useful of the sciences. Its pursuit requires little concentrated thought and no important mathematical knowledge."[59] Around 1800 chemistry was an applied, useful science. Knowledge attained through the "experimental method of chemical analysis" was considered—as Ursula Klein explained—of use in those professions "in which a number of varied substances were employed . . . most especially in pharmacy."[60] In Germany, and later in England as well, chemistry was accordingly introduced into the newly established educational institutions devoted to the practical training of doctors, miners, engineers, and military officers. Having studied chemistry for professional reasons, all their graduates were capable of practicing the "new art" once they had heard of the possibility of making photographic images. As a practical chemist Downing, for example, who had not only passed his examination before the Royal College of Surgeons but also "obtained the practice certificate as apothecary,"[61] was able to communicate photochemical information even more precisely than those who discovered photography and further developed it.[62] Inasmuch as chemistry, the most popular and most useful—though nonmathematical—science, provided the means by which photographs could be produced, the image of Nature as drawing mistress serves as an allegory of both photography and chemistry itself.

This essay is an abridgment of the first chapter of my forthcoming book *Musterbilder*.

1. Hippolyte Gaucheraud, "Beaux Arts: Nouvelle découverte," *Gazette de France* (Jan. 6, National Media Museum, Bradford, 1839): 1; Gaucheraud, "The Daguerotype," *Literary Gazette; and Journal of the Belles Lettres, Arts, Sciences, &c.* (hereafter *Literary Gazette*), Jan. 12, 1839, 28; and François Arago, "Physique appliquée: Fixation des images qui se forment au foyer d'une chambre obscure," *Comptes rendus de l'Académie des Sciences* 8 (meeting of Jan. 7, 1839), 4–6. See also Anne McCauley, "François Arago and the Politics of the French Invention of Photography," in *Multiple Views: Logan Grant Essays on Photography, 1983–89*, ed. Daniel P. Younger (Albuquerque: Univ. of New Mexico Press, 1991), 43–69; and Paul-Louis Roubert, *L'image sans qualités: Les beaux-arts et la critique à l'épreuve de la photographie, 1839–1859* (Paris: Monum/Éditions du Patrimoine, 2006), 36–39.
2. "The New Art," *Literary Gazette*, Feb. 2, 1839, 72–75. This comprises, as noted on 73, "1. Mr. Talbot's Letter. 2. The account of the French Invention. 3. Report of the Royal Institution; and, 4. Report of the Royal Society."
3. "New Art," 75.
4. See Frank A. J. L. James, "Reporting Royal Institution Lectures, 1826–1867," in *Science Serialized: Representations of the Sciences in Nineteenth-Century Periodicals*, ed. Geoffrey Cantor and Sally Shuttleworth (Cambridge, MA, and London: MIT Press, 2004), 75–76.
5. See Maurice P. Crosland, *Science under Control: The French Academy of Sciences, 1795–1914* (Cambridge: Cambridge Univ. Press, 1992), 35.
6. William Henry Fox Talbot, "Some Account of the Art of Photogenic Drawing, or the Process by Which Natural Objects May Be Made to Delineate Themselves without the Aid of the Artist's Pencil," *Athenaeum*, no. 589 (Feb. 9, 1839): 114–17; and in a shorter version in *Abstracts of the Papers Printed in the Philosophical Transactions of the Royal Society of London*, vol. 4, 1837–43 (Jan. 31, 1839), 120–21.
7. William Henry Fox Talbot, *The Pencil of Nature* (London: Longman, Brown, Green and Longmans, 1844), 1–13. See also the essay by Vered Maimon in this volume.
8. James A. Secord suggests that one should rely not only on models from the history of science in descriptions of changing scientific knowledge but also on those of literary and media studies. Secord, "Knowledge in Transit," *Isis* 95, no. 4 (Dec. 2004): 654–72, esp. 666–68. See also my reference to the importance of "philological readings" for the historiography of photography: Herta Wolf, "Pröbeln und Musterbild: Die Anfänge der Fotografie," in *Bildwissenschaft*, ed. Thorsten Hoffmann and Gabriele Rippl (Göttingen: Wallstein, 2006), 112, 124–25.
9. See, for example, Talbot's references, written during the course of his early work on the chemical effects of light, to related studies by William Herschel, Thomas Johann Seebeck, Macedonio Melloni, and Sir Humphry Davy. Notebook L, "London, 8th January 1834," Loan Collection, British Library.
10. Maurice P. Crosland, *In the Shadow of Lavoisier: The "Annales de Chemie" and the Establishment of a New Science* (Oxford: British Society for the History of Science, 1994), 34–35.
11. For more on this, see Wolf, "Pröbeln und Musterbild," 112, 124–25; see also the overview of Talbot's work by Russell Roberts, "Traces of Light: The Art and Experiments of William Henry Fox Talbot," in *Huellas de luz: El arte y los experimentos de William Henry Fox Talbot*, exh. cat. (Madrid: Museo Nacional Centro de Arte Reina Sofia, 2001), 361–80, esp. 364–66; and Roberts, "Specimens and Marvels: The Work of William Henry Fox Talbot," in *Specimens and Marvels: William Henry Fox Talbot and the Invention of Photography* (New York: Aperture Foundation; Bradford, UK: National Museum of Photography, Film and Television, 2000), 5–15. Trial and error and methodical experiment also characterize the evolution of the media, optics, and chemistry that gave rise to photographs; see Herta Wolf, "'Es werden Sammlungen jeder Art entstehen': Zeichnen und Aufzeichnen als Konzeptualisierungen der folografischen Medialität," *Zeitschrift für Medienwissenschaft* 2, no. 3 (2010): 27–41.
12. I have previously explained that the locution "dame Nature" alludes to drawing as an aristocratic skill and virtue and speaks to the iconography of the first British photographs, especially Talbot's. "Art master" or "art mistress" goes back to the Latin *magister*, from which "master" was derived, simply meaning "teacher." Herta Wolf, "'Collections of All Kind Will Be Formed,'" *Photoresearcher*, no. 13 (2010): 54–64.

13. The two techniques had been developed "by two scientific experimenters in different countries, without a knowledge of each other's pursuit." The display of Talbot's work was meant to "establish a date," should each discovery "be proved to be original." "New Art," 74–75.

14. See, for example, Ann Bermingham, *Learning to Draw: Studies in the Cultural History of a Polite and Useful Art* (New Haven and London: Yale Univ. Press, 2000); and Kim Sloan, *"A Noble Art": Amateur Artists and Drawing Masters, c. 1600–1800*, exh. cat. (London: British Museum Press, 2000).

15. "Letter from Sir Chr. Wren to Mr. Treasurer Hawes, November 24, 1694," in *The Wren Society*, vol. 11 (Oxford: Oxford Univ. Press, 1934), 74. See also Sloan, *"Noble Art,"* 104, 108; and Michael Clarke, *The Tempting Prospect: A Social History of English Watercolours* (London: Colonnade Books by British Museum Publications, 1981), 93.

16. *Records of the Royal Military Academy, 1741–1892*, ed. William Jones (London: F. J. Cattermole, 1883), 33, quoted in Luke Hermann, *Paul and Thomas Sandby* (London: Batsford, 1986), 26. See also Bermingham, *Learning to Draw*, 84.

17. "New Art," 75.

18. "New Art," 73–74.

19. Vernon Heath, *Recollections* (London: Cassell, 1892), 48–49. See also Larry J. Schaaf, "'A Wonderful Illustration of Modern Necromancy': Significant Talbot Experimental Prints," in *Photography: Discovery and Invention* (Malibu, CA: J. Paul Getty Museum, 1990), 31–46, esp. 45 nn. 1, 2.

20. See Faraday's Notebook RI MS F4 £, 45, with his "notes relating to the Friday Evening Discourse here on 25 January 1839." In his Notebook RI MS Le 2, Faraday records that Royal Institution discourses were attended by 325 people. Who they were, however, "is impossible to recover." Frank James, e-mail message to the author, Oct. 20, 2010; I am grateful to the editor of Faraday's writings for these references.

21. See James, "Reporting Royal Institution Lectures," 71–72, 74–75.

22. *The Correspondence of Michael Faraday*, ed. Frank A. J. L. James, vol. 2, 1832–Dec. 1840, Letters 525–1333 (London: Institution of Electrical Engineers, 1993), 629. Humphry Davy had also used the term "varnish" to describe the coatings with the help of which he tried to fix shadows on silver chloride and silver nitrate; Humphry Davy and Thomas Wedgwood, "An Account of a Method of Copying Painting upon Glass, and of Making Profiles, by the Agency of Light upon Nitrate of Silver: Invented by T. Wedgwood, Esq. with Observations by H. Davy," *Journals of the Royal Institution of Great Britain* 1, no. 9 (1802): 172.

23. See letter 1204, Faraday to Jean-Baptiste-André Dumas, Sept. 2, 1839, and letter 1207, Faraday to Jean-Baptiste-André Dumas, Sept. 5, 1839, in *Correspondence of Michael Faraday*, 2:607, 610.

24. "New Art," 75.

25. The camera obscura had been referred to as a "machine" nearly eighty years before; see Jean-Antoine Nollet, *Leçons de physique expérimentale*, vol. 5, 2nd ed. (Paris: Chez Hippolyte-Louis Guerin et al., 1758), 534.

26. "M. Herschell [*sic*] a aussi la maladie photogénique, j'ai reçu de lui et de M. Talbot de tristes chlorures, des silhouettes *blanches*, semblables à des vieilles estampes sur lesquelles a passé le coude." (Mr. Herschel, too, has the photogenic illness. I received a letter from him and Mr. Talbot with *white* silhouettes, mirthless chlorides, resembling old stamps which someone has wiped with his elbow.) Letter LIII, April 25, 1839, in *Correspondance d'Alexandre de Humboldt avec François Arago (1809–1853), publiée avec une préface et des notes par Le Dr. E.-T. Hamy* (Paris: E. Guilmoto, 1907), 179 (my translation). The discovery that not only did mercury vapors make "invisible" exposed plates visible but, by such subsequent (chemical) development, the represented objects were reproduced with an extraordinary pictorial quality was, along with a camera that admitted a greater amount of light, the crucial improvement of Daguerre's method over that of Joseph Nicéphore Niépce.

27. "New Art," 73.

28. "New Art," 73. See also Talbot, "Art of Photogenic Drawing," 116.

29. Talbot, "Art of Photogenic Drawing," 114. The second section is entitled "Effect and Appearance of these Images."

30. Davy and Wedgwood, "Copying Painting upon Glass," 171.

31. Jean Senebier, *Mémoires physico-chymiques Sur l'influence de la lumière solaire pour modifier les êtres des trois règnes de la nature, & surtout ceux du règne végétal*, vol. 3 (Geneva: Barthelemi Chirol, 1782), chap. 3, "Action de la lumière solaire sur les précipités d'argent, & sur-tout sur la lune cornée," 192–205. Moreover, in a footnote Humphry Davy and Thomas Wedgwood mentioned similar experiments by Carl Wilhelm Scheele, which had been confirmed by Senebier. They also referred to William Herschel and William Hyde Wollaston's experiments on the workings of light and to Johann Wilhelm Ritter's statements on the reaction of horn silver (silver chloride) to blue and ultraviolet light, published in *Annalen der Physik* 7, no. 4 (1801): 527. Davy and Wedgwood, "Copying Painting upon Glass," 171.
32. Gaucheraud, "Daguerotype," 28.
33. "New Art," 74.
34. "New Art," 74.
35. See, for example, Arago, "Physique appliquée," vol. 8, 4–6; and François Arago, "Le Daguerréotype," in *Comptes rendus de l'Académie des Sciences* 9 (meeting of Aug. 19, 1839), 250–52.
36. Isabelle Stengers, "Ambiguous Affinity: The Newtonian Dream of Chemistry in the Eighteenth Century," trans. Horst Brühmann, in *A History of Scientific Thought: Elements of a History of Science*, ed. Michel Serres (Oxford: Blackwell Reference, 1995), 374.
37. John Frederick William Herschel, "Light," in *Encyclopaedia Metropolitana; or, Universal Dictionary of Knowledge on an Original Plan . . .*, ed. Edward Smedley, vol. 4 (London, 1830), 341–582.
38. Herschel, "Light," 581–82. Ritter wrote in 1801: "On the 22nd of February I also found sun rays on the violet end of the color spectrum, outside of it, revealed with chlorargyrite [silver chloride]. They reduce it even more than violet light itself, and the range of these rays is very great." Johann Wilhelm Ritter, letter to the editor (Ludwig Wilhelm Gilbert), *Annalen der Physik* 7, no. 4 (1801): 527. See also Ritter, "Experiments on the Invisible Rays of the Solar Spectrum," *Journal of Natural Philosophy, Chemistry, and the Arts* 5 (1803): 255–56; and Walter D. Wetzels, "J. W. Ritter and Romantic Physics," in *Romanticism and the Sciences*, ed. Andrew Cunningham and Nicholas Jardine (Cambridge: Cambridge Univ. Press, 1990), 207–8. I am grateful to Jan Frercks, who in an e-mail on May 25, 2004, listed all of Ritter's publications relating to photography.
39. Charles-Guillaume Scheele, *Traité chimique de l'air et du feu: Avec une introduction de Torbern Bergmann*, trans. Baron de Dietrich (Paris: Sous le privilège de l'Académie, 1781), 145, quoted in Herschel, "Light," 581–82. Scheele's book was originally published in Geman in 1777. It is remarkable that Herschel did not quote the English edition, which preceded the one in French: *Chemical Observations and Experiments on Air and Fire: With a prefatory introduction, by Torbern Bergman. To which are added Notes, by Richard Kirwan with a Letter to Him from Joseph Priestley*, trans. John-Reinold Forster (London: printed for J. Johnson, 1780), 92–93.
40. Herschel, "Light," 582. See also John Frederick William Herschel, "Note on the Art of Photography or the Application of the Chemical Rays of Light to the Purposes of Pictorial Representation," in Larry Schaaf, "Sir John Herschel's 1839 Royal Society Paper on Photography," *History of Photography* 3, no. 1 (1979): 47–60; and Herta Wolf, "Die Augenmetapher der Fotografie," in *Neue Vortraege zur Medienkultur*, ed. Claus Pias (Weimar: Verlag und Datenbank für Geisteswissenschaften, 2000), 201–31.
41. Herschel, "Light," 582. Davy and Wedgwood also referred to the elder Herschel. See Davy and Wedgwood, "Copying Painting upon Glass," 171. Interestingly enough, their reference was to the German compilation by Ludwig Wilhelm Gilbert of two of William Herschel's essays, "Untersuchungen über die wärmende und erleuchtende Kraft der farbigen Sonnenstrahlen: Versuch über die nicht-sichtbaren Strahlen der Sonne und deren Brechbarkeit" and "Einrichtung grosser Teleskope zu Sonnenbeobachtungen," in *Annalen der Physik* 7, no. 2 (1801): 137–56. In a footnote on 149, Gilbert listed all the experiments known to him on silver chloride's sensitivity to light, including those of Scheele. This, in turn, was what prompted Ritter to write his letter to Gilbert on Feb. 22, 1801 (see note 38 above). Talbot also referred to the elder Herschel's text in Notebook L, 63, Loan Collection, British Library, likewise in connection with *Annalen der Physik*, by that time (since 1824) edited by Johann Christian Poggendorff (see fig. 47).
42. See Wolf, "Pröbeln und Musterbild," 112, 124–25.

43. Talbot began his account to the Royal Society with the assertion that it was only after obtaining his first "curious results," having "enlarged and corrected" his methods as a result of his "experience," that he "thought of inquiring whether this process had been ever proposed or attempted before." He then referred to the "only definite account of the matter," the report by Davy and Wedgwood. Talbot, "Art of Photogenic Drawing," 114–15.

44. Davy and Wedgwood, "Copying Painting upon Glass," 171, 172.

45. Davy and Wedgwood, "Copying Painting upon Glass," 173.

46. "New Art," 74.

47. Davy and Wedgwood, "Copying Painting upon Glass," 172.

48. "New Art," 74.

49. In order to protect his claim to the discovery, he had been forced "to use all possible expedition in drawing up a statement, and exhibiting specimens to the Royal Society. If I had known that M. Daguerre intended to keep his process so long a secret, I would certainly not have written on the subject in so imperfect and hasty a manner." Talbot, "Photogenic Art," *Literary Gazette*, April 13, 1839, 235–36; that Talbot "exhibited to the [Royal] Society" "portraits of a house and ancient architecture" already can be read in the summary of the meeting of Jan. 31. "Royal Society," *Athenaeum*, no. 588 (Feb. 2, 1839): 97.

50. Charles Toogood Downing, "To the Editor of *The Literary Gazette*, April 8, 1839," *Literary Gazette*, April 13, 1839, 236 (emphasis mine).

51. "Lettre sur la découverte de MM. Niépce et Daguerre," in François Arago, *Œuvres complètes, Mélanges*, ed. J.-A. Barral, vol. 12 (Paris: Gide, 1859), 724–25.

52. See, for example, William Henry Fox Talbot, "Remarks on Chemical Changes of Colour," *London and Edinburgh Philosophical Magazine and Journal of Science* (hereafter *Philosophical Magazine*) 2 (1833): 359–60; and Talbot, "On the Nature of Light," *Philosophical Magazine* 7 (1835): 116.

53. Charles Toogood Downing, *The Fan-Qui in China in 1836–7*, 3 vols. (London: Henry Corburn, 1838).

54. See William Maugham, *The London Manual of Medical Chemistry, Comprising an Interlinear Verbal Translation of the Pharmacopoeia . . .* (London: Whittaker, Treacher, 1831), xv–xvii. After demonstrating such knowledge in an examination before the Royal College of Surgeons, one might be admitted as a member—as was Downing. As for the importance of the *Pharmacopoeia* as a chemistry textbook, see also Ursula Klein, "Not a Pure Science: Chemistry in the 18th and 19th Centuries," *Science* 306, no. 5698 (Nov. 5, 2004): 981–82. Silver salts and their properties are also described in Maugham's manual.

55. See "Bye-Laws of the college, and Standing-orders of the Court of Examiners . . .": "That from, and after, the 1st day of January, 1828, every candidate for the Diploma . . . will be required to produce certificates of having attended one course of lectures on chemistry." I am grateful to Geraldine O'Driscoll, Library and Archive Assistant of the Royal College of Surgeons, for providing me with the bylaws and a list of members of the college in a letter of Aug. 24, 2011.

56. John Frederick William Herschel, *Preliminary Discourse on the Study of Natural Philosophy*, 2nd ed. (London: Longman, Rees, Orme, Brown & Longman and John Taylor, 1833), 70–71.

57. Beaufort was the British Royal Navy cartographer and hydrographer after whom the wind force scale was named in 1906.

58. Even though he claimed to have placed the new art on a sure foundation, Talbot remarked—quite as though acknowledging the unforeseeable workings of the drawing mistress Nature—that "I do not profess to have perfected an Art, but to have *commenced* one; the limits of which it is not possible at present exactly to ascertain." "New Art," 74.

59. Quoted in Walter Faw Cannon, "John Herschel and the Idea of Science," *Journal of the History of Ideas* 22, no. 2 (April–June 1961): 224.

60. Ursula Klein, "Der Chemiekult der Frühromantik," in *Friedrich Schleiermacher in Berlin, 1796–1802*, ed. Andreas Arndt (Berlin: Walter de Gruyter, 2009), 91–92. As the testing schedules for assistant surgeons showed, this was also true of England.

61. See *A List of Persons Who Have Obtained Certificates of Their Fitness and Qualification to Practice as Apothecaries, from August 1, 1815, to July 31, 1840* (London: Gilbert and Rivington, 1840), 59.

62. See "Further Particulars Respecting Photogenic Drawing," *London Saturday Journal* (May 11, 1839): 293–94: "In Mr. Talbot's first communication . . . he omitted to state the means he had used to procure accurately shaded drawings and dark outlines, and to multiply impressions of the same design . . . but, for the benefit of those who are desirous of preparing the paper themselves, we will transcribe the proportions stated to be the best, as given by Mr. C. Toogood Downing, in a communication to the *Literary Gazette*, calculated upon the known chemical qualities of the materials." Another physician who criticized Talbot's imprecise information, and who therefore likewise provided more exact instructions for making photographs, was Golding Bird. See Golding Bird, "A Treatise on Photogenic Drawing," *Mirror of Literature, Amusement, and Instruction* 33, no. 945 (April 20, 1839): 243–44. The journal's title page reproduces, in the form of a woodcut, a photograph of a fern by Bird, a presumption that annoyed Talbot's mother, Lady Elisabeth Feilding. See Lady Elisabeth Feilding to William Henry Fox Talbot, April 30, 1839, Doc. No. 3873, The Correspondence of William Henry Fox Talbot, http://foxtalbot.dmu.ac.uk (accessed Jan. 31, 2012) (original: Talbot Collection, British Library).

Talbot's Art of Discovery

Vered Maimon

William Henry Fox Talbot's two most canonical accounts of photography, "Some Account of the Art of Photogenic Drawing" (1839) and "A Brief Historical Sketch of the Invention of the Art" (1844), differ in tone and emphasis. These differences can, of course, be attributed to the specific circumstances of their composition and publication. The first is a scientific paper presented before the specialized audience of the Royal Society, and the second is an introductory text to *The Pencil of Nature*, intended for a broad audience as part of Talbot's efforts to familiarize the general public with the calotype. Yet the differences between these accounts are not simply circumstantial or rhetorical but pertain to broader historical and epistemological concerns, in particular the collapse of "natural philosophy" as a viable framework for the study of nature in the early nineteenth century. Simon Schaffer has shown that the end of natural philosophy was marked by the emergence of new models of scientific discovery that came out of the shift within the philosophy of science from discovery to justification, as well as the institutionalization of science as a professional practice.[1] Since Talbot's early accounts of photography are also accounts of his methods of discovery, his conceptualization of photography is in fact inseparable from the discovery processes he outlined there. By locating the differences within the premises of British Victorian science, this essay examines the historical discursive framework within which Talbot's early conceptions of photography were made possible. By emphasizing the historical specificity of these accounts and analyzing his early botanical images, it also reconsiders the application of essential forms of intelligibility, namely, the index, to early photography in postmodern theories of photography. Thus, rather than emphasizing continuity within the field of the history and theory of photography, it points to historical differences with regard to the evidentiary status of the photographic image in the early stages of its conceptualization.

In Schaffer's essay "Scientific Discoveries and the End of Natural Philosophy," he argued that "where natural philosophers presented their histories as methods for training practitioners in discovery . . . historians of the sciences from the early nineteenth century separated the disciplined training of scientists from the heroic discovery moment, for which no training was possible."[2] This shift can be seen quite clearly in two of the most important philosophical publications on British empirical science in the early nineteenth century: John Frederick William Herschel's *Preliminary Discourse on the Study of Natural Philosophy* (1830) and William Whewell's *History of the Inductive Sciences* (1837), followed in 1840 by his *Philosophy of the Inductive Sciences*. While Herschel's *Discourse* was written in order to popularize science as an accessible practice to every student of natural philosophy, Whewell's books focus on the unaccountable ways in which actual discoveries were made by highly gifted and exceptional individuals whose work formed the basis for new research programs. As Schaffer stated: "The end of natural philosophy was accompanied by the appearance of models of discovery which appealed to *discipline* and to *genius*, and which have dominated theories of science ever since."[3] Both Herschel and Whewell occupy an important place in Talbot's scientific biography through their connections to Cambridge University, and therefore his accounts, I argue, are informed by their publications, which reveal a complicated view of British science at this historical juncture.[4]

HERSCHEL'S DISCOURSE

In his account in early 1839, Talbot presented his achievement as discovering a method for fixing the discoloration of the image. He described his discovery of the fixing agent as the outcome of a logical process of inductive reasoning grounded in experiments. Talbot then defined photogenic drawing as a "natural magic" and emphasized its relation to physical science and method:

> The phænomenon which I have now briefly mentioned appears to me to partake of the character of the *marvellous*, almost as much as any fact which physical investigation has yet brought to our knowledge. The most transitory of things, a shadow, the proverbial emblem of all that is fleeting and momentary, may be fettered by the spells of our "*natural magic*," and may be fixed for ever in the position which it seemed only destined for a single instant to occupy.[5]

This statement is often aligned with Roland Barthes's claim in *Camera Lucida: Reflections on Photography* that "the realists do not take the photograph for a 'copy' of reality, but for an emanation of *past reality*: a *magic*, not an art."[6] In its status as an emanation from a "real body," the photograph becomes a "certificate of presence," the ultimate proof that something "has been there." Thus, from a phenomenological point of view, Barthes argued, the photograph's "power of authentication exceeds the power of representation."[7] Based on his argument, Carol Armstrong stated that for nineteenth-century practitioners, "photography seemed to offer something *extraordinary*, without precedent, whose uses could only just begin to be imagined. That something was the photograph's unique ability to authenticate. This was a function of its fundamental indexicality."[8]

Yet the use of similar words, such as "magic," does not in any way indicate a conceptual and discursive affinity. Talbot explicitly stated that the phenomenon he discovered is "marvellous," (almost) "as much as *any* fact which physical investigation has yet brought to our knowledge," indicating that the "magic" nature of the phenomenon he discovered is inseparable from the scientific method he applied in his discovery, as he explicitly stated: "This remarkable phænomenon, of whatever value it may turn out in its application to the arts, will at least be accepted as a new proof of the value of the *inductive methods* of modern science."[9] Talbot's conceptualization evokes not only the inductive method but also the philosophical premises informing this mode of investigation within the discourse and practice of natural philosophy. This becomes quite clear when his statement is read in light of Herschel's *Discourse*, which articulates the character and ethos of the natural philosopher's inquiry within Victorian science: "The character of the true philosopher is to hope all things not impossible, and to believe all things not unreasonable. . . . Accustomed to trace the operation of general causes, and the exemplification of general laws . . . he walks in the midst of *wonders*: every object which falls in his way elucidates some principle, affords some instruction, and impresses him with a sense of harmony and order."[10]

Herschel's statement encapsulates the epistemological law of what Michel Foucault, in *The Order of Things*, terms "representation." Natural phenomena are "wonders" because they repeatedly expose, against plain logic, the necessary metaphysical continuity and correspondence between nature and the human mind. Consequently, they simultaneously reinforce theological convictions of the argument of divine design and the independent validity of scientific modes of inquiry that rely on direct experience. Talbot's statement thus indicates the discursive status of the photograph as both a natural phenomenon produced by light and an inductively discovered scientific object produced by an artificial chemical process—a fixed shadow.

The undecided epistemological status of phenomena within natural philosophy as direct objects of observation and as scientific abstractions or instrumental productions is in fact precisely what marks, according to Schaffer, the discursive specificity of natural philosophy as a form of scientific practice. He emphasized that the discourse of natural philosophy was "structured by the dialectic of the anomalous in contrast with the common-sensical."[11] Talbot's formulation simply reinforces the conformity of the photograph, as an object and an "instrumental production," to specific modes of physical investigation and to the philosophical conditions informing them, rather than pointing to its extraordinary character. It is therefore not a subjective phenomenological response to the uniqueness of the photograph as a real trace of the past that constitutes it as magic but the historical and discursive conditions of its formation. The issue is not, as Barthes argued, photography's "mad" status as a "direct emanation from the referent," but rather its reducibility to specific scientific and philosophical forms of intelligibility in which the observation of nature is subjected to theological convictions that cannot be grounded epistemologically or materially. In this regard the photograph is not an ultimate form of proof that an object "has-been-there."[12] Instead it demonstrates the fact that in the 1830s, following the challenges that English empiricism faced from David Hume's skeptical attack on the law of causality and the introduction of Immanuel Kant's transcendental philosophy to England by Romantic poets such as Samuel Taylor Coleridge, the validity of induction can only be based on what was termed "Common Sense" principles.[13] These principles, which ground the metaphysical validity of representation and Herschel's "law of continuity," are, as Thomas Reid argued, "impossible to disbelieve and impossible to prove."[14] After all, Talbot claimed in his account in 1839 that the photograph provides a "new proof" for the value of the inductive method, indicating the necessity of providing a proof in the first place. That is, while his account, like Herschel's *Discourse*, evokes the inductive method in order to promote the moral and cultural value of scientific practice, at the same time it also exposes the fact that its empirical foundations were under attack.[15]

The *Discourse*'s emblematic status in the early nineteenth century derives from its effort to legitimize science within culture as a practice for the cultivation of moral and religious values while also insisting on the independence of scientific inquiry. It thus addresses scientific methodology as a specific neutral field within the philosophy of science, detached from broad epistemological concerns. And while Herschel's formulation of induction is based on empiricist presuppositions that locate the source of knowledge in direct experience, he evaded any in-depth discussion of empiricism as an epistemology. In fact, he saw the *Discourse* as an adaptation and elaboration, following recent discoveries in the physical sciences, of Francis Bacon's *Novum Organum* (1620). Like Bacon, Herschel emphasized that knowledge is pursued collectively for the benefit of mankind as part of a process in which nature is subjected to human will. At the same time he stressed the pursuit of truth as a purely moral and psychological end divorced from practical applications. Herschel also evoked Bacon's theory of prejudices as idols in which any preconceived (social, cultural, linguistic, psychological) idea not grounded in experience is cleared from the mind as a condition for a reliable scientific inquiry. His main interest therefore lay in Bacon's formulation of scientific method as a mechanistic system that regulates the operations of the senses and the mind. Nevertheless, the science historian Richard Yeo has argued that the prevalent description of early nineteenth-century British science as "Baconian" or "inductivist" begs important questions of definition, since these terms did not have a neutral and stable meaning; rather they have been objects of controversy with incompatible interpretations.[16]

WHEWELL'S *HISTORY* AND *PHILOSOPHY*

In Whewell's review of the *Discourse* in 1831, he praised Herschel for providing a clear definition of induction as a regulated process of generalization out of observations, and he compared it to Reid's "very different" definition of induction as "our instinctive belief in the permanent uniformity of the physical laws of nature."[17] In his review Whewell approved of Herschel's definition of induction as a mechanistic method for the discovery of laws whose validity need not be founded on any particular epistemology but on the primacy of experience as the origin of knowledge and the metaphysical argument for divine design. Yet Whewell's views changed during the writing of his *History* and *Philosophy of the Inductive Sciences*. He decided to write his *History* in order to ground his own theory of induction in history, that is, to see whether it succeeded in explaining the emergence of real discoveries in science.[18] While Herschel's *Discourse* was written in order to instruct practitioners of science, Whewell sought to develop a theory of inductive science out of an analysis of the actual history of scientific discoveries. Moreover, while Herschel focused on methodological concerns, Whewell came to realize that these concerns were inseparable from an explanatory theory regarding the sources of knowledge.

Whewell scholars, such as Menachem Fisch and Richard Yeo, emphasize that his *Philosophy* marks a break from traditional British empiricism, and consequently his account of induction and discovery is radically different from Herschel's account in the *Discourse*.[19] For Herschel the role of the mind in discovery is primarily passive and ideas are simply "accumulated sensations," the outcome a gradual and continuous process of generalization from observations; for Whewell the mind is active and there exists a clear qualitative epistemological distinction between ideas and sensations, as he stated:

> But yet a little consideration will show us that an activity of the mind, and an activity according to certain Ideas, is requisite in all our knowledge of external objects. We see objects, of various solid forms, and at various distances from us. But we do not thus perceive them by sensation alone. Our visual impressions cannot, of themselves, convey to us a knowledge of solid form, or of distance from us. Such knowledge is inferred from what we see: —inferred by conceiving the objects as existing in space, and by applying to them the Idea of space.[20]

Whewell showed that there are fundamental ideas, such as space or time, that transcend the conditions of experience. Experience can discover general truths, but it cannot give those truths universality; on the other hand, universal truths, though they borrow their form from ideas, cannot be understood except by the actual study of nature.

Whewell defined discovery as a creative process in which new scientific conceptions and ideas emerge in a discontinuous and unexpected manner. He explained that "the first and great instrument by which facts are . . . combined into important and permanent truths, is that peculiar Sagacity which belongs to the genius of a Discoverer; and which, while it supplies those distinct and appropriate Conceptions which lead to its success, cannot be limited by rules, or expressed in definitions. . . . These are what are commonly spoken of as . . . inexplicable strokes of inventive talent."[21] Discovery, in this sense, is an irreducible imaginative and interpretative process in which scientific concepts are formed not by structurally capturing what has already been found in known facts, but by reading something new into them by the active act of the mind. Whewell's *Philosophy* is structured around a tension between his recognition that discovery cannot be reduced to a set of methodological rules and his efforts to extract such rules from the actual process through which discoveries were made in the past.

It was through his friendship with two of Coleridge's followers, William Rowan Hamilton and Julius Charles Hare, that Whewell became exposed to the more moderate aspects of Coleridge's epistemology and politics.[22] The science historian Geoffrey Cantor explained that Whewell's attitude toward Romanticism changed: initially he was critical of Coleridge's metaphysics, but from 1833 he became sympathetic to the poet's philosophy and modeled his notion of a scientific genius and the intuitive process of discovery on Coleridge's core aesthetic and epistemological terms and ideas.[23] Cantor also pointed out that Whewell's effort to construct a historiographical model that both describes and conceptualizes scientific development was part of a collective effort by a group of Liberal Anglican historians to integrate historical facts and theories. Their historical accounts were defined in opposition both to the rationalist tradition, which emphasized the accumulation of empirical knowledge, and to the speculative tradition, represented in England by Coleridge and William Wordsworth, which emphasized grand historical schemata over facts; nevertheless, they were much closer to the Romantics.[24] This group, inspired by new German scholarship, particularly the work of Barthold Niebuhr, included Thomas Arnold, Thomas Babington Macaulay, and friends of Whewell's in Cambridge, Hare and Newell Connop Thirlwall. All these scholars emphasized the interpretative nature of history writing and the role of the historian's imagination in imposing order and unity on historical facts. Liberal Anglican historians also rejected the idea of progress and accepted the Romantic view of history as "the natural, law-like unfolding of a nation's inner dynamic; an unfolding which did not presuppose progress since a nation finally decays and becomes extinct."[25] In order to oppose a deterministic and utilitarian view of historical development underlining rational historical accounts, they "emphasized the moral and intellectual characteristics exhibited by each nation through its history," as well as the role of creative individuals who both "reflected a nation's values and shaped its history."[26]

While rationalist histories of science, such as Joseph Priestley's *History and Present State of Electricity* (1767), describe the gradual growth of science as marking the progress of the human mind and exclude the achievements of individuals in favor of an "accumulation of facts," Whewell's *History* focuses on the achievements of a gifted discoverer as a "product of his time and as a moral agent."[27] Following the Romantic idea of a poetic genius, Whewell argued that creativity and intuition were crucial to explain the achievements of great discoverers such as Isaac Newton and Johannes Kepler. In this way Whewell shifted the history of science from matter, the "knowledge of the external world," to the mind, "the internal world."[28] Thus, like Herschel, albeit in a different way, Whewell emphasized the moral aspects of scientific practice in order to show that in no way does it undermine natural theology or moral philosophy, but instead it strengthens their premises and convictions. Yet while Herschel conceived of himself mainly as a practicing scientist, Whewell tried to define his role as a metascientific critic at a time when there was no model for such a role within the scientific community. This model existed in the Romantic poet, particularly in Coleridge, who conceived of himself also as a critic, and criticism as a distinct intellectual activity; Whewell adopted this metacommentary position from the cultural realm.[29]

TALBOT'S "ORIGINAL IDEA"

It is quite clear that Talbot's introduction to *The Pencil of Nature*, "A Brief Historical Sketch of the Invention of the Art," is discursively informed by a new model of discovery in which the history of science is described as the history of personal discoveries, and discoverers are considered to be the founders of disciplines and research schools. By this time, five years after his first public announcement, Talbot presented himself as the discoverer of "the principles and practice of Photogenic Drawing," and therefore his account of his discovery now becomes the *history* of the art.[30] Moreover, as I have argued elsewhere, in *The Pencil of Nature* Talbot discussed the role and status of the photographic document in relation to a Romantic form of historicism precisely of the kind employed by Liberal Anglican historians such as Macaulay and Whewell.[31]

In this account the inductive method is never explicitly mentioned, and while Talbot made a number of statements evoking the primacy of experiments, he also emphasized the importance of theory and ideas. Consequently, the account does not present the discovery process itself as a logical one, and its object is not a chemical fixing process for images that can be produced with or without the camera obscura. Instead Talbot focused on the "original idea": fixing the images of the camera obscura, which led to his experiments. He dramatized the specific "heroic" moment of discovery, during a vacation on the shores of Lake Como in 1833, as an inexplicable stroke of inventive power that cannot be explained through any kind of method:

> It was during these thoughts [on the beauty of the images of the camera obscura] that *the idea occurred to me* . . . how charming it would be if it were possible to cause these natural images to imprint themselves durably, and remain fixed upon the paper! Such was the idea that came into my mind. Whether it had ever occurred to me before amid floating philosophic visions, I know not, though I rather think it must have done so, because on this occasion it struck me so forcibly. Such were as nearly as I can now remember the reflections which led me to the invention of this *theory*, and which first impelled me to explore a path so deeply hidden among nature's secrets. And the numerous researches which were afterwards made—whatever success may be thought to have attended them—cannot, I think, admit of a comparison with the value of the *first and original idea*.[32]

Rather than describing a highly systematic and gradual method of discovery modeled on an inductive discovery of a scientific law, Talbot now recounted a story in which an inexplicable stroke of inventive power led to the sudden emergence in his mind of a new idea. Thus what is emphasized is precisely the lack of method as a way to insist on the originality of the discovery. In other words, it is not the mechanical law of continuity that constitutes the reciprocity between nature and human nature responsible for the discovery, but precisely the mind's capacity to transcend experience in a creative way that is epistemologically unaccounted for. What underlines Talbot's new account is an awakening sense of subjectivity together with an awareness of temporality, a certain movement from the exteriority of impressions and sensations into the opaque interiority of the mind's independent capacity of reasoning.

When this personal moment of epiphany is read in light of the history of Victorian science, it is anything but unique. Schaffer has shown that in the new models of discovery, when a single event is given the status of a discovery, a new fact is replicated and given a specific author. Yet historical and sociological studies have shown that these two processes are subject to a series of negotiations inside the scientific community regarding the specific experiments and criteria of adequacy that could lead to the discovery of different phenomena or facts. This suggests, Schaffer claimed, that if "replication and authorship are matters of negotiation, then there is no event which corresponds to an automatic or instant discovery. . . . Subsequently, the story of that process is rewritten. The lengthy enterprise is telescoped into an individual moment with an individual author."[33] Five years after his first announcement Talbot retold the story of his discovery in a way that defined him as the sole originator of the idea, an author, condensing years of research and experimentation into one specific moment.

Talbot's professional and personal contacts with Whewell were more limited than his relations with Herschel, although all of them belonged to what was called the "Cambridge network." Yet the point that needs to be emphasized is that while Whewell's work gave the most sophisticated and comprehensive account of the scientific discoverer as genius and thus indicated the infiltration of Romantic ideas of creativity and genius into scientific accounts, this redefinition of the discoverer was in no way unique at this historical juncture. This Romantic definition also underlined David Brewster's historical and biographical studies *Life of Newton* (1835) and *Martyrs of Science* (1841). Brewster, who was a much closer professional colleague of Talbot's and more involved in his photographic experiments,[34] was part of a different, if not opposed, scientific milieu than Whewell; nevertheless, as Schaffer has shown, Brewster also argued that "the influence of imagination as an instrument of research has, we think, been much overlooked by those who have ventured to give laws to philosophy."[35] By emphasizing the role of the imagination in scientific discoveries, both Whewell and Brewster responded to Kant's definition of genius in his *Critique of Judgment* (1790) as the talent (or natural gift) that gives the rule to art. The works of genius could be copied or become a rule for others to follow, but genius could not be imitated, and therefore science could not be carried out by genius. Based on this characterization and the delimitation of the term "genius" to artists, Kant famously argued that Newton cannot be considered a genius because he presented his discoveries as an outcome of a logical process of reasoning. Reclaiming the imagination and intuition into the realm of science enabled Whewell and Brewster not only to emphasize the religious and moral aspects of scientific practice but also to grant discoverers the status of genius.[36]

Within the British scientific establishment it was Humphry Davy, out of his relations with Coleridge, who presented himself in his chemical, experimental, and philosophical works as a Romantic scientific genius.[37] Talbot was of course familiar with Davy's chemical experimental work and mentioned Davy in his first account as a "distinguished experimenter" and the coauthor, together with Thomas Wedgwood, of the essay "An Account of a Method of Copying Paintings upon Glass, and of Making Profiles, by the Agency of Light upon Nitrate of Silver" (1802), which documented a failed attempt to copy and fix with solar light pre-existing images, including the ones formed in the camera obscura.[38] It was Davy, in his investigations of chemical affinity, who offered a much more "heroic" model of the natural philosopher than the one suggested by Herschel in his *Discourse*. As Christopher Lawrence has pointed out, Davy stated in many of his lectures at the Royal Institution that discovering the animating powers of life or "nature's hidden operations" necessitates a genius. In 1802 Davy publically confronted the Romantic poets' view of chemistry as a lower-order activity when compared with poetry. He argued, "The son of true genius . . . in the search of discovery . . . will rather pursue the plans of his own mind than be limited by the artificial divisions of language." Thus chemistry could "exhibit to men" that system of knowledge "which relates so intimately to their own physical and moral constitution." He also emphasized that when engaging with nature in this way, the mind of the natural philosopher must be active and creative.[39]

FOLDING IMAGES

This necessary relation between the creative power of the discoverer's mind and nature's own creations, which "impelled" Talbot, as he stated, "to explore a path so deeply hidden among nature's secrets," evokes a Romantic mode of inquiry that connects nature and human nature along much more subjective and dynamic lines than the static and mechanical law of continuity. A more explicit evocation of the faculty of the imagination is suggested in his comments about plate IV, *The Open Door*, in *The Pencil of Nature*, in which he suggested the picture's capacity to "awaken a train of thoughts and feelings, and picturesque imaginings" (see fig. 1, p. 2).[40] At this point it is clear that Talbot felt uncomfortable with the association of the calotype with the idea of a "mechanical copy." At stake for him was the need to show that the inseparability of the photograph from nature—the fact that it was produced by "the pencil of nature" and not by the artist's hand—did not simply result in the production of mechanical copies of phenomena. He suggested that nature does not simply trace or copy but in fact introduces difference and variety; that is, nature is an active agent of representation: "It would hardly be believed how different an effect is produced by a longer or shorter exposure to the light, and, also, by mere variation in the fixing process, by means of which almost any tint, cold or warm, may be thrown over the picture, and the effect of bright or gloomy weather may be imitated at pleasure."[41]

The relation between the imagination and nature as an active force underlines Coleridge's aesthetic theory. The importance of his poetic and philosophical project to the conception of photography was noted by Geoffrey Batchen when he pointed out that Coleridge had close personal relations with early "proto-photographers," such as Wedgwood.[42] Batchen showed how Coleridge's insistence on the interactive and dynamic relation between nature and the viewing subject informed the proto-photographers' conception of photography. I locate the importance of Coleridge to the early conceptualization of photography in the way he modeled his definition of the imagination on his essay *Theory of Life*.[43] Against the mechanical view of nature Coleridge suggested a dynamic view of nature as a life force: "*the principle of individuation*," and a "*power* which discloses itself from within as a principle of *unity* in the *many*."[44] Life is a force that reveals itself in a range of different phenomena as that which both divides (such as polarity in magnetism) and connects (such as electricity). Similarly, he defined the imagination in the *Biographia* as an echo of the divine act of creation that "dissolves, diffuses, dissipates, in order to recreate; or where this process is rendered impossible, yet still at all events it struggles to idealize and to unify. It is essentially *vital*, even as all objects (as objects) are essentially fixed or dead."[45] Both the imagination and nature are conceptualized as dynamic forces, ceaseless acts of differentiation that unfold in time.

In light of Coleridge's aesthetic theory, it becomes clear that by grounding his discovery in the inventive powers of the imagination, Talbot also implicitly affirmed the active and differentiating power of nature itself and consequently the aesthetic qualities of the photographic image as an organic form of imitation rather than mechanical copying. This dynamic connection between the imagination and nature explains the importance of the botanical image to Talbot's

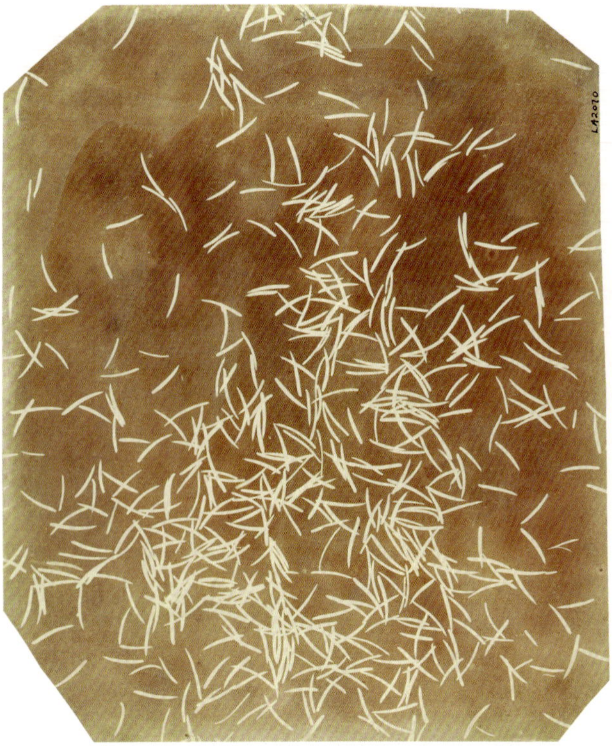

59. William Henry Fox Talbot, *A Cascade of Spruce Needles*, 1839, photogenic drawing, 9 x 7¼ in. (22.7 x 18.5 cm). Talbot Collection, British Library.

conceptualization of photography.[46] The largest category of images in his first major exhibition, in August 1839 at the British Association for the Advancement of Science annual meeting in Birmingham, was of botanical specimens, mostly negatives.[47] While at this point Talbot was unable to produce satisfactory images with the camera obscura, by the time of the production of *The Pencil of Nature* he was able to do so—and still he decided to include a botanical image as well as other images created through contact printing, such as a piece of lace.[48]

When we look closely at some of Talbot's most striking botanical images, it becomes clear that nature is not simply tracing or copying but that something new is formed in the encounter between image and plant. That is, what these images present is not simply an "index" of a plant that tautologically authenticates its existence, "that-has-been," as Armstrong suggested in her analysis of early botanical photographs.[49] Instead it is precisely productivity and ceaseless creation that are emphasized. Consider, for example, *A Cascade of Spruce Needles* (fig. 59), in which it is not the plant's particular arrangement on the paper that creates the image, but rather the different intensities of its needles' distribution on the paper's surface that present the plant as a rhizomatic and extensive diagram of entangled lines and flat, shifting planes. The needles spread out on the paper in a way that

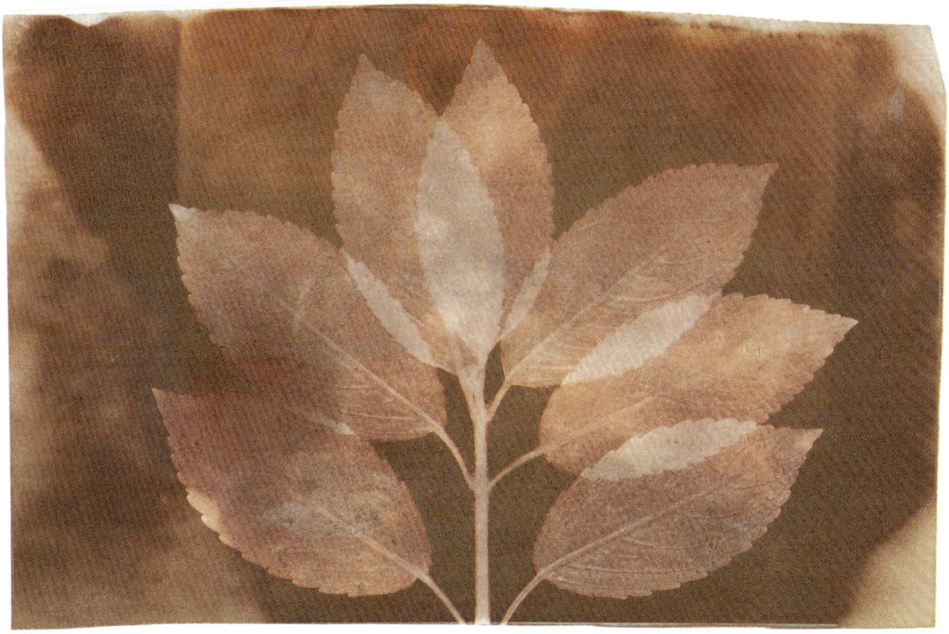

60. William Henry Fox Talbot, *Branch of Leaves of Mercuriàlis pérennis*, 1839, photogenic drawing, 4⅝ x 6¾ in. (11.6 x 17.2 cm). Photographic History Collection, Natural Museum of American History, Smithsonian Institution, Washington, DC.

suggests no starting or end point, bottom or top, for the image but instead a detachable, reversible, and constantly modifiable map. Thus it is not the needles that are delimited by the paper, but rather the paper itself that is articulated and punctuated by their exceeding movement and ceaseless dispersion. The image displays neither a centralized structure nor a clear outline division into parts that can be identified or traced against a fixed table of classification, yet something is formed: an abstract, dynamic map composed of material effects that are at the same time formal signs. In Talbot's camera-less images there is a prevailing sense of unfolding temporality, of a productive encounter with nature's forces that is not simply reproductive or tautological as the spatial concept of the index suggests. While Armstrong showed that botanical photographs failed to function as botanical illustrations because of the essential incapacity of the photography to signify, I interpret this failure in light of the historical premises of Romantic aesthetic theories, which, with their emphasis on the dynamic relations between nature and mind, set botany as a science of classification in opposition to chemistry as the science of forces.[50] Thus what we see in these botanical images is an impossible encounter between a dynamic conception of nature and a static form of representation.

Consider also Talbot's extraordinary *Branch of Leaves of Mercuriàlis pérennis* (fig. 60), in which the image's mode of formation appears inseparable from its specific imagery. In this image the branch is centralized, its stem is cut by the paper, and its leaves' outlined forms are clearly presented, but the leaves also overlap in a manner that creates an internal pattern like a distinct visual rhythm. Another visual pattern is suggested by the irregular shift from dark to light areas and the inconsistent degrees of translucency due to the varied thickness of the leaves and the uneven sensitivity of the paper. The result is that each leaf appears different from the others, while its own specific texture is presented discontinuously with areas that are erased, blocked, or blurred. This inner process of folding in the image hovers between materiality (chemical sensitivity, conditions of luminosity) and signification (patterning and coding). It produces effects that are at the same time *signs* that are inseparable from the depicted object. In this regard the image does not trace or "resemble" the plant but connects different semiotic chains (material, formal, biological, chemical) *within* the image's different states of formation. Signification develops inseparably from the material processes of the image's formation in its encounter with the plant. The prevalence of camera-less photographs in the early years of photography is often cited as proving Barthes's theory of indexicality, in which signification is always presented as necessarily divorced from authentication, that is, from the image's process of production; however, my analysis is meant to show that in the early years of photography, it was impossible to separate these two processes because nature, like the imagination, does not simply trace in a circular manner but re-creates—it functions as a force of both difference and unity. Therefore, the image's material mode of formation functions not as proof for what "has-been," but precisely for the alignment of nature and mind, yet this time not through a metaphysical belief in the mechanical and static principles that govern the two, but rather through their irreducible and dynamic intertwined operations, their ceaseless temporal capacity to introduce change.

By focusing on the epistemological and philosophical premises of Talbot's accounts of photography and his early botanical images, this essay emphasizes the historical conditions within which the conceptualization of early photography took place. This genealogical analysis aims to dislocate historical notions of continuity, which, by attributing essential forms of intelligibility to photography, ignore the highly complex and irreducible ways in which photography's relation to nature and to the real was articulated throughout its history. Photography's philosophical and cultural significance was not always associated with the epistemological quest for evidence and objective truth. And consequently, it was not always its status as a mechanical and automatic copy that defined the specificity of the photographic image. History presents different regimes of truth and forms of intelligibility in which photography's status as an image and document can be highly specific and inconsistent, as this discussion of the philosophical premises of Victorian science indicates.

1. Simon Schaffer, "Scientific Discoveries and the End of Natural Philosophy," *Social Studies of Science* 16, no. 3 (1986): 387–420.
2. Schaffer, "Scientific Discoveries," 387.
3. Schaffer, "Scientific Discoveries," 407 (emphasis in original).
4. Talbot, Herschel, and Whewell were all closely associated with the "Cambridge network." Herschel in particular was closely involved in Talbot's photographic experiments and was himself the inventor of a number of photographic processes. He was also the one who suggested the term "photography" for Talbot's early images. On the relations between the two and Herschel's photographic experiments, see Larry J. Schaaf, *Out of the Shadows: Herschel, Talbot & the Invention of Photography* (New Haven and London: Yale Univ. Press, 1992). Talbot's personal and professional connections with Whewell were less dominant than the ones with Herschel. Talbot's biographer H. J. P. Arnold described Whewell as a friend and mentioned that he was one of the individuals who signed Talbot's certificate of recommendation for the Royal Society. Arnold also mentioned a number of social gatherings surrounding the British Association for the Advancement of Science meetings in which the two participated, as well as a professional scientific correspondence. H. J. P. Arnold, *William Henry Fox Talbot: Pioneer of Photography and Man of Science* (London: Hutchinson Benham, 1977), 65, 80, 82, 313.
5. William Henry Fox Talbot, "Some Account of the Art of Photogenic Drawing; or, The Process by Which Natural Objects May Be Made to Delineate Themselves without the Aid of the Artist's Pencil," in *Photography in Print*, ed. Vicki Goldberg (Albuquerque: Univ. of New Mexico Press, 1981), 40–41 (emphasis in original). Talbot's first photographic process, photogenic drawing, was a "print-out" process that fully depended on solar energy for the production of negative images. His second photographic process, the calotype, was a "developed-out" process in which, after a short exposure time, an invisible image is formed and later developed by chemical means (gallic acid, the chemical agent that Talbot started employing as a developer in 1841). These terms are used by Larry J. Schaaf in his book *The Photographic Art of William Henry Fox Talbot* (Princeton, NJ: Princeton Univ. Press, 2000), 17.
6. Roland Barthes, *Camera Lucida: Reflections on Photography*, trans. Richard Howard (New York: Hill and Wang, 1981), 88 (emphasis in original).
7. Barthes, *Camera Lucida*, 89.
8. Carol Armstrong, *Scenes in a Library: Reading the Photograph in the Book, 1843–1875* (Cambridge, MA: MIT Press, 1988), 12–13 (emphasis added). The status and meaning of the index have become highly debated in recent photography theory. See the collection of essays, comments, and in particular the roundtable discussion in *Photography Theory*, in which it is suggested that the way art historians apply the term "index" to photography theory is based on a misreading of Charles Sanders Peirce's analysis of signs. James Elkins, ed., *Photography Theory* (New York and London: Routledge, 2007). See also Elkins, "What Does Peirce's Sign System Have to Say to Art History?," *Culture, Theory, and Critique* 44, no. 1 (2003): 5–22.
9. Talbot, "Art of Photogenic Drawing," 40 (emphasis added). This understanding of "natural magic" also underlines other scientific publications from this period. See, for example, David Brewster, *Letters on Natural Magic, Addressed to Sir Walter Scott, Bart.* (New York: J. & J. Harper, 1832).
10. John Frederick William Herschel, *A Preliminary Discourse on the Study of Natural Philosophy* (Chicago: Univ. of Chicago Press, 1987), 8, 15 (emphasis added). Herschel was a prominent "gentleman of science," and historians of science consider his *Discourse* to be "the epitome of the nature and ethos of early Victorian science." See Richard R. Yeo, "Reviewing Herschel's *Discourse*," *Studies in History and Philosophy of Science* 20, no. 4 (1989): 541.
11. Simon Schaffer, "Natural Philosophy," in *The Ferment of Knowledge*, ed. G. S. Rousseau and Roy Porter (Cambridge: Cambridge Univ. Press, 1980), 78.
12. The exact term used by Barthes is "that-has-been." Barthes, *Camera Lucida*, 94.

13. The importance of Kant's philosophical revolution and his concept of the organism to Romantic aesthetic theory has been pointed out and analyzed by, among others, M. H. Abrams in his seminal book *The Mirror and the Lamp: Romantic Theory and the Critical Tradition* (London: Oxford Univ. Press, 1953).

14. Richard Olson has shown that Common Sense philosophy was established in Scotland in response to the philosophical work of Hume. Thus, Reid's *Inquiry into the Human Mind* (1764) was conceived to defend the religious beliefs and moral tenets of moderate Scottish Presbyterianism against the corrosive influences of atheistic skepticism on the one hand and necessitarian materialism (as argued by David Hartley and later Joseph Priestley) on the other. Reid's response was based on the notion that these philosophies contradict "certain principles . . . which the constitution of our nature leads us to believe, and which we are under a necessity to take for granted in the common concerns of life." These principles, the "principles of common sense," like causality, are impossible to prove but they must be accepted as unproven ultimate principles, like the axioms of geometry. Richard Olson, *Scottish Philosophy and British Physics, 1750–1880* (Princeton, NJ: Princeton Univ. Press, 1975), 27–30. Herschel's *Discourse* was also indebted to the Scottish school of Common Sense philosophy.

15. Historians of science have shown that the prevalent evocation of scientific methods by "gentlemen of science" in their accounts functioned mainly as a rhetorical tool to promote their discoveries morally and culturally, not to ground their validity epistemologically. See Richard R. Yeo, "Scientific Method and the Rhetoric of Science in Britain, 1830–1917," in *The Politics and Rhetoric of Scientific Method*, ed. John A. Schuster and Richard Yeo (Dordrecht: D. Reidel, 1986), 260.

16. Richard R. Yeo, "An Idol of the Market-Place: Baconianism in Nineteenth Century Britain," *History of Science* 23 (1985): 252.

17. William Whewell, review in *Quarterly Review* 45 (1831): 378.

18. On Whewell's *History* and his theory of induction, see Geoffrey N. Cantor, "Between Rationalism and Romanticism: Whewell's Historiography of the Inductive Sciences," in *William Whewell: A Composite Portrait*, ed. Menachem Fisch and Simon Schaffer (Oxford: Clarendon Press, 1991), 67–86.

19. Menachem Fisch, *William Whewell, Philosopher of Science* (Oxford: Clarendon Press, 1991); and Richard R. Yeo, *Defining Science: William Whewell, Natural Knowledge, and Public Debate in Early Victorian Britain* (Cambridge: Cambridge Univ. Press, 1993).

20. William Whewell, *Selected Writings on the History of Science*, ed. Yehuda Elkana (Chicago: Univ. of Chicago Press, 1984), 149.

21. Whewell, *Selected Writings*, 210–11.

22. On Whewell's relations with Hamilton and Hare and the influence of Coleridge's ideas on his work, see Richard R. Yeo, "William Whewell's Philosophy of Knowledge and Its Reception," and Simon Schaffer, "The History and Geography of the Intellectual World: Whewell's Politics of Language," both in Fisch and Schaffer, *William Whewell: A Composite Portrait*, 175–99, 201–31, respectively.

23. Cantor, "Between Rationalism and Romanticism," 77.

24. Cantor, "Between Rationalism and Romanticism," 75.

25. Cantor, "Between Rationalism and Romanticism," 77.

26. Cantor, "Between Rationalism and Romanticism," 77.

27. Cantor, "Between Rationalism and Romanticism," 80.

28. Cantor, "Between Rationalism and Romanticism," 80.

29. On the role of Whewell as a "critic" of science, see Yeo, *Defining Science*, 70; and Schaffer, "History and Geography," 230–31.

30. William Henry Fox Talbot, "A Brief Historical Sketch of the Invention of the Art," in *The Pencil of Nature* (New York: Da Capo, 1969), n.p.

31. See Vered Maimon, "Displaced 'Origins': William Henry Fox Talbot's *The Pencil of Nature*," *History of Photography* 32, no. 4 (2008): 314–25.

32. Talbot, "Brief Historical Sketch," n.p. (emphasis added).

33. Schaffer, "Scientific Discoveries," 397.

34. On the relations between Talbot and Brewster, see Graham Smith, *Disciples of Light: Photographs in the Brewster Album* (Malibu, CA: J. Paul Getty Museum, 1990). Brewster also wrote one of the first comprehensive accounts of photography, in which he discussed Talbot's discoveries; see Brewster, "Photogenic Drawing, or Drawing by the Agency of Light," *Edinburgh Review* 86, no. 154 (Jan. 1843): 309–45.

35. Schaffer, "Scientific Discoveries," 410.

36. On the relations between science and romanticism in the context of natural philosophy, see Andrew Cunningham and Nicholas Jardine, eds., *Romanticism and the Sciences* (Cambridge: Cambridge Univ. Press, 1990); Stefano Poggi and Maurizio Bossi, eds., *Romanticism in Science* (Dordrecht and Boston: Kluwer Academic, 1994); and David M. Knight, *Science in the Romantic Era* (Aldershot, UK: Ashgate, 1998). On the centrality of the idea of the genius to natural philosophy, see Simon Schaffer, "Genius in Romantic Natural Philosophy," in Cunningham and Jardine, *Romanticism and the Sciences*, 82–98.

37. On Davy's scientific work and his relations with Coleridge, see Trevor H. Levere, *Affinity and Matter* (Oxford: Clarendon Press and Oxford Univ. Press, 1971).

38. Thomas Wedgwood and Sir Humphry Davy, "An Account of a Method of Copying Paintings upon Glass, and of Making Profiles, by the Agency of Light upon Nitrate of Silver," in *Photography: Essays & Images*, ed. Beaumont Newhall (New York: Museum of Modern Art, 1980), 15–16.

39. Christopher Lawrence, "The Power and the Glory: Humphry Davy and Romanticism," in Cunningham and Jardine, *Romanticism and the Sciences*, 213–20. See also Ferdinando Abbri, "Romanticism Versus Enlightenment: Sir Humphry Davy's Idea of Chemical Philosophy," in Poggi and Bossi, *Romanticism in Science*, 31–45.

40. By comparing the photograph to the Dutch school of art, Talbot emphasized that the representation of daily and familiar scenes such as a "casual gleam of sunshine . . . a time-withered oak, or a moss-covered stone may awaken a train of thoughts and feelings, and picturesque imaginings." William Henry Fox Talbot, *The Pencil of Nature* (London: Longman, Brown, Green and Longmans, 1844), n.p.

41. Talbot's statement appears in the *Literary Gazette; and Journal of the Belles Lettres, Arts, Sciences, &c.*, Feb. 13, 1841, 108.

42. Batchen traced the personal relationship and correspondence between Coleridge, Wedgwood, and Davy in order to show how the poet's views on representation informed early inventors' conception of photography. Geoffrey Batchen, *Burning with Desire: The Conception of Photography* (Cambridge, MA: MIT Press, 1997), 60–62, 84–90. Anne McCauley also discussed the common philosophical concerns of Coleridge, Davy, and Wedgwood. McCauley, "Talbot's Rouen Window: Romanticism, *Naturphilosophie* and the Invention of Photography," *History of Photography* 26, no. 2 (2002): 124–31.

43. On the relations between Coleridge's *Biographia* and his *Theory of Life*, see Timothy Corrigan, "The *Biographia Literaria* and the Language of Science," in *Modern Critical Views: Samuel Taylor Coleridge*, ed. Harold Bloom (New York: Chelsea House, 1986). On Coleridge's *Theory of Life*, see Roy Male, "The Background of Coleridge's *Theory of Life*," *Studies in English* 33 (1954): 60–68; see also Alice D. Snyder, *Coleridge on Logic and Learning* (New Haven: Yale Univ. Press, 1929).

44. Samuel Taylor Coleridge, *Hints towards the Formation of a More Comprehensive Theory of Life*, vol. 1 (New York: Harper & Brothers, 1853), 386–87 (emphasis in original).

45. Samuel Taylor Coleridge, *Biographia Literaria*, vol. 1, 1815–17 (London: Oxford Univ. Press, 1962), 202 (emphasis added).

46. On Talbot's botanical images, see Vered Maimon, "On the Singularity of Early Photography: William Henry Fox Talbot's Botanical Images," *Art History* 34, no. 5 (2011): 958–77.

47. For this exhibition in August 1839, Talbot prepared an inventory of about ninety images, of which about fifty were negatives of botanical specimens and only twenty were positive camera images of architectural views and interiors. The inventory is reproduced in *Henry Fox Talbot: Selected Texts and Bibliography*, ed. Mike Weaver (Oxford: Clio Press, 1992), 58.

48. As late as 1844 Talbot still found the negative image of lace or leaves as intelligible as a positive image of these objects. As his stated in his comments about plate XX, *Lace*, in *The Pencil of Nature*: "In copying such things as lace or leaves of plants, a negative image is perfectly allowable, black lace being as familiar to the eye as white lace, and the object being only to exhibit the pattern with accuracy." Talbot, *Pencil of Nature*, n.p.

49. See Carol Armstrong, "Cameraless: From Natural Illustrations and Nature Prints to Manual and Photogenic Drawings and Other Botanographs," in *Ocean Flowers: Impressions from Nature*, ed. Carol Armstrong and Catherine de Zegher (Princeton, NJ, and Oxford: Princeton Univ. Press, 2004), 54–165.

50. In his essay on method Coleridge opposed botany as a science based on artificial classification and on the conception of nature as a sensible product to chemistry as a science of dynamic creativity and pure potentiality. Chemistry is the model for Coleridge's Romantic conception of science and for the philosopher as a complementary figure to the poet. See Samuel Taylor Coleridge, "Essays on the Principles of Method," in *The Collected Works of Samuel Taylor Coleridge*, vol. 4, *The Friend*, ed. Barbara E. Rooke, bk. 1 (Princeton, NJ, and London: Princeton Univ. Press, 1969), 470–71.

“The Caxton of Photography”: Talbot’s Etchings of Light

Larry J. Schaaf

> “Wonderful wonder of wonders!! Vanish aqua-tints and mezzotints. . . . Steel engravers, copper engravers, and etchers, drink up your aquafortis and die! There is an end of your black art . . . the real black art of true magic arises . . . at a few moments’ notice . . . by virtue of the sun’s patent, all nature, animate and inanimate, shall henceforth be its own painter, engraver, printer, and publisher. Here is a revolution in art.”[1]

William Blackwood’s prescient statement was made in March 1839, just weeks after Louis-Jacques-Mandé Daguerre and William Henry Fox Talbot startled the world with their announcements of photography. As a publisher Blackwood immediately grasped that the true significance of the new art would be its influence on the printed page. Unlike Daguerre, Talbot was acutely aware of this as well, for throughout his life he displayed a passion for the world of books and printing. He sited photography firmly within the world of publishing, an attitude especially apparent in his boldly experimental volume *The Pencil of Nature* (1844). Perhaps surprisingly, Talbot spent one decade inventing and perfecting silver-based photography before abandoning it, but three times as long, right up to the end of his life, perfecting a means of expressing nature’s image in time-tested ink on paper. Talbot “died with his boots on,” working on his invention of photogravure until the very end, a quest that moved the photographer John Thomson to call him “the Caxton of Photography.”[2] In 1879, just two years after Talbot’s death, Karl Klíč introduced his photogravure process, based so closely on Talbot’s inventiveness and creativity that it became known as the Talbot-Klíč process.[3] Talbot had initially turned to the salts of silver because of their reaction to light. These were essential for faithfully recording nature’s image, but they proved to be unnecessary and often even fatal for preserving it once recorded. Talbot’s invention of photogravure could be viewed as an improved form of photography, a logical evolution of technology toward the attainment of his original goal.[4] Light-sensitive silver captured nature’s original image; subsequently, robust printer's ink multiplied and preserved it. Some later photographic artists, notably James Craig Annan, agreed with Talbot's vision that photogravure was not just the reproductive art form to which it is relegated by many today but rather the highest form of photographic expression.[5] As one commercial firm later advertised, photogravure could "combine the accuracy of the photograph with the delicacy and strength of steel engraving."[6]

In 1859 Charles Dickens optimistically declared that "photography has become a science, with a literature of her own."[7] He spoke during a period when photography had moved on from the minds of the inventors and the hands of the pioneers into a more public sphere, spawning its own societies and journals. More recently the literature on photography has expanded greatly, including not only monographs and histories but also catalogues raisonnés and other research tools. Unfortunately, studies of the photomechanical processes—those that sought to merge photography with printer's ink—remain sparse.[8] The present essay seeks nothing more than to be a stimulus for further research. In 1902 Talbot's son, Charles Henry, lamented that "if Fox Talbot had lived only a few weeks longer, he would have given his own account of this invention."[9] Although the elder Talbot published very little about his work in photogravure, he created thousands of largely experimental printing plates and proof prints that still survive. His research notes are voluminous, complex, scattered among several collections, and uncatalogued. Very little of this work has been studied in any detail.[10]

The drive to translate photography into printer's ink started immediately in 1839. In fact, the pioneering work of Joseph Nicéphore Niépce in *héliographie* well before that year has at least as much kinship with lithography as it does with the camera.[11] In 1839 Daguerre unwittingly set the stage for the future course of photography. His images were so attractive and detailed as to create a high demand, but they suffered from the drawback of being unique originals made in the camera, unable to be replicated in the same manner as Talbot's negatives. In addition, Daguerre's images were on metal sheets that already looked like printing plates. Some of the first attempts at reproduction employed an original daguerreotype plate as the ground for hand engraving, which of course destroyed the original plate and supplanted nature's image with one largely drawn by the hand of man. Another approach was to etch the original daguerreotype plates chemically, enabling them to capture ink. Armand-Hippolyte-Louis Fizeau was the most successful at this, turning to electrotyping and etching, but his plates were limited in tonality and printing life. An instructive view of these efforts appears in the two volumes of *Excursions daguerriennes: Vues et monuments les plus remarquables du globe*, issued by Noël-Marie-Paymal Lerebours in 1840–43. These included engravings made after daguerreotypes and three plates by Fizeau's photomechanical process (fig. 61).

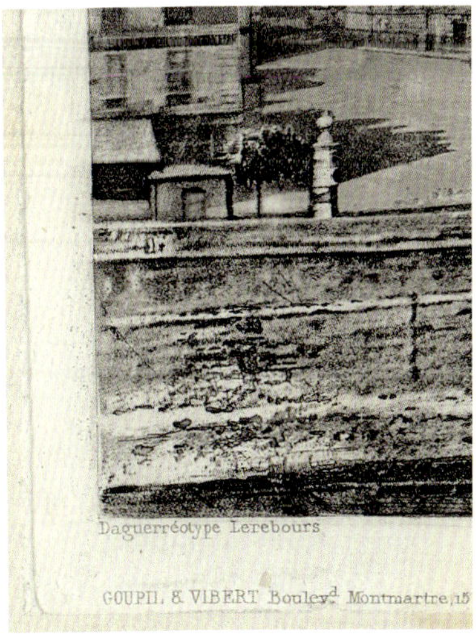

61. Armand-Hippolyte-Louis Fizeau, *Hotel de Ville de Paris* (detail), 1843, print from an engraved daguerreotype plate, hand-retouched, captioned "Daguerréotype Lerebours" and "Gravé par le procédé Fizeau," from an unnumbered plate in Noël-Marie-Paymal Lerebours, *Excursiones daguerriennes.* Hans P. Kraus, Jr., New York. Fizeau was one of many who explored means of translating photographs into printer's ink. Engraving not only destroyed the unique daguerreotype plate but also required extensive handwork to be suitable for printing.

Printing processes fall into three main families, defined by where the ink lies on the printing plate, and all three were explored during the evolution of photography. Examples of relief printing, depending on raised surfaces, include metal type and rubber stamps. Planographic processes, such as lithography, separate the ink into patches on the flat surface of the plate. While Talbot probed different approaches he concentrated most of his efforts on intaglio processes, those that trap ink in wells below the surface of the plate. Instead of engraving a metal plate by hand with a sharp instrument, Talbot etched his plates, replacing the mechanical burin with chemical action, a process later known as photogravure.[12] For Talbot the expressive power of "Nature" had been paramount for photography, an imperative that he carried over to his printing plates. Unlike the majority of his contemporaries he did everything he could to avoid the intervention of the human hand on his plates.[13] Nature drew its own pictures, and natural processes forged its printing plates.

THE TARNISHING OF SILVER

Talbot started on the invention of photography out of a very personal need, motivated by his appalling draftsmanship. He turned to the salts of silver because of their great sensitivity to light, and the results from them were both astonishing and beautiful. Talbot was under no pressure to produce when he made his negatives and prints by hand at Lacock Abbey. Not only did his personal expertise trump many vexing problems, but if impediments arose, he was quite happy to turn his attention to optics or mathematics or cuneiform. He could choose the sunniest of days for printing and toss any prints that were substandard. Water from Lacock Abbey's well was plentiful, and the country house kitchen stood ready to heat it to a useful temperature. The wind in Talbot's intellectual sails was whimsical, for he moved freely and guiltlessly from subject to subject, through either opportunity or inclination.[14]

With the evidence of his own artistic growth, Talbot had every reason to be proud of his productions, having gained a well-founded optimism about the future of the art.[15] In June 1844 he commenced publication of *The Pencil of Nature*, a part book not only issued for its intrinsic worth but intended to demonstrate how photography could aid in the production of books and prints.[16] Lacking a way to reproduce photographs in a printing press, Talbot turned to large editions of silver-based prints, each an original made from his negative and pasted onto a book page.

In order to produce the substantial number of silver prints required, Talbot encouraged the ambitions of Nicolaas Henneman, his former valet, to found the world's first photographic production facility in the bustling town of Reading, on the newly built rail line halfway between Lacock and London. Henneman set up banks of printing frames on racks in the sun, planning to turn out silver prints on a larger scale than had ever before been attempted. Industrializing Talbot's handcrafted process soon exposed its inherent weaknesses, for it was an idea fatally in advance of the technology of the day. Craftsmanship was forced to yield to expediency, and permanency suffered (figs. 62 and 63). Henneman had to meet print quotas in the range of hundreds per week and could not afford the luxury of suspending production on a dull day. Prints made by extended exposure to weak light could never possess the strength of those made in unclouded daylight and were more susceptible to fading. Talbot's friend and scientific supporter John Frederick William Herschel had suggested using sodium thiosulfate (familiar to this day as hypo) to fix Talbot's prints. Hypo could dissolve any remaining active silver salts, allowing them to be washed out of the paper, but it was a fickle ally. If not washed out thoroughly, hypo would destroy the image over time (fig. 64). Copious quantities of clean water were required, and Henneman was at the mercy of a municipal water supply so terrible that it eventually led to a parliamentary inquiry. Reading's erratic water supply, heavily polluted, flowed through ancient wooden pipelines and was often available only a few hours per day. Henneman faced ruinously high fuel costs to heat the large quantity of water required for washing the prints. Talbot's handcrafted quality could not be maintained.

The death knell for printing in silver came early in 1846 when Talbot agreed to supply seven thousand original prints for inclusion in the influential monthly periodical *Art-Union*.[17] Intended to bring photography before a wider public, it instead became a stress test that pushed Henneman's pioneering operation over the brink of failure. Every single print had to be prepared by hand, exposed in sunlight

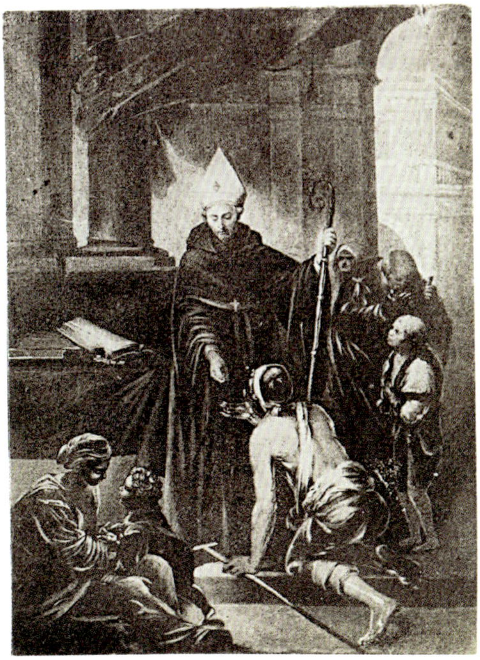

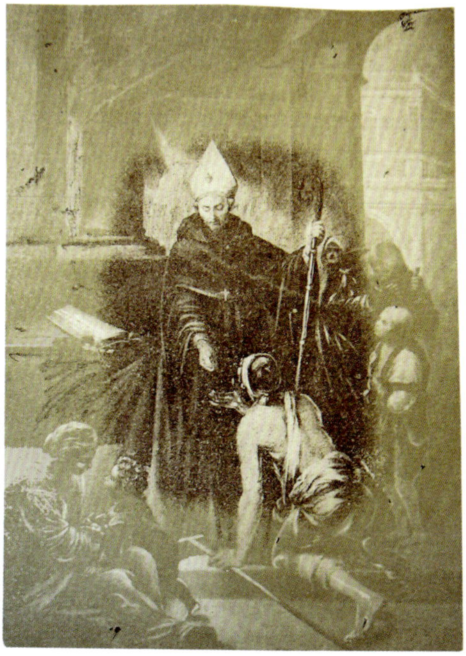

62. Nicolaas Henneman, *Photographic Copy of José Roldán after Murillo, St. Thomas of Villanueva Giving Alms*, 1847, salted paper print from calotype negative, image: 5⅛ x 3⅝ in. (12.9 x 9.2 cm). Talbot Collection, National Media Museum, Bradford.

63. Nicolaas Henneman, *Photographic Copy of José Roldán after Murillo, St. Thomas of Villanueva Giving Alms*, 1847, salted paper print from calotype negative, image: 5⅛ x 3⅝ in. (12.9 x 9.2 cm), in William Stirling, *Annals of the Artists of Spain* (London: John Olivier, 1847), plate 44. George Eastman House, Rochester, NY.

The uncut print in fig. 62 approaches the ideal that Talbot sought and that Henneman was able to achieve. Once it was mounted in a book as in fig. 63 and subjected to environmental stresses, however, fading could be caused by several sources, in this case most likely by the intrusion of sulfur-contaminated air. The image is Henneman's copy of a copy: he photographed the oil on canvas formerly in William Stirling Maxwell's collection at Keir House, Perthshire, copied from the original in the Museo de Bellas Artes, Seville.

under a negative, and then fixed and washed and dried. Under the relentless pressures of production, every available negative was brought into play, but the supplies of reliable sunlight and clean warm water were stretched beyond their limits. This soon evolved into a disaster, spreading hundreds of defective and fading prints throughout the art community. To his further horror Talbot was becoming aware that many of the prints already sold in *The Pencil of Nature* were beginning to fade. He and Henneman had attempted to ratchet up what had been a personal art to an industrial scale.

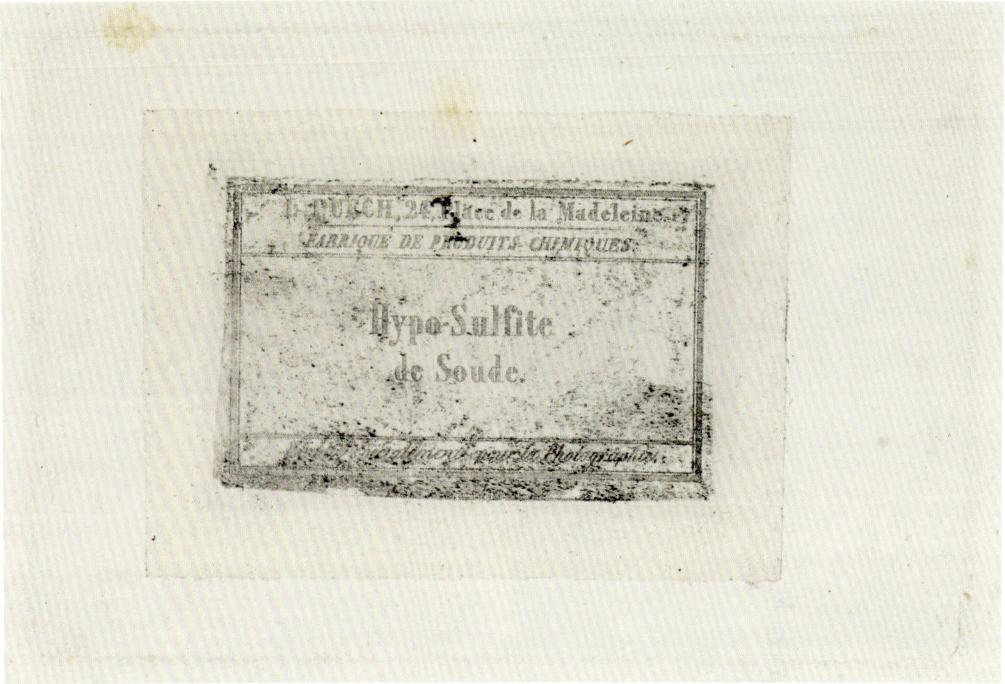

64. William Henry Fox Talbot, *Copy of a Label for Hyposulfite of Soda*, ca. 1858, photoglyphic engraving, 1⅝ x 2½ in. (4 x 6.2 cm). Talbot Collection, National Media Museum, Bradford. Hyposulfite (or "hypo") was the fixer that made silver prints permanent. Improperly washed out, however, it caused the eventual destruction of the image. Talbot's rendering in printer's ink of this friend and foe was undoubtedly a comment on that fact. Given his competition with the French over photography, his unusual choice of a French label for his new process may well have been intentional.

Exacerbating silver prints' original sin was the uncontrollable life that they led after leaving Reading. The binder pasted them onto pages of paper of uncertain quality, using glues preserved with sundry chemicals. Many wound up in London, where the notorious smoke from coal fires put the highly reactive silver in the presence of sulfurous fumes. Although silverware that tarnished under such conditions could be polished, the same chemical reactions destroyed the more delicate photographic image. In the end original silver prints could never be truly permanent, even though subsequent advances in technology would permit others to come closer to this goal than Talbot was able to in the early years of photography.[18] Permanence, the ability to transmit knowledge down through the ages, was the very essence of the book, and *The Pencil of Nature* was an idea ahead of its time. Talbot then sought to marry the veracity of photography with the time-proven productions of the printing press.

THE BEGINNINGS OF PHOTOGRAVURE

The point when Talbot began to think of replacing silver with printer's ink may have been as much a mystery to him as it is unclear to us, for there are scattered references to printing plates throughout his research notes.[19] In 1838, on the very threshold of photography's debut with the public, Talbot recorded a scheme for using the heat of the sun's rays to make a printing plate: "Let a shadow of a plant be thrown on a copper plate covered with wax or something more fusible, so that the heat of the sun may uncover the plate by melting the wax. Then let it be etched with aqua fortis. Some amalgams are fusible at that low temperature, & might answer better."[20] Right from the beginning of 1839, paralleling his work in photography on silver, Talbot recorded various ideas in his notebooks about the production of metal plates.[21] It seems likely that Talbot was influenced by the scanty information emerging from Paris. He knew that Daguerre operated on metal plates, yet he did not know how or even why.

Photography became such a success for Talbot by 1840, in terms of both his aesthetic advances and his discovery of the improved calotype negative, that it is difficult to assess how much importance he himself put on the idea of alternative modes of printing. With photography on silver becoming more troublesome by the late 1840s, however, one manuscript document of his stands out. On November 28, 1847, Talbot drafted a "Proposed method of transferring Photography to Steel Engraving."[22] Not a very practical process (it involved selectively corroding a plate through the action of electricity), nonetheless the direction of its stated intent was clear. That goal was to be manifested in a patent application five years later. Looking back at this period, his son concluded that although "the precise date at which Talbot first turned his attention to the photographic etching of metal plates cannot be exactly determined . . . his family remember that he was actively engaged in experiments in 1852"; indeed, it is from this date that the documentary record clearly outlines Talbot's progress.[23]

PHOTOGRAPHIC ENGRAVING (1852)

Talbot's first clearly identifiable photogravure process was laid out in his patent No. 565, entitled "Improvements in the Art of Engraving"; he filed the preliminary specifications on October 29, 1852, and finalized them on January 24, 1853. The patent outlined the production of a photographic image on a metal plate and then the use of this image as a resist, a pattern that either blocked or allowed the chemicals to pass, thus controlling the etching in each area. Having done their work in faithfully recording nature, all the photographic chemicals were then removed, enabling the resulting plate to be printed in a conventional printing press using standard printer's ink. Talbot restricted his description to steel plates, mentioning in passing that zinc plates or lithographic stones might also be used, but notably he made no reference to copper plates at this time.

The photographic part of this process was a refinement of Mungo Ponton's discovery in 1839 that potassium bichromate could be used to make an image.[24] Once the blank steel plate was carefully cleaned, Talbot coated it with a warm solution of gelatin and potassium bichromate in water, spreading it evenly with a glass rod. The plate was then dried with gentle heat, leaving a thin orange film of bichromated gelatin. This operation had to be conducted in reduced light, for the plate was now sensitive to light. An object such as a leaf, or a photographic positive, was superimposed on the plate under a sheet of glass, and the sandwich was placed in daylight for a period between half a minute and five minutes or longer. When removed from the light the plate would exhibit a very weak image, remaining orange where the object blocked the light, but having turned a muddy green where light exerted its action. This greenish image represented areas of the gelatin that had been tanned, or hardened by the action of the bichromate under the influence of light. This hardening raised the melting point of the gelatin in the exposed areas, thereby making it less permeable to water. A brief wash in warm water removed the unhardened areas of gelatin, leaving a photographic image on the plate. Dark areas of the original were rendered as bare metal, whereas the lighter areas were protected by a film of hardened gelatin. Talbot then etched this with a solution of platinic chloride brushed on the plate.[25] Within a minute or two the acidic solution ate away the steel in the areas lacking the protection of a gelatin layer, leaving shallow pits in the surface of the steel wherever there had been dark areas in the original. A vigorous scrubbing removed the remaining gelatin and any residual chemicals. At this point photography's work was done. It had performed its magic of transferring nature, and the resulting plate could be printed commercially just like any conventional engraving.

This "photographic engraving" process (as it was termed in the popular press) showed great promise in Talbot's quest to harness nature as his etcher, in much the same way as he had previously sought its aid in his sketching through the invention of photography. Results with fine opaque objects such as grasses were particularly striking. Talbot had invented the first photogravure process.

THE "PHOTOGRAPHIC VEIL" AND THE HALFTONE DOT

As promising as it was, Talbot's photographic engraving process had serious shortcomings, poorly reproducing intermediate tones and failing in larger areas of high density. Part of the difficulty was that the gelatin began hardening from the outer surface, not always reaching down to the surface of the plate. When the exposed plate was washed some of the middle tones were undercut and lost. Other than employing thin layers and cautious washing, there was little Talbot could do about this (the resolution of this problem would come decades later with the use of carbon transfer tissue). Physics worked against him, for another mechanical effect came into play. The first step in printing an intaglio plate is to cover the entire surface with ink. The excess is then removed with a cloth, or more commonly wiped off with a blade called a "doctor." Fine lines retain their ink, but in broader areas of tone, surface tension causes the ink to stick together, pulling it out of the trough and leaving only the ink clinging to the periphery. Thus, broad dark areas are reproduced by a strong outline surrounding a weak inner expanse (fig. 65a).[26]

In order to overcome this problem, Talbot suggested applying a familiar aquatint ground, although there is little evidence that he practiced this with any regularity. By dusting the etched plate with a resin powder, and then melting it into the surface, he could create a fine granularity that would retain the ink. But this innovation was eclipsed by one of his experimental observations. In 1853 he published a refinement of photographic engraving, stating that "if a piece of black gauze or crape is the object selected for representation, it produces an engraving of itself which is marvellously accurate. But when two folds of the gauze are laid across each other obliquely, then the resulting engraving requires a lens in order to separate from each other and distinguish clearly the lines belonging to the two portions of the gauze. Now, if this engraving be printed off, the result offers to an eye at a moderate distance the appearance of an uniform shading" (figs. 65 and 66).[27] Talbot had invented the essence of the halftone dot (fig. 67).[28] By breaking up large tonal areas into small dots, he created a way for a single ink of standard density to provide the appearance of intermediate grays, without the problem of the ink being dragged out of large areas. The eye would fuse these masses of fine dots into a uniform tonality. Talbot placed his sensitized plate under two or three folds of fine cloth for the first exposure. Then, without processing, he made a second exposure under a leaf or a photographic positive. Wherever this original did not protect the plate, the light obliterated the pattern of the cloth. The previous image of the cloth was retained, however, in those areas where the leaf blocked the second exposure to sunshine. Talbot explained that "the image of the leaf itself is still covered with a network of innumerable lines. Now, let this be etched in the

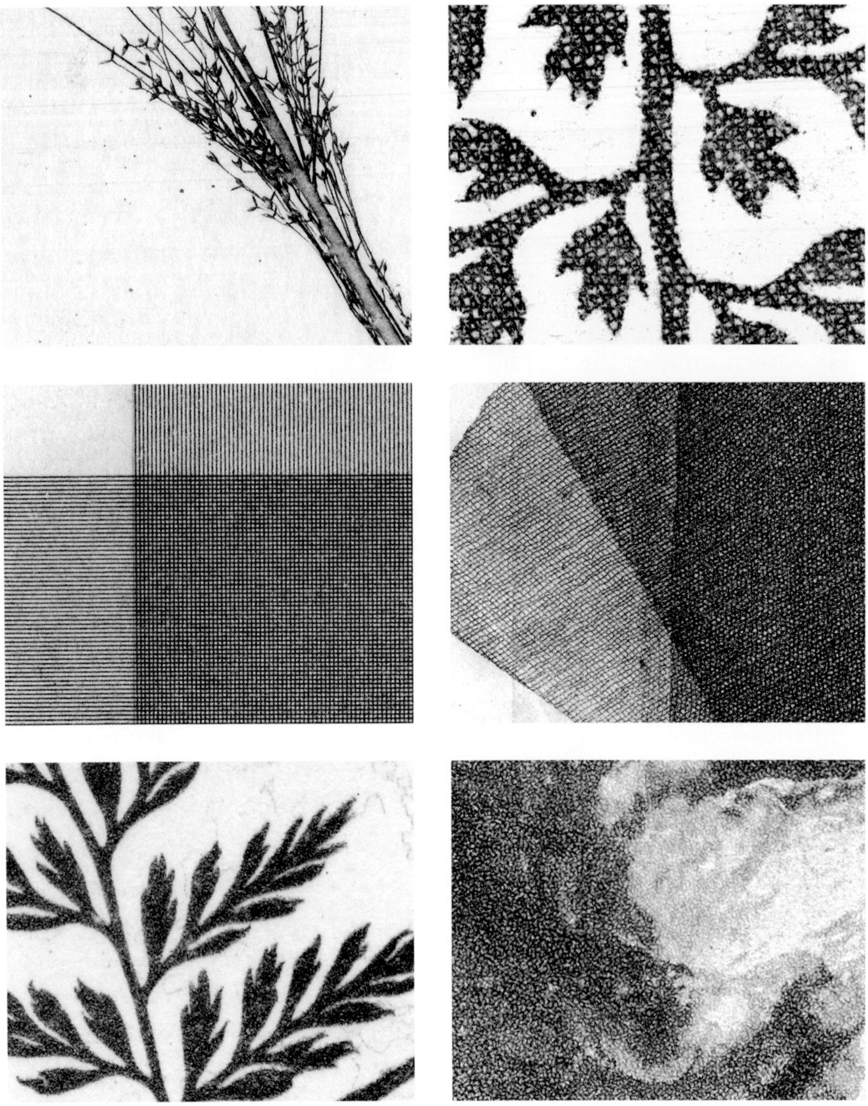

65. Magnifications of Various Effects in Talbot's Photogravure Plates. Hans P. Kraus, Jr., New York. Clockwise from top left: *a. Graminée du midi de la France — Three Stems of Grasses* (detail), probably 1852–57, photographic engraving. In Talbot's original photographic engraving of 1852, broad areas of deep tonality were simply a ditch in the plate; when the excess ink was wiped off, capillary attraction pulled the ink out of these recesses, leaving it only on the edges. *b. Foliage, with Superimposed Pattern of Gauze*, ca. 1852–57, photographic engraving from a steel plate. Superimposing the image of a finely woven cloth broke up these areas, allowing the ink to remain; at viewing distance the eye merged this into one tone. *c. Three Sheets of Gauze, Crossed Obliquely*, ca. 1852–57, photographic engraving. Various layers of fine cloth, arranged at angles, produced more complex and sometimes more successful veils. *d. Source of the Arveyron from beneath a Glacier near Chamouni*, 1860s, photoglyphic engraving from an original photograph by an unknown photographer. Liquid aquatint graining, invented by Talbot, produced a less detectable grain pattern; this was the basis of "Photo-sculpsit." *e. Truncated Fern*, probably 1858 or later, photoglyphic engraving. This shows Talbot's application of conventional aquatint graining. *f. Veil, Engine-Ruled Lines, Crossed at Right Angles*, possibly July 29, 1859, ink on paper, possibly from a photographically reproduced plate. Talbot also attempted mechanically ruled screens to achieve the same effect.

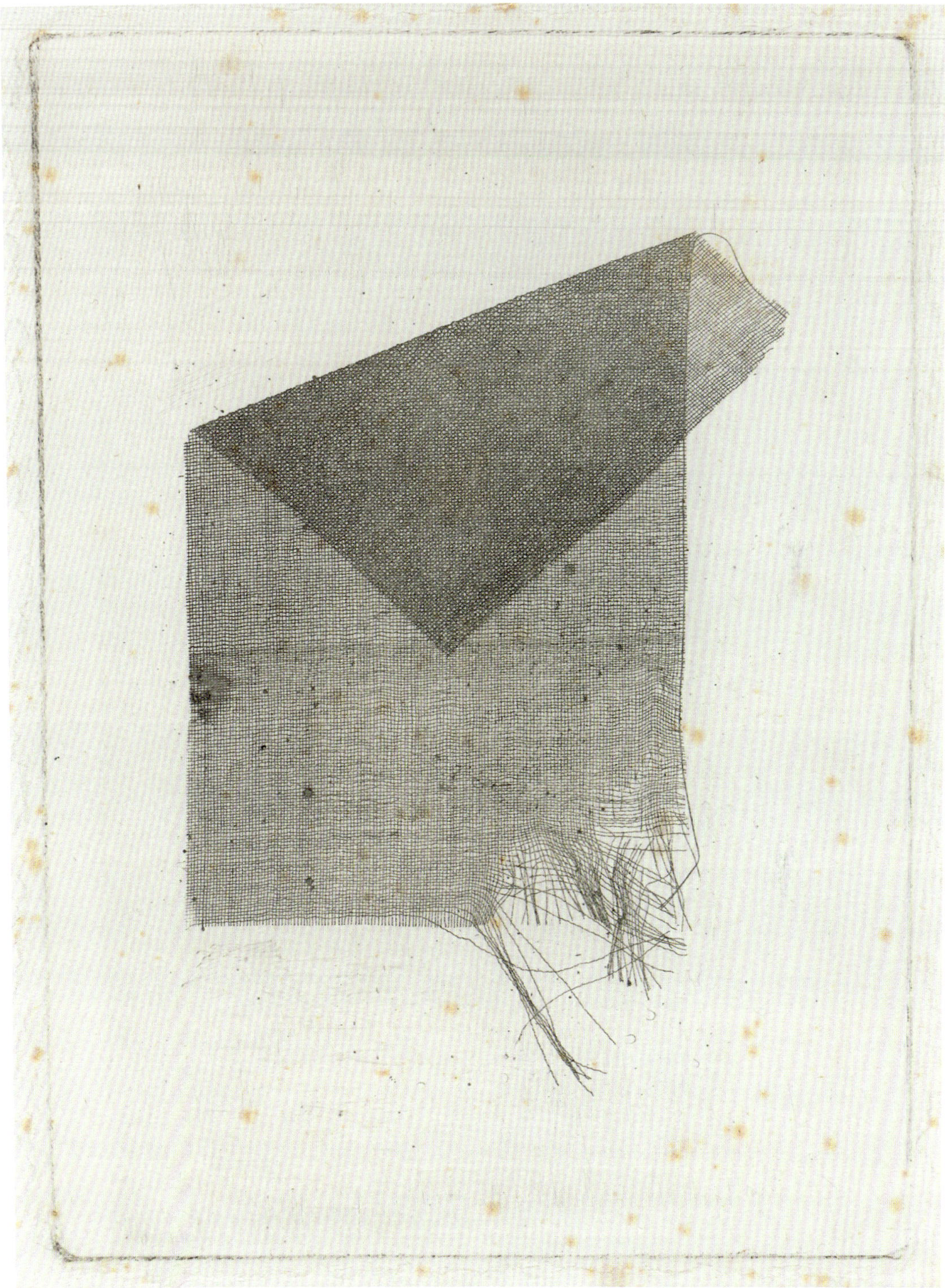

66. William Henry Fox Talbot, *Folded Piece of Muslin*, ca. 1852, photographic engraving, plate: 4 x 2⅞ in. (10.2 x 7.2 cm). State Library of Victoria, Melbourne. It was observation of the tones created by the weave of the cloth that inspired Talbot to use this as a photographic veil.

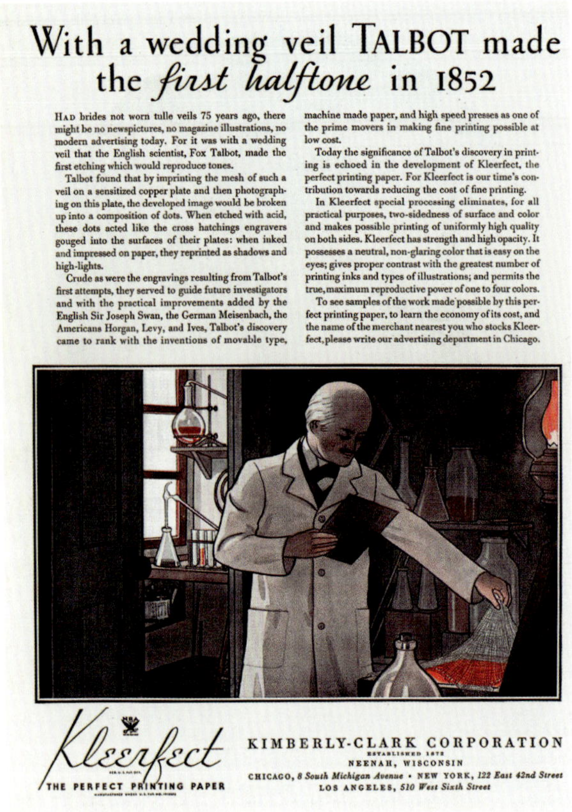

67. Artist unknown, *With a Wedding Veil Talbot Made the First Halftone in 1852*, advertisement from an unidentified trade journal, ca. 1952. Private collection. A century after Talbot's critical breakthrough, many in the printing industry still recognized his seminal contributions to their daily work. Although the story of the wedding veil may have been apocryphal, Talbot's discovery of the halftone dot (initially produced with a fine cloth screen) was the basis of most photographic reproduction throughout the twentieth century.

way already described, and let the resulting etching be printed off. The result is an engraving of the leaf which when beheld by the eye at a certain distance appears uniformly shaded, but when examined closely is found to be covered with lines very much resembling those produced by an engraver's tool, so much so that even a practical engraver would probably be deceived by the appearance. This crape arrangement I call a *photographic veil*."[29]

In addition to common gauze, Talbot investigated various finely woven specialty cloths and mechanically ruled lines and screens.[30] "These things, which I believe have not been heretofore used in the fine arts, I would denominate photographic screens or veils," he wrote in his patent specification.[31] Although an aquatint ground could also be used, this was tricky and had to be applied to each plate, whereas Talbot pointed out that a single photographic veil could serve for numerous plates. The noted photohistorian and printing historian Edward Epstein observed that "Talbot was not only the first to use potassium bichromated gelatin on metal (steel), but he was also the first to employ both the term and the use of 'photographic screens.' Historically this lays the foundation of the present-day reproduction processes."[32]

These early photographic engravings have several distinguishing characteristics. As with Talbot's earliest contact negatives made by his photogenic drawing process, their subject matter was dominated by botanical specimens and pieces of fabric. For a brief period at the beginning, he also produced positives from his own earlier calotypes. In many ways his camera-less original images are the most seductive to us now, tying his new invention back to his first one, but Talbot soon became caught up in creating the perfect printing plate more than in making totally new images. He acquired his own printing press, but he clearly saw his contribution as being in the making of the plates, and most of these he had printed by commercial printers such as George Barclay and Thomas Brooker.[33] He soon began relying on glass positives made by other photographers, typically French, and had no concerns about the authorship of the image. Weak inking in broad areas of tonality is a characteristic of his earliest plates. The presence of a photographic veil most likely dates a proof to these early years.

PHOTOGLYPHIC ENGRAVING (1858)

A new patent signaled Talbot's next major advance in photogravure, a related set of innovations and practices that had implications not only for photogravure but for broader areas of the printing industry. Filed on April 21, 1858, and finalized on October 14, 1858, No. 875 carried the same title as his patent filed in 1852: "Improvements in the Art of Engraving." In it he declared, "to the processes here described I have given the name of photoglyphic engraving (from the Greek for carving with light)." Talbot explained that "the new art has received the name of Photoglyphic Engraving an appellation suggested by the analogy of Hieroglyphic and some other words—such an engraving can be called simply a Photoglyph if a single term to express it should be found requisite or convenient."[34] For photoglyphic engravings he repeated the initial steps of his first process, coating a steel, zinc, or now a copper plate with bichromated gelatin and exposing it under an object or a positive photograph. Every experimenter from Niépce through to Talbot had washed away portions of the resist, leaving the plate exposed for etching. But in a radical departure Talbot realized that not only was this unnecessary, but in fact it was "injurious to the beauty of the result. . . . Much more beautiful engravings are obtained upon plates which have not been washed, because the more delicate lines and details of the picture have not been at all disturbed."[35] He dusted the exposed plate with resin and melted this into the plate over an alcohol lamp. Contrary to expectation this did no damage to the gelatin. He claimed that "this process may be called the laying of an aquatint ground upon the gelatine, and I believe it to be a new process."[36] This aquatint ground replaced his photographic veils, which had seemed so promising. Perhaps Talbot's goal of enlisting nature as his etcher informed this decision. Talbot's son, Charles Henry, later felt the veils were "discarded, probably because it was a desideratum that the *structure* of the ground should not appear."[37] The random granularity of resin grains was a phenomenon of nature; the artificial pattern of man-made cloth was not.

In a second major contribution Talbot described his discovery of a startling new etchant, which he called "perchloride of iron" (ferric chloride). Engravers commonly employed nitric acid, putting up with its noxious fumes and its deleterious effects on their clothing. Talbot's first etchant was more gentle but relied on expensive platinum. The real advantage of the iron chemical agent was his discovery that it could be adjusted in strength to allow control over the tonalities in etching. His new etchant was rapidly adopted by engravers throughout the printing industry and is commonly used by photogravure artists to this day. Counterintuitively, stronger solutions penetrated the gelatin more slowly, thus having milder effects. Talbot generally worked with three strengths: a saturated solution of ferric chloride, a solution of one part of water added to five or six parts of the ferric chloride solution, and one diluted fifty-fifty. Talbot would start etching with a strong solution and then move on to more diluted (more active) ones as needed. This could even be done selectively in areas of unusual density, such as the shadows of buildings. Modern practitioners will keep several different strengths of etchant at hand (loosely called Baumé, after a measure of specific gravity), using each as needed for a particular plate.

While none of Talbot's photogravures were offered for commercial sale, his photoglyphic engraving process of 1858 was sufficiently practical to make its way into actual publications. His steel plates could hold up to printings in the thousands, although there was no demand at the time for such a quantity. Steel plates were more difficult to etch, however, and were subject to rusting. One solution was to use multiple, nearly identical copper plates, replacing each as it wore out after perhaps two hundred prints. In the first widespread demonstration of his

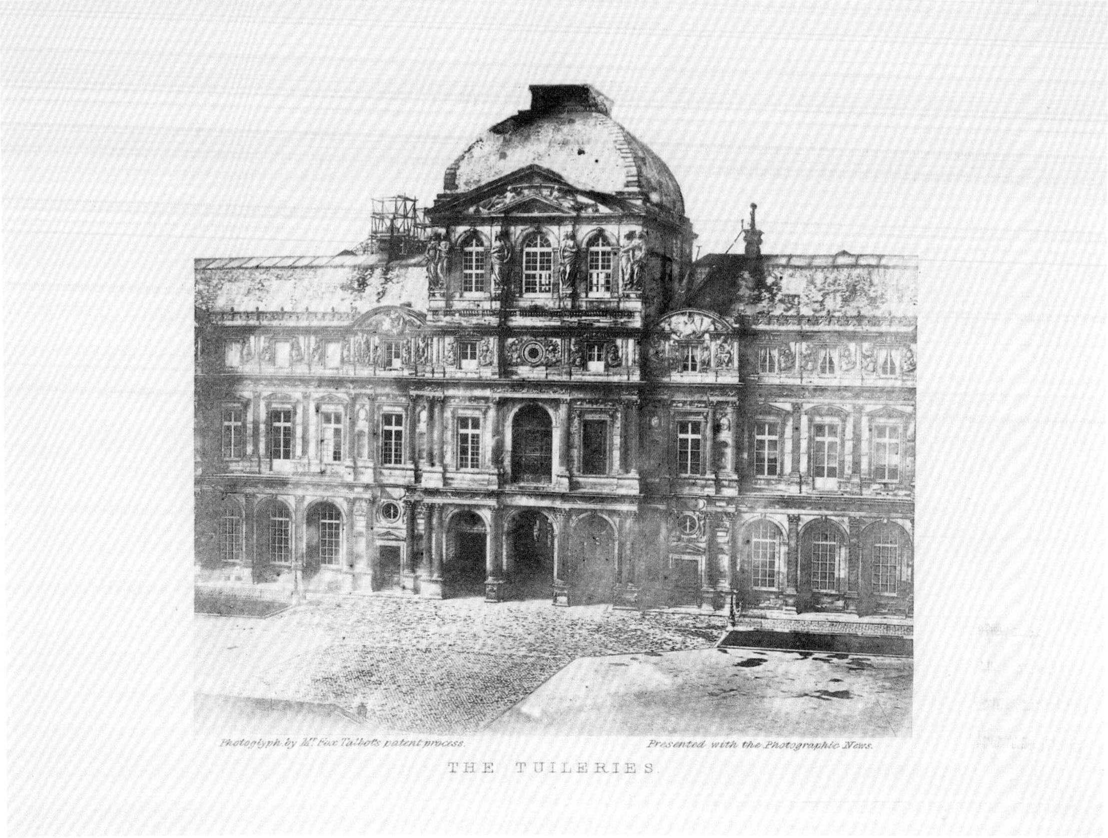

68. William Henry Fox Talbot, *The Tuileries*, photoglyphic engraving, plate: 7½ x 10 in. (19 x 25.3 cm), presented with *Photographic News*, Sept. 16, 1859. Gernsheim Collection, Harry Ransom Center, University of Texas at Austin. William Crookes, the editor, was a great supporter of Talbot's photogravure investigations and issued a number of examples in the journal. They were just short of being commercially practical, but they clearly indicated the direction that publication would take.

process, multiple plates were used for different original photoglyphic engravings in each copy of the *Photographic News* on two occasions in 1858 and 1859 (fig. 68).[38] Talbot published a botanical example in a scientific journal in 1863, noting that "if this art had been invented a hundred years ago, it would have been very useful during the infancy of botany, when communications with distant countries were so difficult. It would have been easy for botanical travellers . . . to have taken with them a small printing-press, and to have worked off a small edition of fifty copies of each engraving they made. . . . It would have greatly aided modern botanists in determining the plants intended by these authors, whose descriptions are frequently so incorrect that they are like so many enigmas, and have proved a hindrance and not an advantage to science."[39]

Rather than using multiple copper plates, or a single but more problematic steel one, Talbot turned to the French process of *aciérage*, in which a coating of iron was plated onto the finished copper plate. Usually but incorrectly called steel-facing, this plating of iron was sufficiently robust to bolster the plate in printing, but it did not destroy fine detail. Best of all, the printer could see when the iron was wearing thin and apply new coatings indefinitely.

The eccentric astronomer royal for Scotland, Charles Piazzi Smyth, afraid neither of experimenting nor of pushing boundaries, had long been an admirer of Talbot's inventions and became the biggest promoter of photoglyphic engraving. He asked Talbot to produce a large photoglyphic plate to illustrate part of Smyth's official report on his expedition in 1856 to Tenerife, in the Canary Islands (the other photographs were represented by albumen prints). The report was published in 1863; 250 copies were sent to the worldwide list of the Royal Observatory, Edinburgh, and an additional 250 photoglyphs were distributed at that year's meeting of the British Association for the Advancement of Science (fig. 69). As Smyth astutely observed in his introduction, "To the inventor alike of photography and photoglyphy, it must be comparatively indifferent by which of his two methods these unusual Teneriffe landscapes are introduced into this book, though to readers in a future century it may make a great difference; for the photoglyph must last as long as the paper it is printed on, but the photograph may go the way of some of those beautiful specimens exhibited last year at the International Exhibition, and which faded before the eyes of the nations then assembled."[40] In fact, Talbot was awarded a medal for photoglyphic engraving at this same International Exhibition of 1862, in London (fig. 70).

Examples from the period of photoglyphic engraving tend to be made from glass positives donated by other photographers (most often French), and the most common size was a half-stereo plate. These usually exhibit an aquatint ground, normally easily visible under magnification. Experimental examples from the early 1860s often have a very faint three-digit number scratched into the plate. Reproduced in reverse just outside the image area, these hairline numbers correlate to surviving Talbot notes, and they not only make it possible to date the plate but also often allow precise identification of his experimental procedure.[41]

"PHOTO-SCULPSIT" (1866)

By the late 1850s Talbot's publishing activity and public visibility had moved heavily toward cuneiform and linguistic study, away from photography and other physical sciences.[42] It is clear from his correspondence and notes, however, that his interest in photogravure remained central to his intellectual activities. In 1859 one of his daughters married and moved to Scotland. From that time Talbot and the

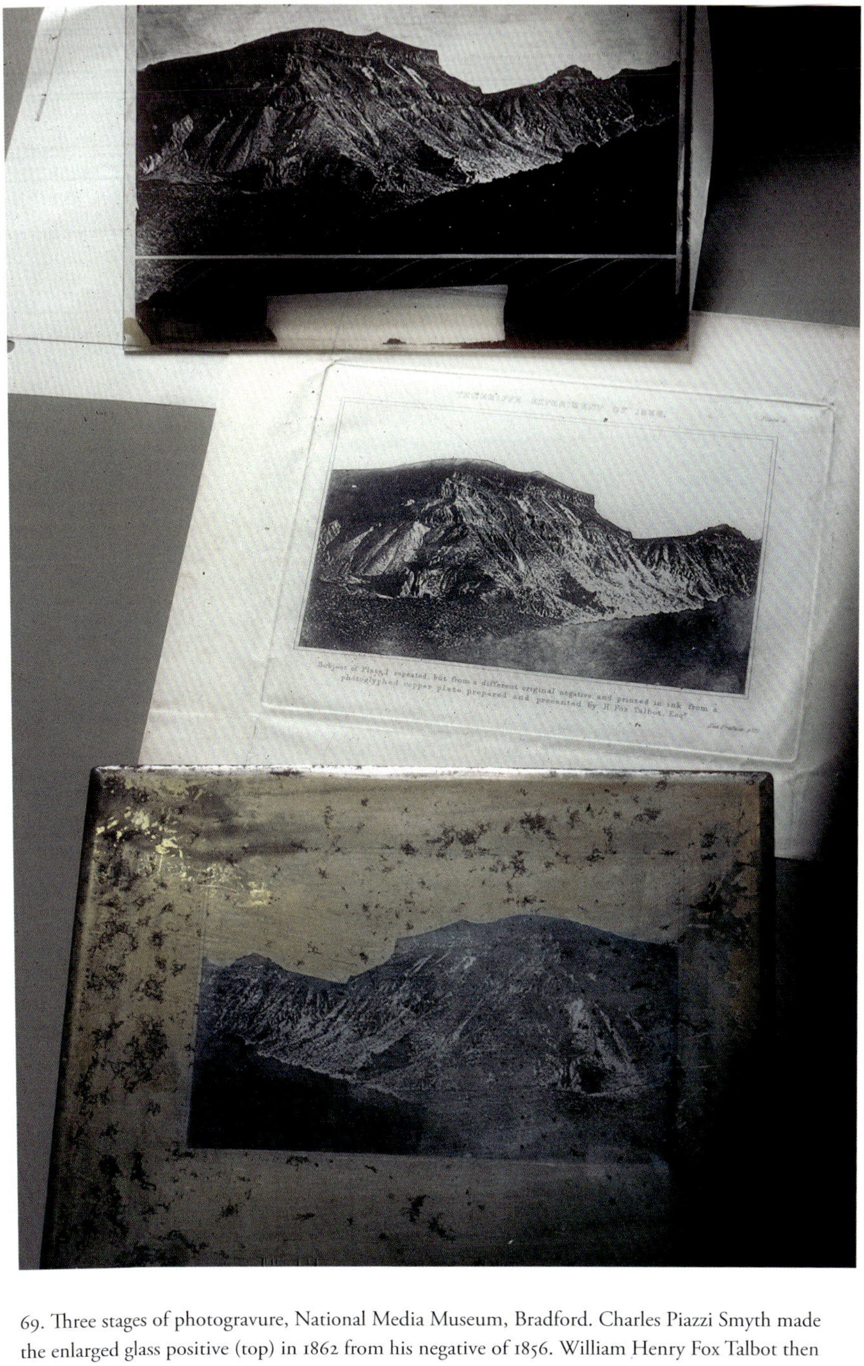

69. Three stages of photogravure, National Media Museum, Bradford. Charles Piazzi Smyth made the enlarged glass positive (top) in 1862 from his negative of 1856. William Henry Fox Talbot then used this to make a steel photoglyphic engraving plate (bottom), which produced the prints for Smyth's *Astronomical Observations Made at the Royal Observatory, Edinburgh. Vol. XII for 1855–59* (Edinburgh: Neill, 1863) (middle).

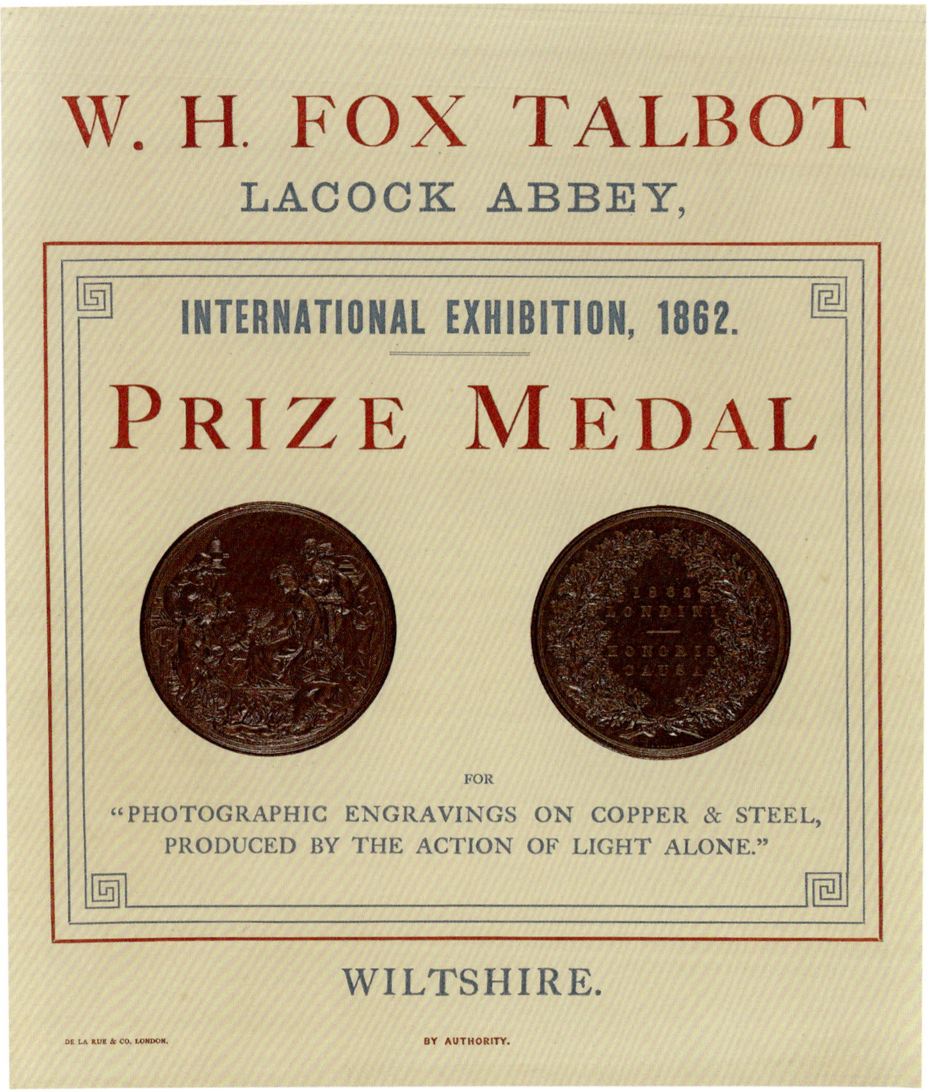

70. Medal from International Exhibition of 1862, London, "For Photographic Engravings on Copper & Steel. Produced by the Action of Light Alone," cast wax on card, 13 x 11⅜ in. (33 x 29 cm). J. Paul Getty Museum, Los Angeles. During his lifetime Talbot actually received more recognition for his work in photogravure than for his original invention of photography.

rest of the family stayed for extended periods there, primarily in and around the capital, Edinburgh, which provided a renewed stimulus for his photogravure work. Previously, much of his research had been conducted at Lacock, supplemented by occasional visits to printers in London. In Edinburgh he was more actively involved in scientific meetings and even used the laboratories at Edinburgh University. Being resident for extended periods in an important center of printing activity provided both industrial resources and a pool of talented craftsmen. At first the Talbots rented in various places in New Town, but from June 1861 to December 1863 they resided at Millburn Tower, conveniently on the rail line to Glasgow near Gogar, just outside Edinburgh. The entire family took an interest in Talbot's advances in his new art, and in 1861, shortly before the move there, his daughter Rosamond wrote excitedly about Millburn Tower: "What would please you, Papa, beyond anything is the existence of a large empty room, separated from, though close to the house, with three south windows looking out on the garden, and a good fire-place, just the place for your engravings."[43] From 1864 until 1871 the Talbots frequented the house they purchased in Great Stuart Street, right in the center of Edinburgh.

Talbot's greatest advance in photogravure, the crowning achievement of his researches and a potentially commercially viable process, took place back at Lacock. Sadly, it is also the most sparsely documented part of his work, not defined by a patent. The publishers of a translation of Gaston Tissandier's *History and Handbook of Photography* realized that the original had too strong a French bias, and in 1877 they asked Talbot to write an appendix to balance the narrative from a British perspective.[44] As expected, he related the story of his conception and invention of photogenic drawing and his subsequent great advance in the calotype process. It is revealing of Talbot's own view of his accomplishments that the entire third section was to be devoted to his new work on photoglyphic engraving. Talbot arranged to supply a thousand copies each of two different photoglyphic engravings; each surviving copy of the Tissandier edition has two authentic Talbot originals bound in and inscribed "H. F. Talbot photo-sculpsit."[45] On September 12, 1877, the publishers received a letter from Talbot saying that he had not been feeling well, but that "the third part is in preparation & will complete the Appendix."[46] Tragically, within a week he had died.

Charles Henry, Talbot's son, conducted an intensive search for the draft manuscript of that final part but never succeeded in locating it. Although he lived separately, Charles had paid close attention to his father's work and had been the senior Talbot's confidant. He completed the third section himself the best he could, revealing publicly for the first time what he knew of his father's last process. Charles wrote that the standard method of producing the aquatint ground "was uncertain and troublesome, and was superseded by a much better and very ingenious method, discovered since the enrollment of the specification in 1858, and never yet published. Common resin and camphor are dissolved in chloroform." When this volatile solution is poured on an exposed plate, "the chloroform immediately evaporates, leaving a film of resin and camphor on the surface of the gelatine. The plate is then warmed over a spirit lamp, which causes the camphor to evaporate, leaving the resin in minute particles adhering to the surface of the gelatine. This method insures a much more even distribution of the resin than the former."[47] Charles remembered much later that "there were occasional small explosions of the camphor vapour during the plate-warming." These did not seem to hurt the plate, however. He further noted, "Also, my father's latest practice was to employ, not steel, but copper plates, and to have them afterwards coated with steel."[48]

In any case the prints from this third period can be identified under magnification. Swirl marks from the liquid aquatint are visible, and the structure of the ground is more wormlike than the sandy appearance of the grounds in 1858. In 1866 the artistic printer Thomas Brooker told Talbot, "I have this day forwarded to you 3 proofs of each of your last 3 plates. Permit me to say I think they are the most successful of any you have done."[49] Most strikingly, the best of Talbot's photogravures from this period approach the standard of quality of later productions by artists such as James Craig Annan (fig. 71).

In a later letter Charles revealed that his father's final process "was within my knowledge, though not published by him. He must have devised it after taking out his last patent, and would not allow me to divulge it, under the impression, I suppose, that it might possibly be included in some future patent." Recalling the bitter patent fights of the 1850s that had severely damaged his father's reputation, Charles ruefully observed that "he seems never to have entirely lost his faith in patents, though they proved to him simply a source of disappointment. Under his photo-engraving patents, I believe, no license was ever granted, and only one application for a license was ever received."[50]

The origin of the term "Photo-sculpsit" (sculpted by light) for this later process (or more likely, family of later processes) remains a mystery.[51] It does not appear anywhere in Talbot's correspondence and thus far has not been traced in any of his research notes.[52] Charles opined that it was conceived in 1877 for the new edition of Tissandier, writing that "the word *photosculpsit* was probably introduced on the plates, in the absence of instructions from Mr Talbot as to the lettering. He invariably called his invention *photoglyphic engraving*, and would not be likely to suggest the above word, which is open to the objection that photosculpture has been adopted to denote another and a totally different process."[53] It is

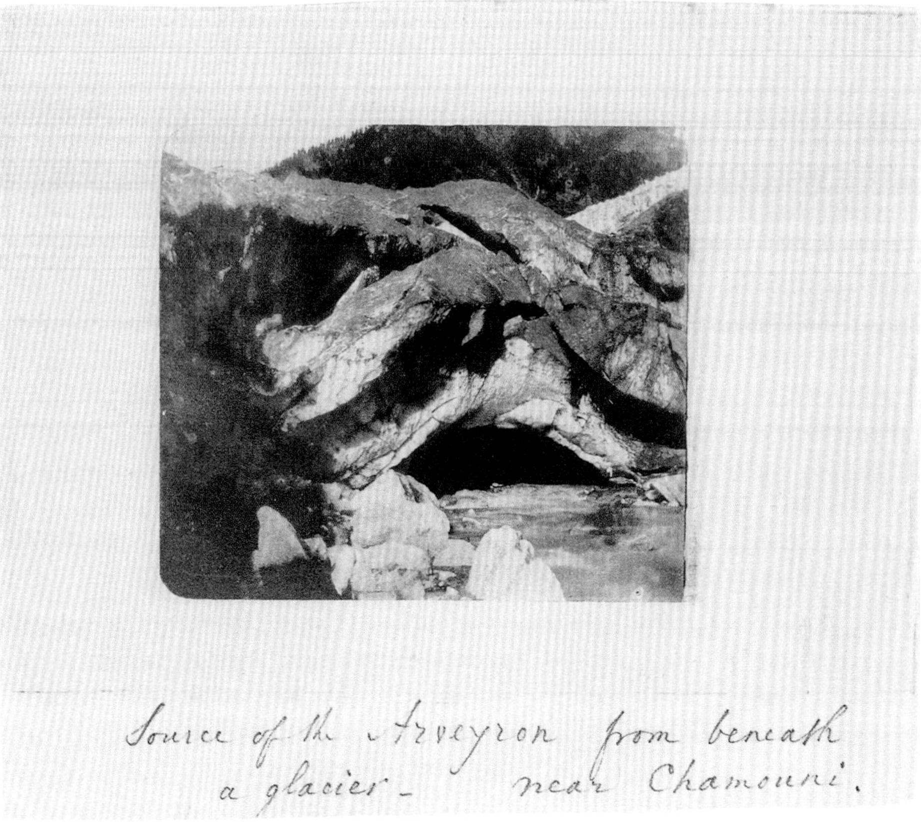

71. William Henry Fox Talbot, *Source of the Arveyron from beneath a Glacier near Chamouni*, ca. 1865, photoglyphic engraving, possibly Photo-sculpsit, titled by Talbot in pencil on recto, plate: 3¾ x 4⅞ in. (9.4 x 12.5 cm). Hans P. Kraus, Jr., New York. Halfway through his three decades of research in photogravure, Talbot had already progressed to the point where some of his plates were outstanding in tonality and uniformity, the equal of later commercial fine-print productions. Here he used a resin dissolved in solvent for the ground.

possible, however, that the term dates back to 1866, to a never-realized project that was the conceptual companion of *The Pencil of Nature*. On August 29, 1866, Talbot wrote to the engraver Ferdinand Joubert, "I have thoughts of publishing a small work illustrated by specimens of this invention—it may extend to 8 or 10 numbers, (3 plates in each number)—they will not be retouched by hand in the smallest degree—I am aware that a skilful engraver might improve them, but my present object is, to show what can be done without any difficulty by the process itself."[54]

These advanced plates took on a new life, in the form of a promotion conducted in 1900 by the *Process Photogram*. Charles Henry Talbot loaned three original copper plates that had been prepared by his father. Readers who collected twelve coupons over the course of the year could select one of the three hundred prints to be made from each plate. Up to one hundred of these prints were to be made on "fine India paper" and sold for five shillings in aid of "The Fox Talbot

Memorial Fund," established to restore the chancel of the Church of St. Cyriac in Lacock, where Talbot was once a lay rector.[55] These were to be printed by the Art Reproduction Company in London, and the plates were *Church of St. Maurice at Vienne in France*; *Portal at St. Trophimus, Arles*; and *Avignon*. Long thought to have been part of an unidentified publication, these three have engraved titles and dates and are numbered "I.," "II.," or "III."; they are the only prints known to carry the term "Photo-sculpsit" outside of those in Tissandier. So far it has not been possible to determine if the inscriptions were done within Talbot's lifetime, perhaps in 1866 or later, or were added in 1900. On April 4, 1901, Charles Henry Talbot presented the Science Museum, London, with four plates and four prints made by his father. At some point long past, some of these inventory numbers appear to have been confused. At least one is a steel-faced (*aciérage*) copper plate entitled *Portal of St. Trophimus, Arles, July 14, 1866*, with the plate number "II." This is probably one of the three plates and prints later shown in the Loan Exhibition at the Victoria and Albert Museum in 1905.[56] It is also a rare case where a fragment of Talbot's private research notes can be matched to a specific plate. On July 14, 1866, he "returned to *normal* aqua, viz. 1 alc + 1 ether. but the solids are a little diminished, via. only ¾ of their *normal* quantity. Also omitted to heat the image. Made thus a good engraving of *Portail St Trophime* in 5' sun latish." Talbot observed that it "prints beautifully." He found that this was an "*Exceedingly simple process* & takes the least time to execute—Also is less capricious & more dependable than others." Like many of the hints within Talbot's photoglyphic work, this raises as many questions as it answers, teasing us with another mystery: "This process seems identical with the *Pheasant process*. . . ."[57] Since many of the chemicals for early experimenters came from the kitchen, there was a tradition of culinary influences on photography, but presumably this reference was not to substances extracted from game birds but rather to a photoglyphic engraving of them. It remains to be discovered.

In 1859 Francis Beatty of Dublin observed that he had "from time to time tried all the processes for engraving of photographic pictures, including the daguerreotype: and, as a practical engraver and a photographer of upwards of twenty years' practice, I consider the photoglyphic process far superior to all others. . . . I consider that this, the last of the many inventions and improvements of Mr. Fox Talbot, crowns the capital of photography."[58] A short time later he called photoglyphy "one of the most extraordinary of the many remarkable inventions of the present age," saying that Talbot was a man "to whom photography owes nearly all its beauties . . . the results to science, art, and education, which must flow from the practical development of this new process . . . cannot be easily foreseen. . . . Science . . . has called into action the finger of nature, ever faithful and true, and inscribed upon her production, not the words 'painted by Nicholl, and engraved

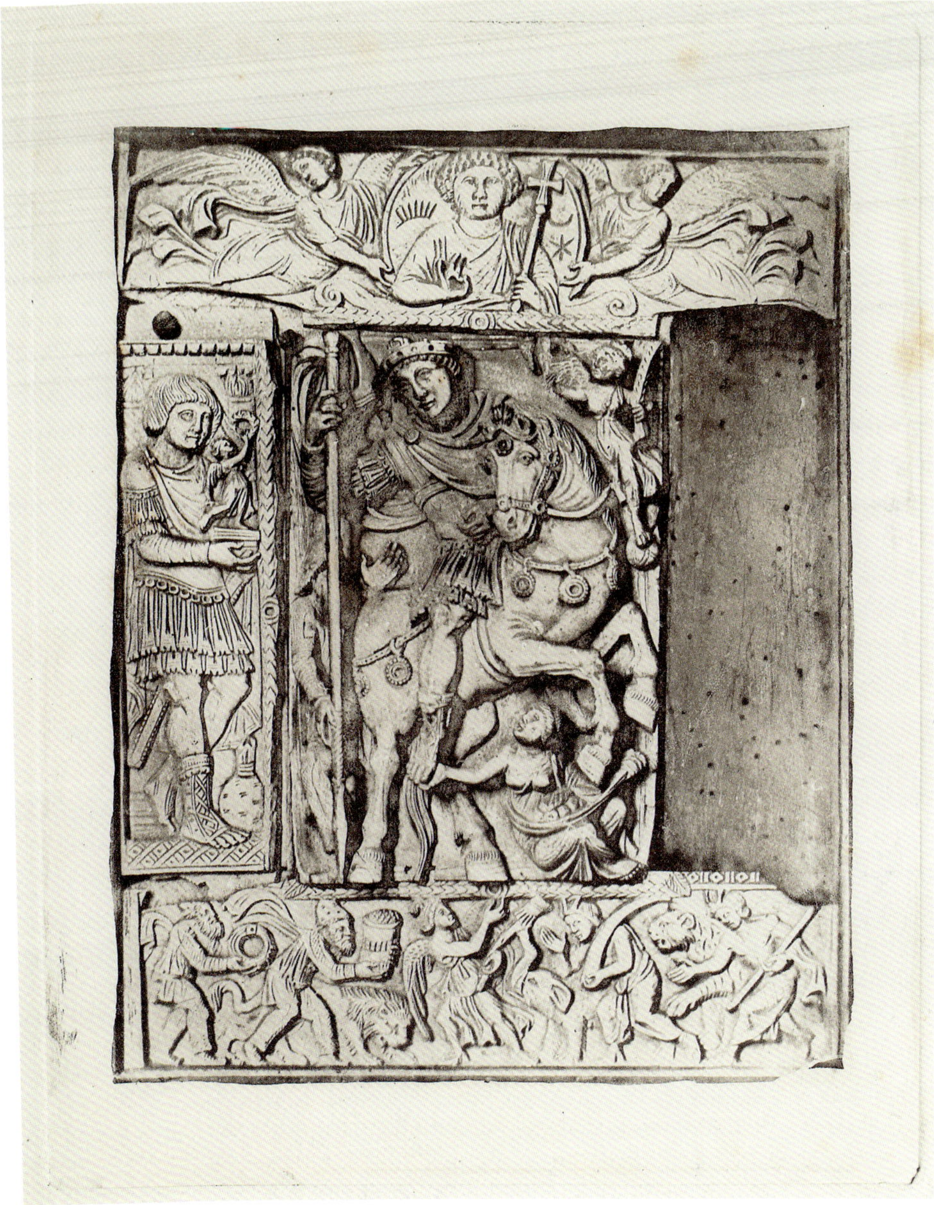

72. William Henry Fox Talbot, *The Barberini Ivory*, summer 1865, photoglyphic engraving, plate: 8 x 5⅞ in. (20.2 x 15 cm). National Gallery of Canada, Ottawa. Talbot produced several photogravures for use in actual publications. Alexander Nesbitt, a friend of Talbot's wife, had the original photographed in Rome and hoped to use Talbot's photoglyph to illustrate his article for the journal. *Archæologia.* The publications committee decided, however, that the negative was too small, and this plate was not published.

by Beatty,' but 'painted by light, and engraved by chemistry.'"[59] More than anyone else at the time, Beatty understood that Talbot's real goal from the start was simply to make nature's images permanent.

When Talbot announced the first of his photogravure work in 1852, he modestly claimed that he "recently had the good fortune to advance another step in the path of photographic discovery" and expressed the hope "that what I have accomplished will prove of great practical utility."[60] Talbot's last research note related to photoglyphic engraving shows that he was actively experimenting with it in the last year of his life. Dated October 25, 1876, it records a process leading to a "superior *Veil*"—an improved halftone dot.[61] For Talbot photogravure had been the ultimate progression of his original invention of photography. Nature not only would do his sketching for him but would then etch his printing plates, transforming these silver sketches into editions rendered in permanent—and beautiful—printer's ink (fig. 72). As one contemporary put it, Talbot "has made Apollo his own engraver."[62]

1. "New Discovery—Engraving, and Burnet's Cartoons," *Blackwood's Edinburgh Magazine* 45, no. 281 (March 1839): 383–84.
2. Thomson first applied this appellation in "Photo-Engraving," *The Year-Book of Photography and Photographic News Almanac* (1883), 86. His paper entitled "Geographical Photography" for the meeting in 1906 of the British Association for the Advancement of Science gave it wider circulation. Other authors later applied the term to Walter Woodbury for his Woodbury-type process.
3. This can be seen in fascinating magnified photographs that vividly demonstrate the similarity between Talbot's etching, Klíč's commercial method, and twentieth-century rotogravure when it was at the height of its glory. See A. J. Bull and Herbert Mills Cartwright, "Fox Talbot's Pioneer Work in Photo-Engraving," *Photographic Journal* 77 (May 1937): 307–13.
4. My interest in this area has its roots in my "Piazzi Smyth at Teneriffe: Part 2, Photography and the Disciples of Constable and Harding," *History of Photography* 5, no. 1 (1981): 27–50. In this I explored the attempts by Smyth and Talbot to use photoglyphic engraving in a practical publication and included Smyth's observations on the differences between conventional photographic prints and photogravure. I have also been influenced by the feedback to a lecture of mine: "'Another Step in the Path of Photographic Discovery'—William Henry Fox Talbot & the Invention of Photogravure," delivered in March 1995 at the International Photogravure Colloquium, Institute for Research in Art, Tampa, Florida. In this combined historical and practical workshop, I was fascinated to observe how modern photogravure artists were using Talbot's exact methods, all the while having no idea that the inventor of photography was also the inventor of their working techniques. Although these lectures were never published, this informed my text in *Sun Pictures Catalogue Twelve: Talbot and Photogravure* (New York: Hans P. Kraus, Jr., 2003). The present essay is built upon these. An excellent guide to modern practice, recording the personal experiences of a photogravure artist, is Johan de Zoete, *A Manual of Photogravure: A Comprehensive Working-Guide to the Fox Talbot Klíč Dust-grain Method*, trans. Christopher Harrisson (Haarlem: Joh. Enschedé en Zonen, 1988).

5. See William Buchanan, *The Art of the Photographer: J. Craig Annan, 1864–1946* (Edinburgh: National Galleries of Scotland, 1992); and Larry J. Schaaf, *Sun Pictures Catalogue Eleven: St. Andrews and Early Scottish Photography, Including Hill & Adamson* (New York: Hans P. Kraus, Jr., 2002), 79–80. The latter also discusses the closely related carbon print (70–78).

6. Advertisement for the N.Y. Photo-gravure Company's *Sun & Shade: An Artistic Periodical*, in *The American Annual of Photography and Photographic Times Almanac for 1892*, 62.

7. Charles Dickens, "Photographic Prints," *All the Year Round: A Weekly Journal* 1, no. 7 (June 11, 1859): 162.

8. Photomechanical processes present many challenges to the researcher, for much of the developmental work was rooted in industry and many of the approaches were proprietary. Substantial contributions were made by individuals who took secrets to their graves. Frequent duplications of efforts were masked by the concoction of proprietary names for common processes.

9. Charles Henry Talbot, "Some Account of Fox Talbot's Process of Photographic Engraving," *Penrose's Pictorial Annual* 8 (1902–3): 9.

10. The largest collection of Talbot's printing plates and photomechanical prints is held at the National Media Museum, Bradford, transferred from the Science Museum, London. In the 1930s Alexander Barclay, a curator at the Science Museum, actually had more interest in photomechanical processes than in photography when he approached Matilda Talbot about donating her grandfather's work. Proportionately he selected more of Talbot's work in ink than he did the work in silver. Barclay intended to publish on this subject but never completed any significant text. The first notable analytical publication in this area was done by A. J. Bull and Herbert Mills Cartwright. See Bull and Cartwright, "Fox Talbot's Pioneer Work," 307–13. This was later expanded on by Eugene Ostroff, curator of photography at the Smithsonian Institution. See his "Etching, Engraving & Photography: History of Photomechanical Reproduction," *Journal of Photographic Science* 17, no. 1 (1969): 65–80; and Ostroff, "Photography and Photogravure: History of Photomechanical Reproduction," *Journal of Photographic Science* 17, no. 4 (1969): 101–15. Of the standard histories of photography, perhaps the most useful indicator for photomechanical work is the venerable Josef Maria Eder, *The History of Photography*, trans. Edward Epstein (New York: Columbia Univ. Press, 1945). Epstein was a printer and printing historian, and his influence on Eder, a close personal friend, is evident. Sadly, wartime conditions forced a separation of the two colleagues and truncated their contributions.

11. After being intrigued by the new art of lithography, the French inventor Niépce began experimenting with light-sensitive materials in 1816 and was successful in making images by using several methods. He is most famous for the camera view on pewter that he took to England in 1827 and left with Francis Bauer, a botanical illustrator at Kew Gardens. Niépce entered into partnership with Daguerre but died before photography was publicly announced. It was Bauer's passionate protest over Niépce's treatment by Daguerre that brought his name before the public in 1839.

12. Photogravure (sometimes photo-gravure) was not Talbot's favored term, but it later became the standard designation for this family of processes. It appears to have entered the language with the short-lived London studio of the Grand Atelier des Arts in 1869, which advertised permanent cabinet portraits in "photogravure"—although what process it actually used is unknown. The word was popularized by Henri Rousselon's work for the French firm of Goupil & Cie starting in 1872.

13. For Talbot the retouching of photographic negatives and prints was anathema. This contrasted, for example, with the attitude of the artist and photographer David Octavius Hill. See Larry J. Schaaf, "Science, Art and Talent: Henry Talbot and Hill & Adamson," *History of Photography* 27, no. 1 (2003): 13–24.

14. This is clearly reflected in his research notebooks, where each page can present a totally new field of thought (or indeed, each section of one page); see the introduction by Mirjam Brusius and Chitra Ramalingam, and the essays by Anne Secord, June Barrow-Green, and Eleanor Robson, in this volume. Talbot was always this way. At the age of twenty-one, he wrote to his stepfather about the relaxing effects of multiple interests: "There is nothing like having two or three different pursuits for se délasser." William Henry Fox Talbot to Charles Feilding, April 8, 1821, Doc. No. 922, The Correspondence of William Henry Fox Talbot, http://foxtalbot.dmu.ac.uk (accessed Jan. 31, 2012), hereafter "Correspondence" (original: Talbot Collection, hereafter "TC," British Library, hereafter "BL").

15. His growth can be seen in the largely chronological arrangement in Larry J. Schaaf, *The Photographic Art of William Henry Fox Talbot* (Princeton, NJ: Princeton Univ. Press, 2000).

16. See the discussion of this in Larry J. Schaaf, *Introductory Volume to the Anniversary Facsimile of H. Fox Talbot's "The Pencil of Nature"* (New York: Hans P. Kraus, Jr., 1989), 40–42.

17. These were published in *Art-Union* 8 (June 1846). In the July issue the editor explained that "the whole of the Talbotypes issued with the June Number of the ART-UNION were taken from the actual objects they represent; they were, strictly, copies from NATURE; in no case had a print been made use of for the purpose of transfer. It is needless to state that prints or drawings may be easily multiplied by this process; but Mr. Talbot, in selecting examples for our Journal, carefully omitted all specimens of that class—confining himself entirely to SUN-PICTURES FROM NATURE." *Art-Union* 8 (July 1846): 195.

18. The production of large runs of silver-based photographic prints for publication was later accomplished with some success in France by Louis Désiré Blanquart-Evrard. Considered to be the "Gutenberg of photography" by the pioneering French photography critic Francis Wey, in 1851 Blanquart-Evrard opened the Imprimerie Photographique at Lille. It was one of the first truly industrialized photographic operations, using specially milled paper, careful manufacturing, temperature control, and precise timing to turn out up to three hundred prints a day from each negative. Rather than relying on the sun, Blanquart-Evrard made a highly sensitive paper that would respond to a brief exposure and could then be developed and subsequently toned with gold. Some contemporaries considered the ashen tones of these prints acceptable, but many regarded his prints as cold and inartistic compared to the warmer tones of traditional salted paper prints. Blanquart-Evrard employed a large staff but kept his process proprietary, sharing it only with his business partner, Thomas Sutton of Jersey. After Blanquart-Evrard's death, and after it was no longer commercially viable, Sutton published the only known account of the process: "Description of the Printing Process Employed by M. Blanquart-Evrard at his Establishment at Lille," *British Journal of Photography* 19, no. 634 (June 28, 1872): 308–9.

19. Many of Talbot's other research interests are preserved in series of notebooks, but the majority of his known documentation on photogravure takes the form of loose sheets of notes. There are hundreds of these pages, jumbled and little understood. The majority are in the Talbot Collection at the National Media Museum, Bradford, but important ones are preserved in various public and private collections.

20. Notebook O, 1836 (date begun), entry dated Jan. 21, 1838, Add. MS 88942/1/195, TC, BL.

21. See the facsimile of his two most critical notebooks in Larry J. Schaaf, *Records of the Dawn of Photography: Talbot's Notebooks P & Q* (Cambridge: Cambridge Univ. Press, 1996).

22. Add. MS 88942/1/350, TC, BL.

23. Talbot, "Some Account of Fox Talbot's Process," 9.

24. While modern usage among chemists favors dichromate over bichromate, the historical term will be found in most literature on the subject and has been employed throughout this text. Although Ponton pioneered the use of bichromate to make images, he had no idea of the role that it could play in hardening substances such as gelatin. He merely soaked plain paper with the bichromate and exposed it to light, yielding a weak image but no physical effect that could be further applied. Mungo Ponton, "Notice of a Cheap and Simple Method of Preparing Paper for Photographic Drawing, in Which the Use of Any Salt of Silver Is Dispensed With" (paper read before the Royal Scottish Society of Arts, May 29, 1839), *Transactions of the Royal Scottish Society of Arts* 1 (1839): 336–38.

25. When Dr. D. A. Spencer, president of the Royal Photographic Society, heard this, he "wondered why with all the chemicals available on his shelves he chose this very expensive one." Spencer chaired the Dec. 22, 1936, meeting in which A. J. Bull presented his lecture entitled "Fox Talbot's Pioneer Work in Photo-Engraving." See Bull and Cartwright, "Fox Talbot's Pioneer Work," 313.

26. This visual effect will be familiar to those who used early xerographic copy machines, the inability of which to maintain an electrostatic charge across a broad area caused a similar outlining effect.

27. William Henry Fox Talbot, "Photographic Engraving," *Athenaeum*, no. 1331 (April 30, 1853): 532–33. See also Talbot to *Athenaeum*, April 25, 1853, Doc. No. 6762, Correspondence (original: TC, BL).

28. While Talbot's procedure differed from modern halftone practice, the conceptual basis was the same. In 1897 the printing historian William Gamble pointed out that "it matters not whether the screen be of gauze, or of ruled lines, or of geometrical stipple, or irregular stipple, there are underlying principles common to all. . . . The object in all cases is to break up the photographic image into a grain. . . . The credit of the idea of breaking up the tones by means of a screen undoubtedly belongs to Fox Talbot, who . . . suggests fully what was afterwards adopted in the practice of the half-tone process." Gamble, "History of the Half-tone Dot," *Photographic Journal* 21, no. 6 (Feb. 6, 1897): 126–27.

29. Talbot, "Photographic Engraving," 532–33. See also Talbot to *Athenaeum*, April 25, 1853, Doc. No. 6762, Correspondence (original: TC, BL).

30. One instance is the group of samples he obtained from the gauze maker J. Hennecart of Paris in 1854. See J. Hennecart to Talbot, Feb. 24, 1854, Doc. No. 7016, Correspondence (original: TC, BL). Talbot's screen was actually too fine for the printing techniques of the day. In modern times Harold White enlarged Talbot's screen somewhat and successfully used it to print a copy of an Antoine Claudet daguerreotype portrait of Talbot. The successful experiment is reproduced in Harold White, "A Note on W. H. Fox Talbot and Photo-engraving," *Journal of the Printing Historical Society* 13 (1978/79): 64–66. White's creation was so convincing that it was mistakenly published as an original Talbot in Hubertus von Amelunxen, *Die aufgehobene Zeit: Die Erfindung der Photographie durch William Henry Fox Talbot* (Berlin: Verlag Dirk Nishen, 1988), 57.

31. William Henry Fox Talbot, "Specification of a Patent Granted to William Henry Fox Talbot, of Laycock Abbey, in the County of Wilts, Esquire, for Improvements in the Art of Engraving," filed October 29, 1852, and issued April 29, 1853, in *The Repertory of Patent Inventions, and Other Discoveries and Improvements in Arts, Manufactures, and Agriculture, Enlarged Series—Vol. XXII. July–December 1853* (London: Alexander Macintosh, 1853), 62–63.

32. Edward Epstein, "William Henry Fox Talbot," *Photo-Engravers Bulletin* 23, no. 11 (June 1934): 29.

33. These craftsmen also supplied practical experience and critiques for Talbot's work. Little is known about George Barclay, but he was a master engraver as well as printer. Thomas Brooker was a printer on copper and steel plates and "Artistic Printer" (a commercial category of printer). The best record of Talbot's interaction with these and other printers is contained within their letters and invoices; see Correspondence.

34. Talbot to William Crookes, Sept. 16, 1858, Doc. No. 7695, Correspondence (original: TC, BL). Talbot was clearly responding to what he had learned from the reaction to his original term of "photogenic drawing": Herschel, among others, criticized its inflexibility. See Larry J. Schaaf, *Out of the Shadows: Herschel, Talbot & the Invention of Photography* (New Haven and London: Yale Univ. Press, 1992), 28–30.

35. William Henry Fox Talbot, "Specification of a Patent Granted to William Henry Fox Talbot, of Laycock Abbey, in the County of Wilts, Esquire, for Improvements in the Art of Engraving," filed April 21, 1858, and issued October 14, 1858, in *The Repertory of Patent Inventions, and Other Discoveries and Improvements in Arts, Manufactures, and Agriculture, Enlarged Series—Vol. XXXII. July–December 1858* (London: Alexander Macintosh, 1858), 365.

36. Talbot, "Specification of a Patent" (1858), 365.

37. Talbot, "Some Account of Fox Talbot's Process," 13.

38. One of seven half-stereo-sized plates from negatives by the Parisian photographers Clouzard & Soulier was included in *Photographic News* 1, no. 10 (Nov. 12, 1858). A larger print of the Tuileries was included in the issue of Sept. 16, 1859.

39. William Henry Fox Talbot, "Photoglyphic Engraving of Ferns; with Remarks," *Transactions of the Botanical Society of Edinburgh* 7 (June 1863): 568–70. This was not the journal's first photoglyph, however. Charles Piazzi Smyth included one entitled "Young Dragon Trees, near Orotava, Teneriffe" to illustrate his article "On the Manner of Growth of Dracaena Draco in Its Natural Habitat, as Illustrating Some Disputed Points in Vegetable Physiology," *Transactions of the Botanical Society of Edinburgh* 1 (March 1859): 250–61, pl. 4.

40. Charles Piazzi Smyth, *Astronomical Observations Made at the Royal Observatory, Edinburgh. Vol. XII for 1855–59* (Edinburgh: Neill, 1863). For a more complete discussion of Talbot's experiments in this period, see Schaaf, "Piazzi Smyth at Teneriffe," 27–50.

41. This is the largest intact set of notes that has been located. It is held by the Talbot Collection, National Media Museum, Bradford.
42. See the essay by Eleanor Robson in this volume.
43. Rosamond Talbot to Talbot, March 9, 1861, Doc. No. 8331, Correspondence (original: TC, BL). Millburn Tower, with its rich history, is now a listed property, but so far it is lacking the blue plaque that should mark Talbot's photogravure researches.
44. Gaston Tissandier, *A History and Handbook of Photography*, rev. ed., ed. J. Thomson (London: Sampson Low, Marston, Searle, & Rivington, 1878); Talbot's appendix is on 345–82.
45. A reproduction of a map of the area surrounding Penicuik, Scotland, is bound opposite 373, and a Walter Woodbury "View in Java" is bound opposite 376.
46. A zinc plate facsimile of this letter is reproduced in Tissandier, *History and Handbook*, 367. See also Talbot to Sampson Low, Marston, Searle & Rivington Co., Sept. 1877, Doc. No. 839, Correspondence (original: TC, BL).
47. Tissandier, *History and Handbook*, 372.
48. Charles Henry Talbot, letter of Jan. 29, 1893, in "Fox Talbot's Photoglyphic Process," *British Journal of Photography* (Feb. 3, 1893): 76–77.
49. Thomas Brooker to Talbot, June 28, 1866, Doc. No. 9102, Correspondence (original: TC, BL).
50. Talbot, "Some Account of Fox Talbot's Process," 13.
51. The term "Photo Sculpsit" gained some currency in the early twentieth century—perversely, however, for a craftsman-manipulated plate that was exactly the opposite of the unretouched translation of the photograph sought by Talbot. This could lead to considerable confusion. Antony Griffiths described them as "hand photogravure" and explained that "the craftsman who prepared a plate often had his name engraved underneath the image, followed by the word *photosculpsit*, sometimes abbreviated to *Ph. Sc*." Griffiths, *Prints and Printmaking* (Berkeley: Univ. of California Press, 1996), 123–24. Sir Kenneth Clark, discussing Herbert Percy Horne, *Alessandro Filipepi, Commonly Called Sandro Botticelli, Painter of Florence* (London: G. Bell & Sons, 1908), wrote that "Emery Walker's photo-sculpsit reproductions have exactly the right balance for the printed page—something that has never been done again. He does therefore seem to me to be the last craftsman who applied himself to the new medium of photography." Quoted in Brooke Crutchley, preface to *Emery Walker: Some Light on His Theories of Printing and on His Relations with William Morris and Cobden-Sanderson*, by Colin Franklin (Cambridge: privately printed, 1973). At the end of the list of plates in Horne's book, the statement is made that "with a few exceptions, all the pictures have been photographed for this book by Mr. Emery Walker, who has also made and printed the plates." Some of the plates are inscribed "Walker & Boutall ph. sc." or "Walker & Boutall Ph. Sc."
52. As mentioned previously, these are voluminous, complex, scattered among several collections, uncatalogued, and, for the most part, unexamined.
53. Manuscript note in Charles Henry Talbot's hand, collection of the Royal Photographic Society, now in the National Media Museum, Bradford. The process he referred to, a mechanical technique for producing busts, was invented by François Willème and patented in France in 1860. Talbot would have been sensitive about terminology, for objections were raised immediately in 1839 about his term "photogenic drawing," with his friend Sir John Herschel promptly suggesting that the word "photography" (in parallel with words such as "telegraphy") would be more suitable.
54. This letter is known to survive only in draft form and has contractions. A verbatim transcription is Ferdinand Jean Joubert de la Ferté to Ferdinand Jean Joubert, Aug. 29, 1866, Doc. No. 9128, Correspondence (original: TC, BL). Talbot's draft copy is in the TC, BL.
55. "An Interesting Relic," *Process Photogram* 7, no. 74 (Feb. 1900): 61. The idea of photographers supporting the fund was first raised by G. E. Brown in *Nature* 57, no. 1469 (Dec. 23, 1897), and then frequently in the photographic press.

56. They were the same titles as in the *Process Photogram* issues and may have been the same plates. In the exhibition in 1905, Talbot's examples were displayed in the same bay as Joseph Nicéphore Niépce's heliogravure of 1827, the last time that relic was publicly displayed until the photo historian Helmut Gernsheim rediscovered it in the 1950s. See the *Catalogue of the Loan Exhibition of Process Engraving Held at the Victoria & Albert Museum, South Kensington, 1905* (London: Victoria and Albert Museum, 1905). David A. Hanson slightly confused this story in his entry for "St. Maurice" in the internal catalogue of his collection at the Sterling and Francine Clark Institute, Williamstown, Massachusetts: "The printing plates are owned by the British Museum and were exhibited, along with the three pictures, at one of the photomechanical exhibits at the turn of the century."

57. Two-sided manuscript sheet in Talbot's hand, dated July 11, July 13, July 14, and July 14, [18]66. Talbot Collection, National Media Museum, Bradford. While other notes from 1866 have been identified, this sheet seems to be an isolated survivor for this month.

58. Francis Beatty, letter to the editor, *Photographic Journal* 6, no. 92 (Dec. 15, 1859): 111.

59. *Northern Whig, Belfast*, Dec. 28, 1859.

60. Talbot to *Athenaeum*, April 4, 1853, Doc. No. 6749, Correspondence (original: TC, BL). See also "Photographic Engraving," *Athenaeum*, no. 1328, April 9, 1853, 450–51.

61. Talbot Collection, National Media Museum, Bradford.

62. "Fine-Art Gossip," *Brighton Gazette*, Oct. 28, 1858, 7.

INSTITUTIONS AND NETWORKS

Bel and the Dragons: Deciphering Cuneiform after Decipherment

Eleanor Robson

To Assyriologists William Henry Fox Talbot is best known as the man who instigated the "decipherment event" of 1857, which led to widespread acceptance that cuneiform script really could be read. Assyriologists know little, if anything, of his other accomplishments and largely ignore the rest of his thirty-year Assyriological career. Most historians of science and photography, meanwhile, pass over it in silence.[1]

My aim here is not to rehabilitate Talbot as a great Assyriologist, for he was not one, particularly, although neither was he a complete incompetent. Rather, I shall revisit that decipherment event and its subsequent reception, in order to question what it means to say that a complex script such as cuneiform has been deciphered. For, as we shall see, even two decades later Talbot's notebooks reveal the complex and circuitous ways by which he arrived at translations of cuneiform texts. I will focus on just three of his hundred or so Assyriological notebooks, in which he prepared an edition of a piece, published in 1876, that Talbot called "The Fight between Bel and the Dragon," and which is known to modern scholarship as part of Tablet IV of the Babylonian Epic of Creation, *Enūma Eliš* (literally, "when above").[2] I will use this analysis to conclude with some preliminary reflections on what Assyriology meant to Talbot, and how it related to the other great intellectual passions of his life.

TALBOT AND DECIPHERMENT

In 1839 Talbot published a short book entitled *The Antiquity of the Book of Genesis*, in which he sought to trace the biblical work's influence on other preclassical civilizations, based on allusions to it in ancient texts.[3] Within just a few years, however, completely new ways of addressing human origins and biblical historicity had opened up, thanks in particular to the discovery of the ancient city of Nineveh. The long-ago capital of the Assyrian Empire, on the banks of the Tigris

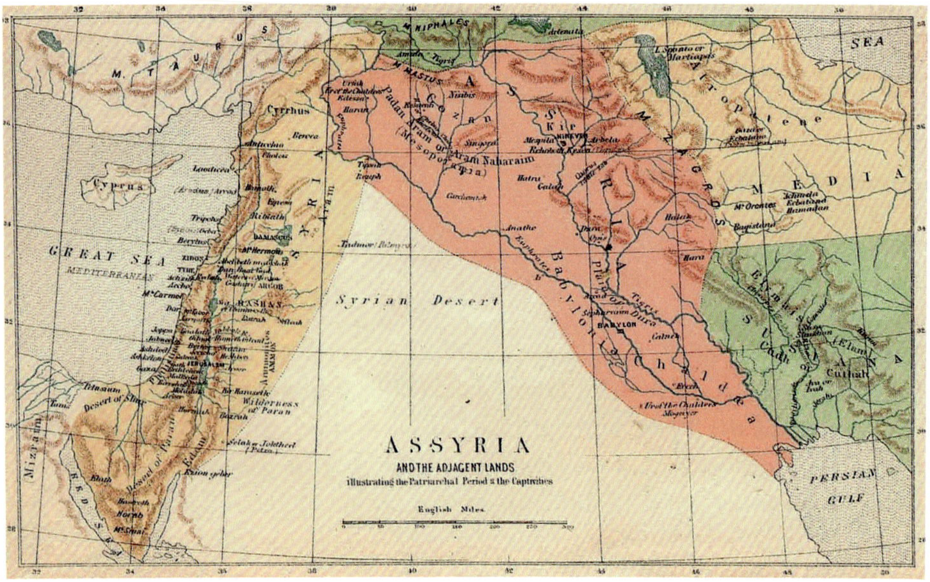

73. Map of the ancient Middle East, American Bible Society, 1880.

River in northern Mesopotamia (modern Iraq) and the erstwhile home of the prophet Jonah, began to yield spectacular finds of sculptures and inscriptions. Not surprisingly Talbot took a keen interest in Assyrian antiquities as soon as they began to appear in the collections of the British Museum in the late 1840s.[4] He was particularly fascinated by the as-yet-undeciphered inscriptions, written in a script known as cuneiform.

The cuneiform, or wedge-shaped script of the ancient Middle East, had first come to Western academic attention in the early nineteenth century, through the discovery of inscribed monuments in western Persia and then the import of smaller objects from the eastern Ottoman Empire (fig. 73).[5] Many men worked on its decipherment, all over Europe. By the 1840s it was clear that there were in fact three cuneiform scripts. Old Persian and Elamite, which had both been used by the Persian court in a few restricted contexts during the mid-first millennium BCE, had been relatively easy to decipher. Assyrian cuneiform, however, which was proving to be much more widespread and enduring, was also by far the hardest to crack. It comprised at least six hundred different signs in several different categories: syllables; logograms, representing whole words; and determinatives, which classified words into different groups.[6] In certain circumstances different signs could substitute for one another, so there were often many legitimate spellings of a single word. Even more confusingly, almost all signs could assume different functions and different pronunciation depending on the context.

By mid-century one particular genre of cuneiform text, the narrative account of Assyrian battles and conquests, was starting to make good sense. Multiple versions of the same inscriptions, brought back from the Assyrian royal cities by British and French explorers starting in the late 1840s, allowed for systematic study of variations in the use of signs. The fact that they depicted events and kings already known from the later books of the Old Testament, albeit often from a very different viewpoint, also allowed for external corroboration. In 1850 and 1854, respectively, the explorer Austen Henry Layard and his erstwhile assistant Hormuzd Rassam shipped two particularly large hauls of tablets to the British Museum from Nineveh. Amounting to well over twenty-five thousand pieces, they became known collectively as the Kouyunjik Collection, after the mound in which they were discovered, or King Assurbanipal's Library.[7] Coincidentally or not, it was at about this point that Talbot started to get heavily involved.

Talbot's first two Assyriological notebooks are from 1853, following his purchase of a copy of Layard's *Inscriptions in the Cuneiform Character from Assyrian Monuments* (1851).[8] The notebooks trace his first attempts at translating portions of the annals of the Assyrian kings Sennacherib, Esarhaddon, and Assurbanipal, translations that were privately published in 1856.[9] Thereafter, until his death in 1877, Talbot filled at least two notebooks a year—often ten or more—on Assyriological matters and published his research prolifically (fig. 74).

In the mid-1850s, however, he was still working hard to establish himself as a member of the Assyriological community. In April 1854 he wrote rather self-effacingly to Samuel Birch—then an assistant in the British Museum's Department of Antiquities—mentioning his first, failed foray into promotion of his work at the British Association for the Advancement of Science.[10] A month later Talbot sent a letter to the keeper of the Department of Antiquities, Edward Hawkins, requesting copies of the photographs of newly arrived Kouyunjik tablets that the museum was then preparing, but he received no reply.[11] A year or so later he began a correspondence with Edwin Norris, secretary of the Royal Asiatic Society and earlier decipherer of Elamite, in which they compared translations and recommended readings to each other, and which was to continue until Norris's death in 1872.[12] Just a few months later Talbot wrote more confidently, and at length, to the editor of the *Journal of Sacred Literature and Biblical Record*—a letter that was published in the January 1856 edition of that journal.[13] This first Assyriological publication is important to our story for two reasons: because it shows Talbot's commitment to convincing skeptics that cuneiform could indeed be read; and because it describes the decipherment method on which he continued to rely until

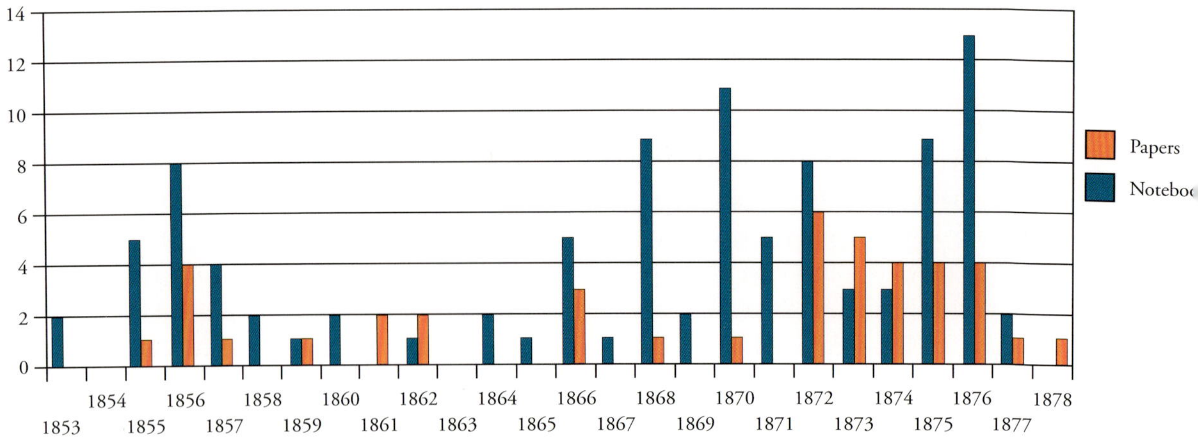

74. William Henry Fox Talbot's Assyriological output, 1853–78. Compiled by the author.

the end of his life, namely, comparative etymology, especially with Hebrew. We shall return to the second point later, but for now let us concentrate on the first.

Talbot sympathetically laid out the skeptics' objections: first, that the decipherers were guilty of circular reasoning, because their interpretation of the script depended on their understanding of the language, and vice versa; second, that it was impossible to choose between rival interpretations, some of them radically different from the others. He then gave a long account in favor of the rigorous methods of Henry Creswicke Rawlinson and Edward Hincks, their independent discovery of key points, and their critical engagement with each other's work.

Talbot sent a copy of the paper to Norris and, via him, to Rawlinson, as soon as it was published.[14] But it seems that Talbot did not consider his own testimony, as a newcomer to the field, to be sufficient proof of the validity of Rawlinson's and Hincks's work. So in September or October 1856 he wrote again to Norris, requesting membership in the Royal Asiatic Society and proposing a public experiment that he said would confirm beyond doubt that cuneiform really could be read.[15] Although Talbot's letter no longer survives, the initial formulation can be inferred from Norris's enthusiastic reply. It seems to have been that one or more independent translations should be made of some soon-to-be-published Annals of Sennacherib and then publicly compared. In that same letter, however, Norris reported

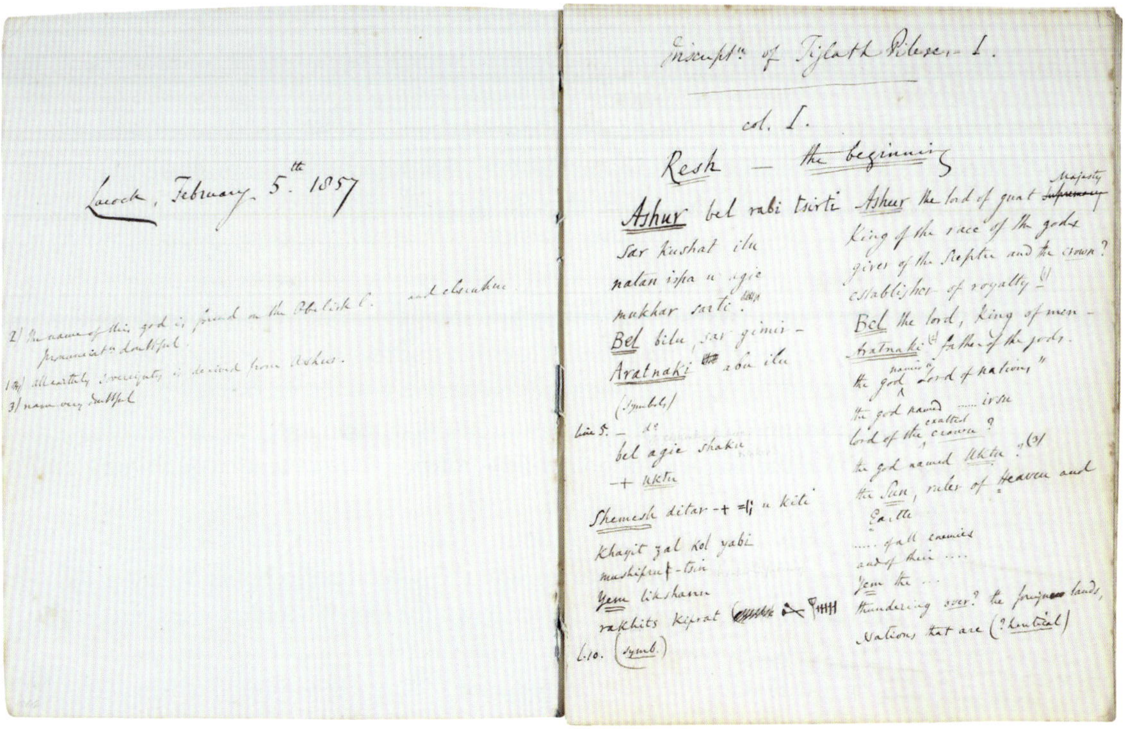

75. William Henry Fox Talbot, first page of Talbot's draft edition of Annals of Tiglath-pileser I, Notebook "Egyptian Archaeology / Inscription of Tiglath-Pileser I (back to front)," 1856–57. Add. MS 88942/1/146, Talbot Collection, British Library.

that Rawlinson "thinks it would be a more satisfactory proof of the truth of the decipherment, if you should take up something which is quite new, such as the Annals of Tiglath-Pileser," which Rawlinson was then preparing for publication.[16] This became the definitive format of the exercise. Behind-the-scenes preparations were long and convoluted, with Norris acting as "go-between at least half-a-dozen times" before declaring in mid-January 1857 that "the red-tape (metaphysical) is all unwound" and that the inscription was ready for Talbot to work on.[17]

Talbot worked fast and intensively—there are three surviving notebooks of his work in progress, one dated to February 5, 1857—and was finished within six weeks (fig. 75).[18] He formally wrote to the Royal Asiatic Society in early March, enclosing his translation. Norris replied on the 13th that he would keep the "sealed packet" locked away until a meeting six days hence, when it would be compared with Rawlinson's freshly published version; "I have not the smallest doubt of their agreement in all essential particulars," the ever-optimistic Norris concluded.[19] But in late April Norris was reporting further delays. Rawlinson wanted to know how Talbot had laid out his translation, whether in paragraphs or line by line, so as to follow the same method and thus facilitate comparison. And by this time Hincks and the French Orientalist Julius Oppert also wished to participate, pushing the date of the public scrutiny back yet further.[20]

Norris fully expected Talbot to be invited to the formal opening, but as it turned out the four translations were scrutinized by a judging committee in a closed session on May 20. Norris secretively and disappointedly reported back to Talbot the following day that "now that it is done, nothing seems likely to come of it." He lamented, "To me, and I may say, to Sir G. Wilkinson, the result was perfectly satisfactory as regarded the real basis of the decipherment; but the others wished to enter into the linguistic proof while admitting they were unacquainted with any Semitic language!"[21] This is hardly surprising, given that apart from the Egyptologist Gardiner Wilkinson, the committee comprised the classical historians George Grote and Henry Hart Milman, and the natural philosopher William Whewell, who "could not stay long."[22] Never down for long, though, by the end of the letter Norris was promising Talbot to show the papers to the society's president, the Sanskritist Horace Hayman Wilson, and adding, "His opinion will have great weight, and I am sure it will be favourable."

The committee did somehow manage to deliver a report to Norris by the end of the month, however, and he presented it to a public meeting of the society in early June. But again there were delays, and the report finally appeared in the *Journal of the Royal Asiatic Society* four years later, perhaps to coincide with the publication of Rawlinson and Norris's cuneiform text that would enable others to verify the accuracy of the translations.[23] Nevertheless, the committee did "agree in bearing testimony to a very remarkable concurrence in the translations compared," despite the fact that "many lines . . . of difficult construction, or uncertain meaning, ha[d] been left untranslated."[24]

And that, it appeared, was that: cuneiform was a real script, writing a real language, Assyrian, that really could be read, even if details were still to be worked out. If histories of script decipherment even mention the 1857 event—and several of them do not[25]—then it is as the final chapter of the story, after which work on cuneiform moved from the challenges of decipherment to the routines of translation.[26] But, as we shall see, it was not nearly as simple as that.

Although the decipherment committee had judged Talbot's work not to be quite of the same quality as Rawlinson's and Hincks's, it is clear that his peers considered him a full-fledged member of the Assyriological community.[27] Even the mean-minded Egyptologist Wallis Budge called him "no mere amateur Orientalist."[28] His notebooks of the post-decipherment decades are thus a unique window into the interests and methods of high Victorian Assyriology.

Throughout the 1850s and 1860s much effort was expended in working through the twenty-five thousand or so tablets and fragments from Nineveh and other Assyrian cities that Layard, Rassam, and others had deposited in the British Museum. With the aim of disseminating as much raw data as soon as possible, it became—and to a large extent remains—standard Assyriological practice to publish volumes of cuneiform texts, whether as hand-drawn copies or typeset facsimiles, without alphabetic transliteration or translation. Other scholars were then free to pick and choose the most amenable or interesting texts for more detailed study and edition, for the most part having to take on trust the reliability of the published representations of the inscriptions. From 1864 Talbot also had a set of large-format photographs of the first few hundred tablets from the British Museum's Kouyunjik Collection. We know he worked from these at least some of the time because their margins and reverses are covered in his penciled notes.[29]

In 1861 Rawlinson and Norris put out their first volume entitled *Cuneiform Inscriptions of Western Asia* (still known to Assyriologists today simply as "Rawlinson"), featuring almost exclusively narrative historical accounts including the Tiglath-pileser inscription that had been used for the decipherment event of 1857 (fig. 76).[30] Many of these inscriptions were in fact composite copies, compiled from several fragmentary manuscripts, whose aim was to re-create as large a portion of the "original" composition as possible. The text of the Tiglath-pileser inscription, for instance, was actually compiled from inscriptions on four different objects (fig. 77). Soon the authors' series moved into the library tablets of ancient Nineveh, with volume 2 (1866) treating ancient word lists and bilingual reference works (of the type that Assyriologists know as lexical texts), which proved to be essential tools for increasing Assyriologists' confidence in their understanding of the script.[31]

The third and fourth volumes appeared in 1870 and 1875, respectively.[32] Largely put together by the young assistant keeper George Smith but still known familiarly as "3 and 4 Rawlinson," they finally gave access to the mythology and scholarship of the Assyrian Empire. Even with the aid of the lexical material published in volume 2, much of this proved frustratingly intractable: full of technical terminology for concepts that were deeply alien to its modern readers, couched in densely

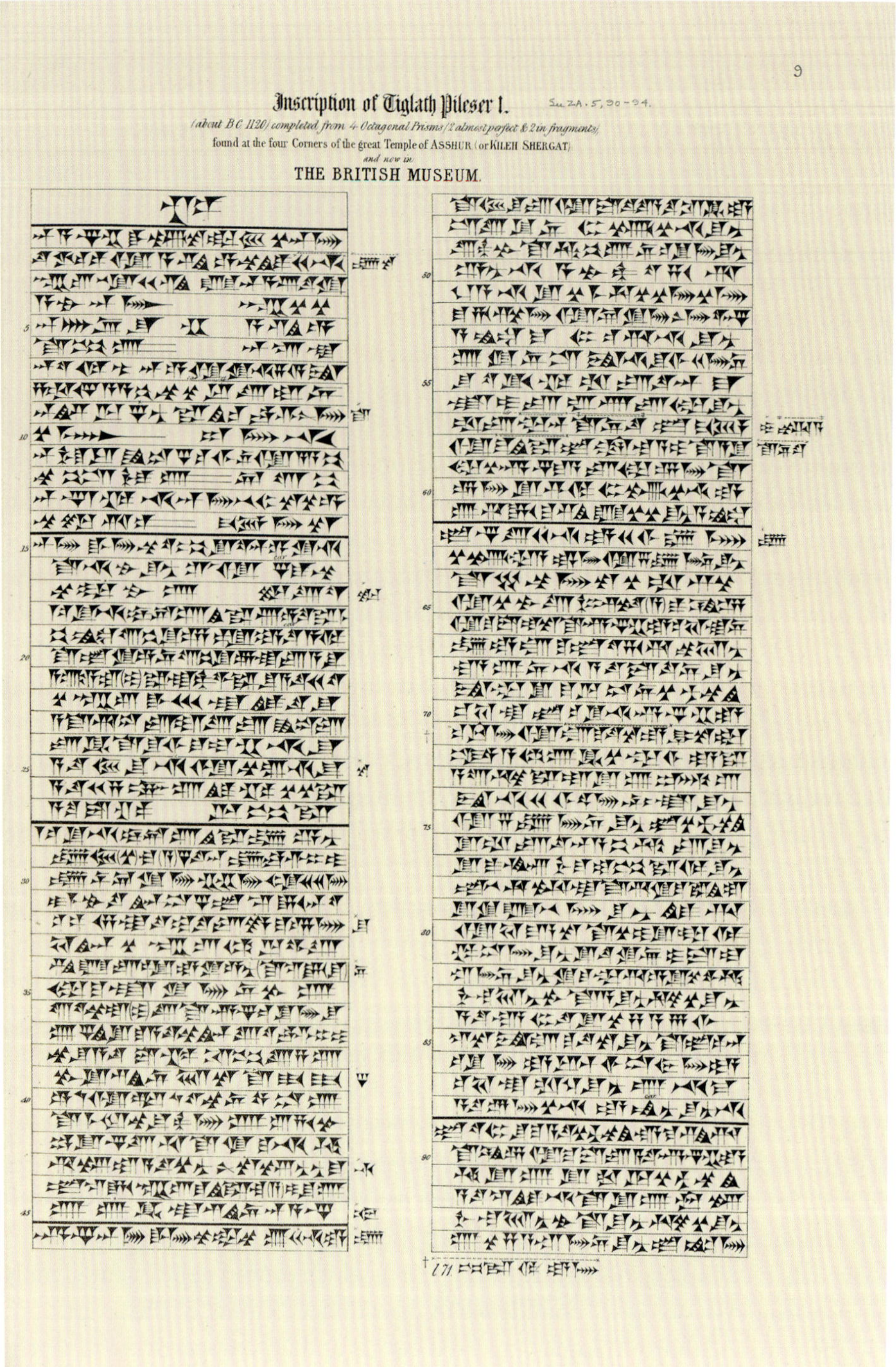

76. Annals of Tiglath-pileser I (Assur, ca. 1050 BCE), lithograph used for decipherment, in Henry Creswicke Rawlinson and Edwin Norris, *Cuneiform Inscriptions of Western Asia*, vol. 1, *A Selection from the Historical Inscriptions of Chaldaea, Assyria, and Babylonia* (London: British Museum, 1861), plate 9. Babylonian Collection, Yale University Library.

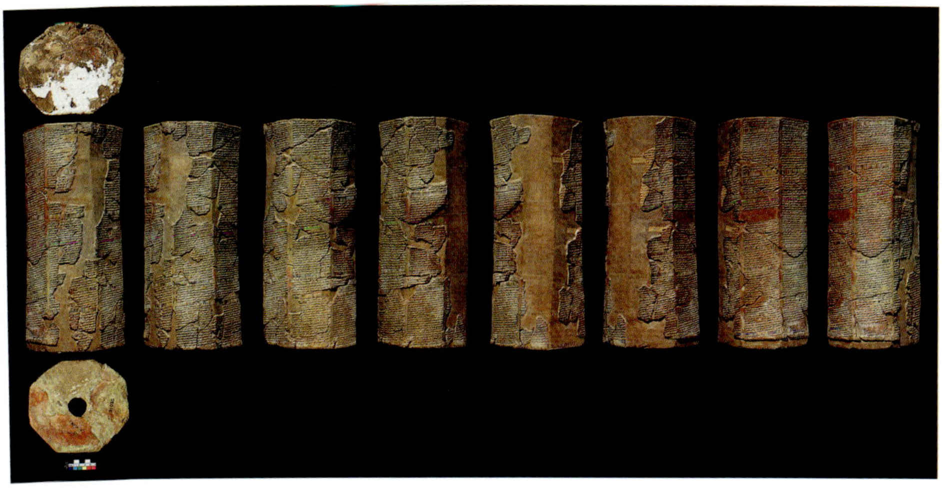

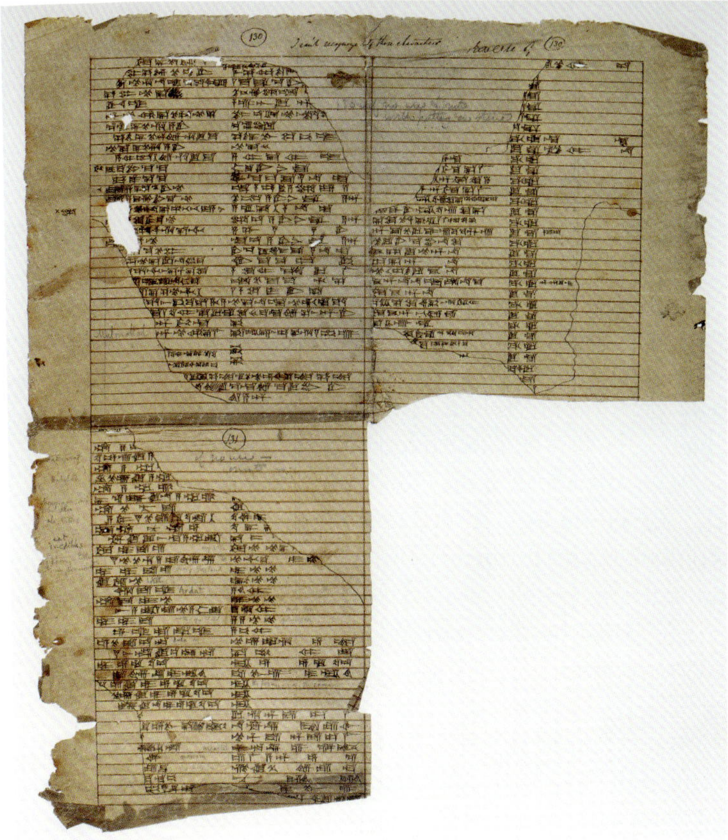

77. Modern photograph of prism K 1621a, one of the four sources used by Rawlinson to compile his lithograph of the Annals of Tiglath-pileser. British Museum.

78. William Henry Fox Talbot, "I can't recognize half of these characters!," drawing in ink on tracing paper of two tablets from the Kouyunjik Collection, date unknown. Add. MS 88942, Talbot Collection, British Library (partially catalogued).

logographic script whose interpretations were highly genre-specific. At the top of a beautifully executed scale drawing of two Sumerian-Akkadian bilingual works, now among Talbot's papers in the British Library, he wrote plaintively, "I can't recognize half of these characters!" (fig. 78).[33] This was not an isolated problem. As we shall see, although the 1857 event had established the principles by which cuneiform script operated, reading it was still a major challenge.

BEL, THE DRAGON, AND BIBLICAL ARCHAEOLOGY

Talbot's article "The Fight Between Bel and the Dragon, and the Flaming Sword Which Turned Every Way" is in many ways typical of his output at the time. It appeared in volume 5 of the *Transactions of the Society of Biblical Archaeology* (June 1876), one of twenty-three papers by him to appear in that journal. As part of a London-based group of Assyriologists and Egyptologists around Keeper Samuel Birch at the British Museum, Talbot had been a founding member of the Society of Biblical Archaeology in 1870 and helped to keep it financially afloat.[34] Its aim was to promote "the investigation of the Archaeology, Chronology, Geography, and History, of Ancient and Modern Assyria, Arabia, Egypt, Palestine, and other Biblical Lands, the promotion of the study of the Antiquities of those countries, and the preservation of a continuous record of discoveries, now or hereafter to be in progress."[35] In fact, there was no "biblical archaeology" to speak of at the time, the Crimean War having put an end to explorations in 1856. In 1872, however, Smith discovered an Assyrian parallel to the Flood story of Genesis 6–7, publicized very effectively as the "Chaldean Deluge" at a meeting of the society that December. This encouraged the hunt for further parallels to the earliest books of the Old Testament and culminated in Smith's collected translations, *The Chaldean Account of Genesis*, in 1876. Sponsored by the *Daily Telegraph* and the British Museum, Smith also resumed Assyrian excavations with foolhardy and fortuitous trips to Nineveh in 1873–75.[36]

"Bel and the Dragon" appeared just before Smith's book came out, having been read to a meeting of the society in March of that year.[37] It was planned as the third of a projected three-part series, beginning with "The Revolt in Heaven"; the never-published "Creation of Man" (which Talbot later renamed "The Fall of Man") was to be the intermediate part, even though Talbot was not entirely sure that all three parts were related.[38] But he thought it "likely that the Dragon was the instigator of the revolt, and likewise the Tempter who seduced mankind, although I cannot find that this is distinctly stated. Therefore Bel in his vengeance destroyed the Dragon."[39]

"Bel and the Dragon" opens with a short introduction to "one of the most striking narratives of the Chaldean mythology," and a translation that begins:

1. [broken]
2. . . . . and with it his right hand he armed.
3. His flaming sword he raised in his hand.
4. He brandished his lightnings before him.
5. A curved scimitar he carried on his body.
6. And he made a sword to destroy the dragon,
7. which turned four ways; so that none could avoid its rapid blows.
8. It turned to the South, to the North, to the East, and to the West.[40]

Here already is Bel—his name appears later—and here already is the dragon, under attack. Bel's sword, Talbot noted, "resembles so strongly the Sword of the Cherubim in Genesis which 'turned every way, to keep the way of the Tree of Life,' that the same celestial weapon must surely be intended"—that is, "the burning sword" by which the Cherubim kept sinful mankind out of the Garden of Eden (Gen. 3:24).[41]

In fact, this passage of *Enūma Eliš* features no dragons but does include a cosmic battle between the heroic god Marduk—here called Bel, meaning "lord" in Babylonian—and Tiamat, the primordial chaotic, female sea. In a recent translation:

1. [He nocked the arrow, he fixed the bowstring for it.]
2. [He raised the *miṭṭu*-weapon], he made his right hand grasp it,
3. He hung the bow and the quiver at his side.
4. He put lightning in front of him,
5. He filled his body with blazing flame.
6. He made a net to encircle the innards of Tiamat,
7. He mustered the four winds so that none of her might escape.
8. South wind, north wind, east wind, west wind.[42]

Bel's weapons include a mace, a bow and arrow, a net, lightning, and four plus seven different winds—but no flaming swords. So where did Talbot conjure the dragon and the sword from, and why? The article continues with a typeset cuneiform text with interlinear alphabetic transliteration and translation, then finally a line-by-line commentary. Three of Talbot's surviving notebooks contain his preparatory work on the paper. Together they enable us to reconstruct, to some extent, the means by which he produced his edition.

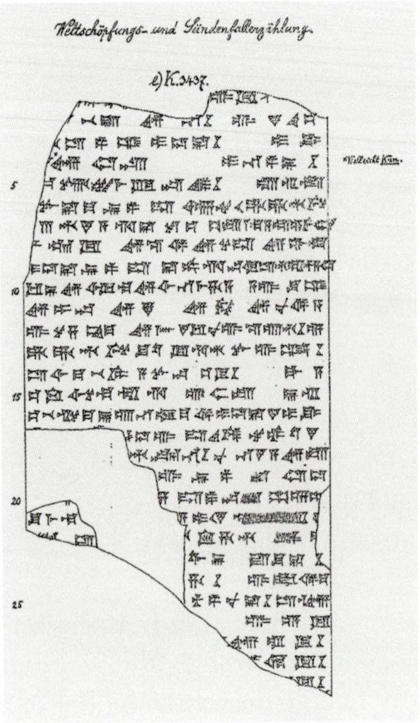

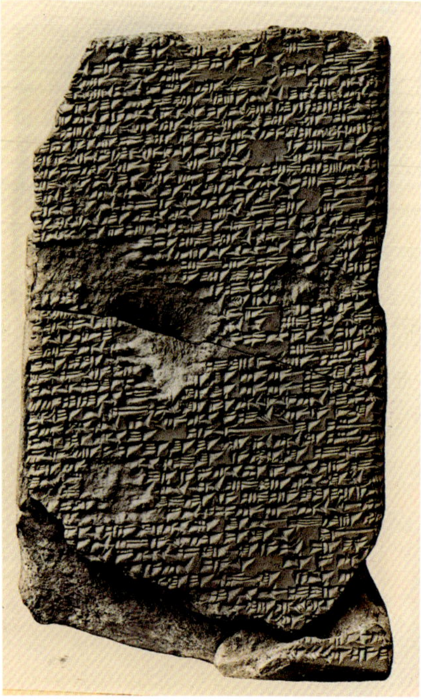

SOURCES AND METHODS

Talbot's primary source was not an actual cuneiform tablet, even though the Kouyunjik Collection tablet concerned, K 3437, was housed in the easily accessible British Museum.[43] Nor was it a photograph, even though by this time he was able to receive photos of tablets directly from the museum's contracted photographer Stephen Thompson and was still pushing for photography in the Assyriological field.[44] Talbot, like many Assyriologists then and now, considered photographs "insufficient to work from."[45] Thus, Talbot worked first from a scale drawing that had just been published by the German Assyriologist Friedrich Delitzsch, as plates 44–45 of his *Assyrische Lesestücke* (1876; fig. 79). This novel and important work was an Assyriological reader, designed to enable students to teach themselves to read Assyrian in cuneiform script.[46] When Smith presented a newly discovered fragment to the Society of Biblical Archaeology in November 1875, which was joined to Delitzsch's piece (fig. 80), Talbot incorporated that too.[47] As was standard British Museum practice at the time, Smith's representation of the tablet was in typeset cuneiform (fig. 81), reconstructed from an alphabetic rendering, rather than an autograph copy such as Delitzsch defended.[48]

The very earliest drafts of Talbot's paper do not survive. One of the three extant notebooks contains a working draft of the introduction, translation, and edition of the obverse of the tablet; the other two contain line-by-line reading notes for

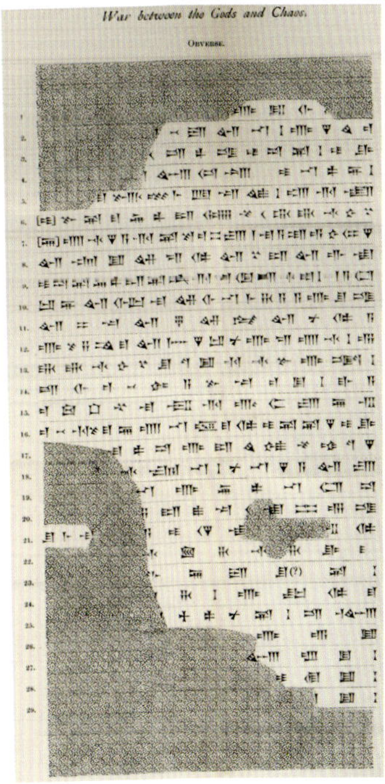

79. Friedrich Delitzsch, autograph copy of *Assyrische Lesestücke* (Leipzig, 1876), plate 44, representing British Museum tablet K 3437 (Nineveh, ca. 650 BCE). Author's copy.

80. Unknown photographer, Tablet K 3437 after fragment was joined to tablet, undated. Room 641, British Museum.

81. George Smith, "On Some Fragments of the Chaldean Account of the Creation," *Transactions of the Society of Biblical Archaeology* 4 (1876): plate 5, lithograph. Yale University Library.

each side of the tablet.[49] The multiple revisions to both translation and notes show how Talbot struggled with almost every word. Here, for instance, is the draft translation on pages 1–2 of the first notebook, which reveals that the "flaming sword" in line 3 was a late arrival (fig. 82):

> Line 1. [broken] . .
> 2. . . . . and with ^ [some weapon] it his right hand he armed
> 3. his [strong] ~~[sharp]~~ ispatu —flaming! sword? ^ he raised in his hand ~~he raised high~~?
> 4. he ~~held his~~ ~~shook waved his~~ brandished his lightnings before him
> 5. ~~His~~ With a ~~crooked~~ curved scymitar he ~~girt around~~ protected his body
> 6. ~~He made~~ (K) ~~his~~ a ~~sabre weapon? ready~~ ~~sulmu~~ crooked? ~~to~~ ^ to destroy & he held a sword ~~attack~~ the dragon
> 7. ~~He took [his position] at the 4 cardinal points that she might not~~ It turned 4 ways so that ~~she might not escape its revolving blades~~ ~~or, He placed~~ dodged ~~himself at the~~ none c[d.] escape elude dodge its ~~turning~~ ragmi rapid blows
> 8. ~~[it turned]~~ to the South, to the North, to the East, ^ & to the West [~~it turned~~] it turned

And on the facing, left-hand page is this revision to line 6:

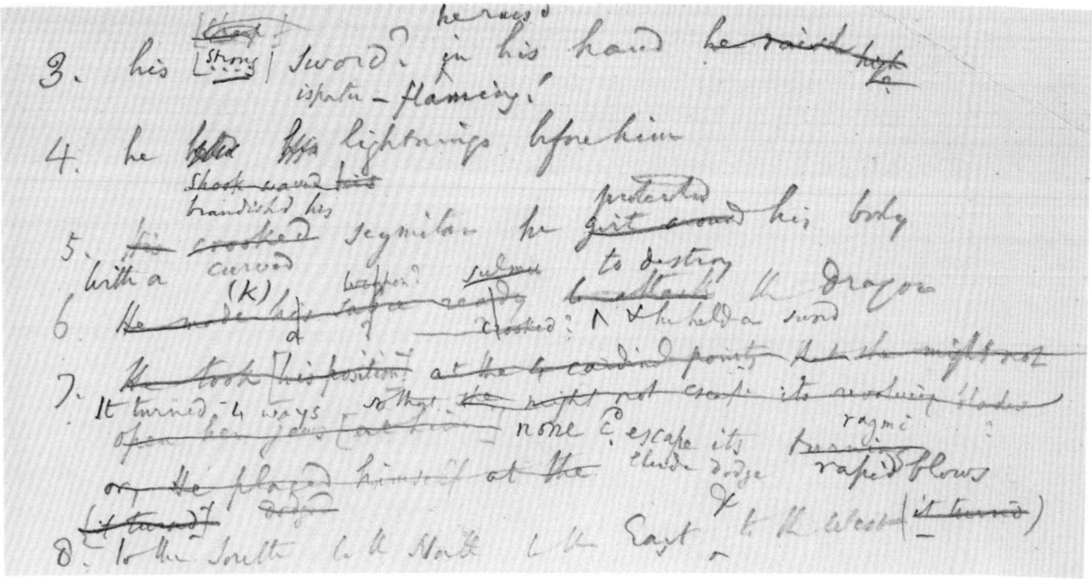

82. William Henry Fox Talbot, draft translation of "Bel and the Dragon," lines 3–8, Notebook "The Fight between Bel and the Dragon," 1876. Add. MS 988942/1/121, Talbot Collection, British Library.

(K) he made ~~that crooked weapon~~ his weapon ~~ready ?~~ perfect to attack the dragon

This was a matter not just of finding the right nuances for the text but of making fundamental decisions about the meanings of almost every word. But how was he to do that? He and Norris had both made substantial efforts toward constructing a glossary of Assyrian a few years earlier: in three long articles for the *Journal of the Royal Asiatic Society*, Talbot had collected notes on key Assyrian nouns and verbs; meanwhile, Norris had reached the letter N in the third volume of his *Assyrian Dictionary* before he died in 1872.[50] But both works necessarily focused on the terminology of the Assyrian kings' annals, full of smiting, pillage, and booty. The newly interpreted mythological works were in a completely different linguistic register, with an entirely new vocabulary.

Talbot wrote barely a word about paleography, orthography, morphology, syntax, or any other linguistic aspects of the ancient text. Perhaps not surprisingly, given his long-standing interests, it was comparative etymology that was his primary focus. The obvious target was Old Testament Hebrew, being roughly contemporary with Assyrian and with a long tradition of translation behind it. In almost every line of his notes, in both draft and publication, he compared newly discovered Assyrian words with similar-sounding Hebrew ones. More rarely he drew on "Chaldean Syriac" (that is, Iraqi Aramaic), Arabic, and/or ancient Greek.

As for reading the script itself, by the mid-1870s it was rare to come across completely new cuneiform signs. Talbot could quite confidently—if often erroneously—

assign a reading to almost every sign in the text. But which reading he chose often depended on whether he could find a Hebrew cognate for the resulting word. So, for example, in line 6 he read the first word (which is now understood as *issi*, "she shouted") as *gissi*, "she sprung," choosing the value *gis* over *is* (and other values he knew, such as *iz*), because it "seems to be from גוז [the Hebrew root gimel vav zayin] transit celeriter (Schindler)."[51]

The only problem was that Talbot was not a confident Semitist. He rarely gave a Hebrew root without a Latin translation from "the Lexicons" (for example, p. 9). He never named his sources explicitly, but occasional references to "Sch" or "Schindler," "Ges" or "Gesenius," enable us to identify his core reference works. The former is Valentin Schindler's *Lexicon Pentaglotton* (1612), published in London in 1634.[52] The latter must be one or more of three very closely related versions of a work by Wilhelm Gesenius: his *Lexicon Manuale Hebraicum et Chaldaicum in Veteris Testamenti Libros* (1834) in the original Latin or, more likely, in the parallel German version (for on p. 19, note to reverse line 7, we read "Gesenius renders it *drehen*").[53] Another possibility is the English translation by Samuel Prideaux Tregelles, which went through several London editions from the 1830s to the 1870s, as sometimes in Talbot's notes the Hebrew is translated directly into English: "Sulmu, to slay (?) Heb. שלם [shin lamed mim] to make an end of: to finish, or consume" (p. 9).[54]

A SWORD AND THREE DRAGONS

Comparative etymology is clearly at work in Talbot's rejection of the "strong" or "sharp" sword in favor of a "flaming" one. The key phrase is in obverse line 3, where the word that he restored without comment as *kakku*, "a sword," is entirely missing (and which we now know means the more generic "weapon"). In fact, parallel manuscripts enable us to restore the signs ^{giš}PAN here: the logogram PAN for *qaštu*, "bow," preceded by the semantic determinative giš for "wooden objects."[55] But if line 3 of the tablet had been complete, Talbot would still have had his "flaming sword," for the same sequence reappears in his line 5. Here he read *gizzir*, "scimitar" (legitimately taking ^{giš} as *giz* and understandably confusing the very similar-looking signs PAN and *zir*). He noted on page 9 that *gizzir*'s literal meaning is "Cutter, from Heb. גזר [gimel zayin resh], to cut asunder" (fig. 83).

The notebooks give us some further insight into how he came to that translation. On page 4 of the first notebook, he wrote:

5. Samsir a scimitar — no! [*added later*]

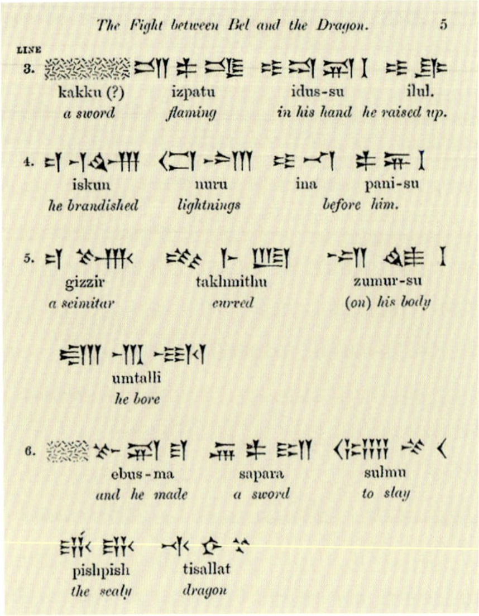
The Fight between Bel and the Dragon. 5

LINE

3. kakku (?) *a sword* izpatu *flaming* idus-su *in his hand* ilul. *he raised up.*

4. iskun *he brandished* nuru *lightnings* ina pani-su *before him.*

5. gizzir *a scimitar* taklimithu *curved* zumur-su *(on) his body* umtalli *he bore*

6. ebus-ma *and he made* sapara *a sword* sulmu *to slay* pishpish *the scaly* tisallat *dragon*

83. William Henry Fox Talbot, "The Fight between Bel and the Dragon, and the Flaming Sword Which Turned Every Way," *Transactions of the Society of Biblical Archaeology* 5 (1876): 5. Yale University Library.

Here he correctly decided against a misreading of the first sign as *sam* (which is too visually complex) but kept *sir* = *zir*. The second notebook contains the note in its almost final form, with a mistaken further biblical reference removed:

> 5. gizzir a sword, lit^{ly} Cutter fr Heb גזב gizir to cut asunder. ~~A gizir of iron is ment in 2 Sam xii~~.

Talbot left his translation of *sapatu* as "sword" in line 6 unannotated; we now know it means "net."[56]

Finally, let us look at *izpatu*, "flaming," which Smith already understood to mean *išpatu*, "quiver."[57] Talbot devoted a very long note to justifying this translation, admitting that it "appeared to me of uncertain meaning at the time when I transmitted this paper to the Society" (p. 8). He gave three inconclusive (and largely erroneous) parallels from other Assyrian texts but concluded, most importantly, that the etymology "may be derived from a Hebrew and Arabic root שב [shin bet] 'to burn' whence the Hebrew שביבא [shin bet yod bet ayin] 'flames,' found in Daniel iii, 22, 'the flame of the furnace slew those men' " (pp. 8–9).

Here then, Talbot privileged syllables over determinatives and logograms, and he entangled etymological and cultural similarity to produce a text that fulfilled readers' thirst for biblical parallels: parallels for the very moment, indeed, in which Adam and Eve were expelled from Paradise.

Talbot's dragon proves to be rather more complicated, not least because it turns out that no fewer than three dragons are involved in his interpretation. The first is from Revelation 12:3, in which "there appeared a dragon in heaven, having seven heads," which was defeated and expelled by Saint Michael. In his introduction to "The Revolt in Heaven," Talbot already noted that "it is not unworthy of remark that the Chaldean dragon had seven heads, like that spoken of in Revelations," referring to the text "2 R[awlinson] 19" (p. 350 with note 2).[58] There is no mention of a seven-headed monster, however, either in Talbot's rendering of "Bel and the Dragon" or in *Enūma Eliš*'s depictions of Tiamat as rendered by modern translations. And "2 R 19" is a manuscript of a myth about the warrior god Ninurta, who carries a mace "which slays like a seven-headed serpent."[59]

The second dragon is from Daniel 14, also known in English as "Bel and the Dragon," which Protestants now consider part of the Apocrypha but was originally included in the King James Version. Talbot did not mention this dragon directly, but as it is so transparently the source of his article's title, we must also address it here. In "Bel and the Dragon," written in Aramaic in the second century BCE, Daniel is a Jewish captive in the court of King Astyages in Babylon (thus setting the narrative in the sixth century BCE).[60] Here, however, the dragon is not Bel's foe but his companion, which Daniel (not Bel) slays by feeding it until it bursts (Dan. 14:23–27):

> And in that same place there was a great dragon, which they of Babylon worshipped. And the king said unto Daniel, Wilt thou also say that this is of brass? Lo, he liveth, he eateth and drinketh; thou canst not say that he is no living god: therefore worship him. Then said Daniel unto the king, I will worship the Lord my God: for he is the living God. But give me leave, O king, and I shall slay this dragon without sword or staff. The king said, I give thee leave. Then Daniel took pitch, and fat, and hair, and did seethe them together, and made lumps thereof: this he put in the dragon's mouth, and so the dragon burst in sunder: and Daniel said, Lo, these are the gods ye worship.

This dragon clearly is not Tiamat, but neither is it entirely the invention of the Aramaic author. Marduk-Bel's elegant *mušḫuššu*-dragon is widely attested in textual and material remains from Babylon of the sixth century BCE, from lapis lazuli cylinder seals to the famous glazed bricks of the Processional Way (fig. 84).[61]

The third and final dragon appears in what Talbot called "the representations of the combat on Babylonian cylinders in Mr Smith's Chaldean Genesis, p. 62, &c." (p. 2; fig. 85). This image, also featured on Smith's front cover (fig. 86), shows a hybrid lion-eagle doing battle with a god clutching lightning bolts. It is a drawing of a stone relief taken by Layard in 1850 from the warrior god Ninurta's temple in

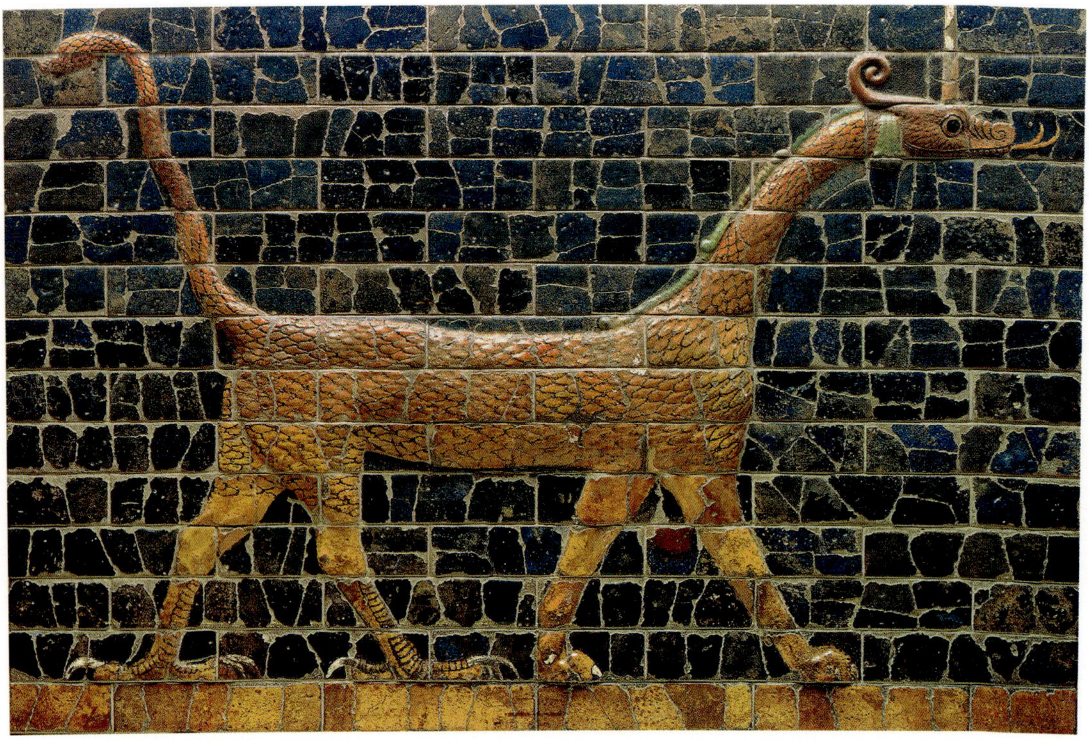

84. Artist unknown, Marduk's *mušhuššu*-dragon from the Processional Way in Babylon, 652–604 BCE, terra-cotta glazed and molded bricks, 45½ x 65¾ in. (115.6 x 167 cm). Detroit Institute of Arts.

the Assyrian city of Nimrud for display in the British Museum (fig. 87).[62] Although the front-cover version is emasculated, the version inside the book clearly shows the monster's snake-headed penis. It cannot therefore be a representation of the female "dragon" in Talbot's text. In fact, this is the evil Anzu bird, who stole the gods' Tablet of Destinies and was eventually defeated by the warrior god Ninurta.[63] Anzu is now known in many incarnations, from the third millennium BCE on, but the pertinent version of his story is told in the *Epic of Anzu*. This work was known to Talbot and his contemporaries as fragments of *The Sin of the God Zu*. Smith's very preliminary translation in *The Chaldean Account of Genesis* already draws attention to Anzu's lion-eagle hybridity.[64]

In short, Talbot's dragon was not created by etymological means; Talbot had nothing to say about his reading of the word "tisallat" (which, since Smith, we read as the name Tiamat).[65] Rather, a mixture of biblical, Assyrian, and other ancient Near Eastern imagery, both visual and textual, was synthesized into a single super-monster.

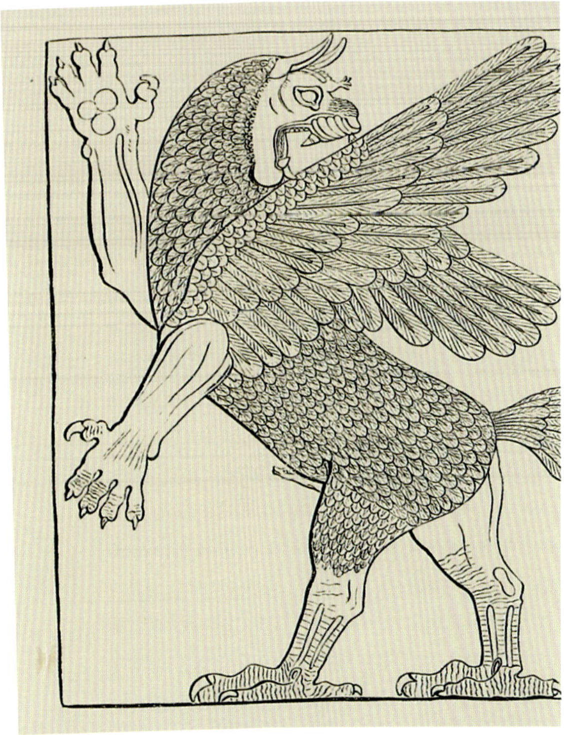

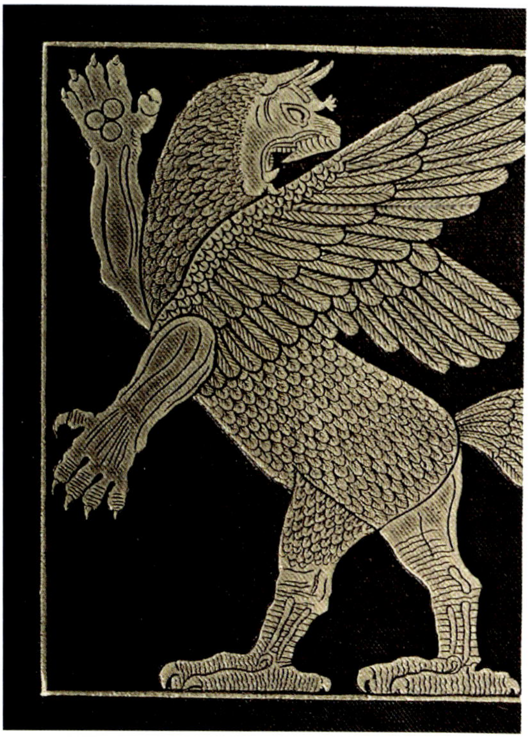

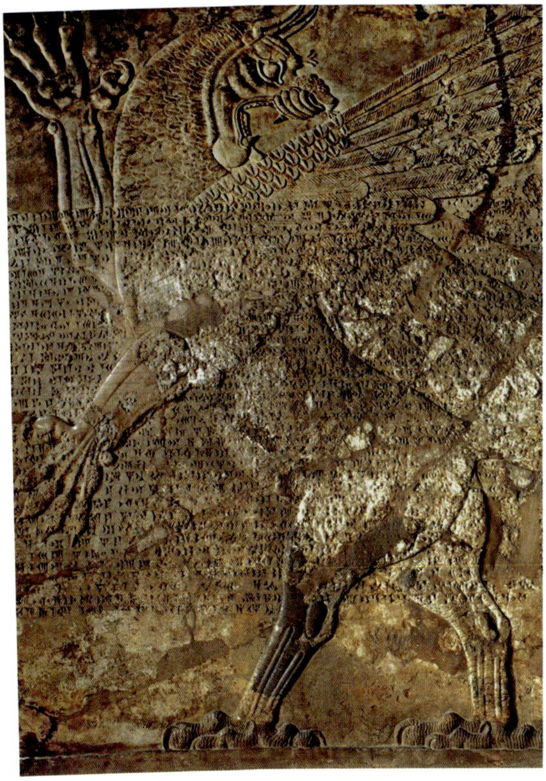

85. Artist unknown, "Fight between Merodach (Bel) and the Dragon," in George Smith, *The Chaldean Account of Genesis* (London: Low, Marston, Searle, and Rivington, 1876), facing p. 62.

86. George Smith, *The Chaldean Account of Genesis,* 6th ed. (London: Low, Marston, Searle, and Rivington, 1879), cover illustration.

87. Unknown artist, *Anzu Bird*, bas-relief at the entrance of the god Ninurta's temple in Nimrud (detail), 865–860 BCE, gypsum. British Museum.

CONCLUSIONS

The great decipherment event of 1857 was not the heroic revelation of the truth of cuneiform that it is often portrayed as being. Despite being highly managed by Norris and Talbot, it was in fact a near failure whose outcome took years to publish. And while it eventually convinced most of the intellectual community that the decipherers had not been deluding themselves, in that the workings of cuneiform had been established in principle, it did not make the work of Assyriologists much easier in practice.

As we have seen, Talbot had to negotiate with many intermediaries in order to access ancient Assyrian texts, even twenty years after decipherment: not the tablet itself or even a photograph but an interpretative drawing, or yet a typeset lithograph reverse-engineered from someone else's transliteration. And even supposing that that representation was sufficiently reliable, it was still not possible to make translations directly from Assyrian cuneiform to alphabetic English: they had to be routed through at least two other languages. Further, although the existence and role of semantic determinatives and logograms in cuneiform were understood in theory, it was rare for them to be taken into account in actuality; syllabic readings were almost always preferred.

It may seem surprising, even shocking, that Talbot spurned the photograph in his Assyriological work. But cuneiform tablets are notoriously resistant to photography, if the aim of the photograph is completely legible representation of the inscription.[66] Unlike most writing technologies, which produce legibility through color contrast on a two-dimensional surface, cuneiform is essentially a series of shadows formed by depressions into a cambered, often very convex, surface. The act of reading entails holding it in one's hand and constantly turning it against a low-raking light source to catch the full range of shadows. A static tablet under a static light is rarely fully legible, even when it is in perfect condition with no surface flaking, cracks, mottling, or other visual "noise." A few expert specialist photographers—such as Roger Fenton in Talbot's time, and some museum employees today—have had the necessary skill, but until the advent of new techniques in digital photography in the early twenty-first century, black-and-white scale drawings remained the preferred means of Assyriological communication of the textual content of tablets.

By situating their endeavors within the frame of biblical archaeology—indeed, by creating that frame—British Assyriologists hindered as much as helped themselves. On the one hand they attracted great attention, praise, and funds for their work from an establishment and public still eager for external corroboration of the Old Testament's veracity. On the other, they were so intent on finding parallels,

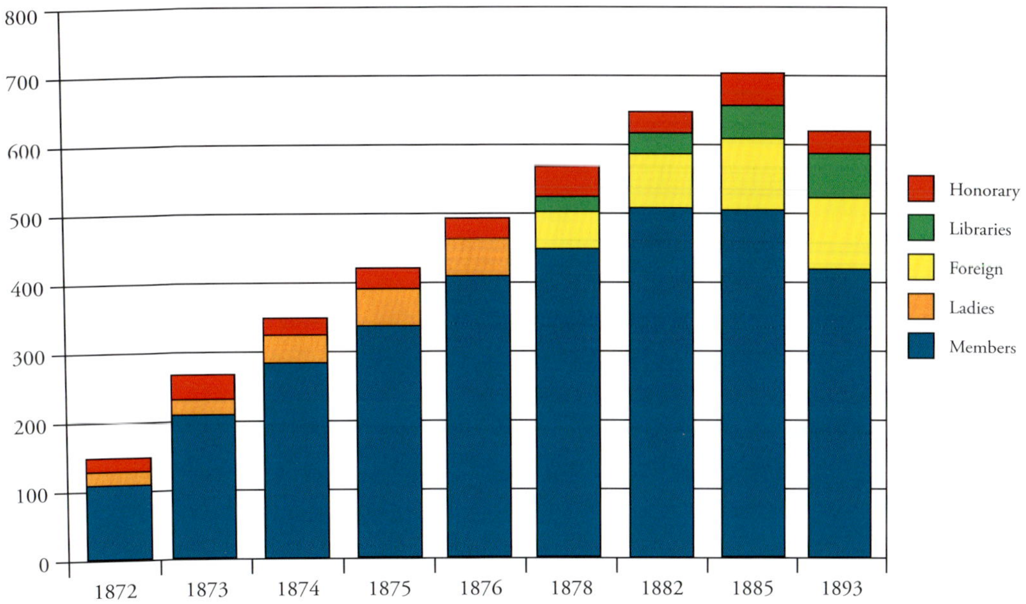

88. Membership of the Society of Biblical Archaeology, ca. 1872–95, from membership lists printed in *Transactions of the Society of Biblical Archaeology* 1–9 (1872–93). Compiled by the author. Note that membership categories changed in 1878.

both linguistic and conceptual, with biblical scripture that dragons and flaming swords were created that would take many decades to correct.

In fact, the Society of Biblical Archaeology and its journal were primarily vehicles for promoting the philological study of Assyriology and Egyptology; even in the first volume only about a fifth of the articles' titles allude to the Bible. Nevertheless, the society's membership far outstripped, in both quantity and quality, what could have been expected of a more narrowly defined organization (fig. 88) and counted Prime Minister William Gladstone among its (honorary) vice presidents. But by the time the society petered out in the late 1880s, it had served its purpose: Assyriology was flourishing as an academic discipline across Europe and the United States, and, in Britain at least, it was beginning to establish an independent identity that had little to do with the Bible.

Talbot has largely been written out of the history of Assyriology, except as the instigator of—yet not always a participant in—the decipherment event of 1857.[67] But for twenty years thereafter, and especially in the 1870s, he was a key member of the British Assyriological community, admitted as much for his demonstrated knowledge as for his social standing. While his translations were always superseded, often very quickly, by others in the field, his prompt and prolific publication habits at least gave his peers a starting point from which to work. Perhaps most importantly, his involvement in the fledgling Society of Biblical Archaeology, both intellectual and financial, helped to secure an audience and status for British Assyriology that began to set it on a professional footing.

Finally, we can now see Talbot's quarter-century love affair with Assyriology as the ultimate intellectual challenge that he had been seeking all his life. Like the study of mosses, mastery of cuneiform required an all-absorbing attention to minute differences of visual detail. The endless series of underdetermined linguistic problems to be solved, within the elegant constraints of Semitic linguistics, surely spoke to Talbot's mathematical training. In this he also drew more obviously on his long-standing fascinations with comparative etymology, ancient history, and the origins of Judeo-Christian civilization. Cuneiform gave him an endless supply of raw material to work on, along with the gratifying cycle of public delivery and speedy publication within the camaraderie and (mostly) friendly rivalry of the tiny community of experts. Assyriology provided, in short, a limitless source of apparently unattractive and intractable problems of the sort that Talbot had relished since he was a child. From this perspective it is clear that Talbot's Assyriology was far from an irrelevance, an old man's hobby with which he idled away the quiet decades between his great invention and his death. On the contrary—indeed, to be deliberately contrarian—one could even argue that photography was just a phase he went through on the way to finding his true vocation.

I am very grateful to Katrina Dean, John Falconer, and their colleagues for their generous assistance at the British Library, and especially to Mirjam Brusius: for the pleasure of working with her on cataloguing Talbot's Assyriological notebooks and on her PhD, and for the many ways in which she has enriched my understanding of early Assyriology.

1. For example, "Talbot spent the last twenty-five years of his life developing and perfecting an effective photogravure process." Malcolm Daniel, "William Henry Fox Talbot (1800–1877) and the Invention of Photography," in *Heilbrunn Timeline of Art History* (New York: Metropolitan Museum of Art, 2004), http://www.metmuseum.org/toah/hd/tlbt/hd_tlbt.htm.
2. William Henry Fox Talbot, "The Fight between Bel and the Dragon, and the Flaming Sword Which Turned Every Way," *Transactions of the Society of Biblical Archaeology* 5 (1876): 1–21. The standard modern edition is Philippe Talon, *The Standard Babylonian Creation Myth: Enūma Eliš* (Helsinki: Neo-Assyrian Text Corpus Project, 2005). The best complete modern English translation is by Benjamin Foster in *Before the Muses: An Anthology of Akkadian Literature* (Bethesda, MD: CDL Press, 2005), 436–86.
3. William Henry Fox Talbot, *The Antiquity of the Book of Genesis: Illustrated by Some New Arguments* (London: Longman, Orme, Brown, Green and Longmans, 1839). David Gange presented a paper on this topic, "The Antiquity of Genesis and the Origins of Civilisation," at the interdisciplinary conference "William Henry Fox Talbot: Beyond Photography," held at the Centre for Research in the Arts, Social Sciences, and Humanities, University of Cambridge, June 24–26, 2010, but he was unable to contribute to this volume.

4. This is evident, in part, from correspondence addressed to him. For example: "If you had not chanced to come that day I might have been ages without knowing of that curious transaction & translation of the Assyrian Inscriptions!" Harriet Georgiana Mundy to William Henry Fox Talbot, Aug. 17, 1844, Doc. No. 2443, The Correspondence of William Henry Fox Talbot, http://foxtalbot.dmu.ac.uk (accessed Jan. 31, 2012), hereafter "Correspondence" (original: Talbot Collection, hereafter "TC," British Library, hereafter "BL"). For another: "What are you occupied with at present—are the Assyrian antiquities from Niniveh very curious." William Strangways to Talbot, April 24, 1848, Doc. No. 6137, Correspondence (original: TC, BL).
5. See, for example, Cyrus Herzl Gordon, *Forgotten Scripts: Their Ongoing Discovery and Decipherment*, rev. ed. (New York: Basic Books, 1982); and Maurice Pope, *The Story of Decipherment: From Egyptian Hieroglyphics to Maya Script*, 2nd ed. (London: Thames & Hudson, 1999).
6. For a convenient introduction to cuneiform, see Eleanor Robson and Karen Radner, *Cuneiform Revealed* (2009), http://oracc.org/saao/knpp/cuneiformrevealed/.
7. On Layard's discovery of 1850, see Mogens Larsen, *The Conquest of Assyria: Excavations in an Antique Land, 1840–1860* (London: Routledge, 1996), 262–64; on Rassam's of 1854, see David Damrosch, *The Buried Book: The Loss and Rediscovery of the Great Epic of Gilgamesh* (New York: Henry Holt, 2007), 110–13.
8. Hatchard & Co. (publisher) to Talbot, March 19, 1853, Doc. No. 6317, Correspondence (original: TC, BL); and Austen Henry Layard, *Inscriptions in the Cuneiform Character from Assyrian Monuments*, ed. Edward Hawkins (London: Harrison and Son, 1851).
9. The three were rulers of Assyria in the seventh century BCE. For a convenient introduction to this period, see Eleanor Robson and Karen Radner, *Knowledge and Power in the Neo-Assyrian Empire* (2007), http://oracc.org/saao/knpp/. There is a bibliography of Talbot's Assyriological works in *Transactions of the Society of Biblical Archaeology* 6 (1878): 556–59.
10. Talbot to Samuel Birch, April 25, 1854, Doc. No. 7140, Correspondence (original: Department of the Middle East, British Museum, hereafter "BM").
11. Talbot to Edward Hawkins, May 19, 1854, Doc. No. 6965, Correspondence (original: Department of the Middle East, BM). Talbot eventually received a set of photos of about 250 tablets in 1864, once he was well established as a "professional" Assyriologist. His long battles with the British Museum over photography of tablets and other inscribed artifacts is documented in detail in Mirjam Brusius, "Preserving the Forgotten: William Henry Fox Talbot, Photography and the Antique" (PhD diss., Univ. of Cambridge, 2011; Akademie Verlag, forthcoming, and English edition in preparation); see also the essay by her in this volume.
12. Talbot to Edwin Norris, May 15, 1855, Doc. No. 6055, Correspondence (original: TC, BL).
13. Talbot to Henry Burgess, Oct. 1855, Doc. No. 786, Correspondence (original: published); and William Henry Fox Talbot, "On the Assyrian Inscriptions," *Journal of Sacred Literature and Biblical Record* 2 (1856): 414–25.
14. Edwin Norris to Talbot, Feb. 15, 1856, Doc. No. 7226, Correspondence (original: TC, BL).
15. Edwin Norris to Talbot, Oct. 3, 1856, Doc. No. 1186, Correspondence (original: TC, BL).
16. Tiglath-pileser I reigned 1114–1076 BCE. Rawlinson's suggestion was made ostensibly in order to avoid accusations of collusion by "determined objectors" but also, it seems, because he was fearful of Talbot's preempting his volume, expected in July 1857; see Antonio Panizzi to Talbot, Jan. 12, 1857, Doc. No. 92, Correspondence (original: TC, BL); and Brusius, "Preserving the Forgotten," 157. It finally appeared as Henry Creswicke Rawlinson and Edwin Norris, *Cuneiform Inscriptions of Western Asia*, vol. 1, *A Selection from the Historical Inscriptions of Chaldaea, Assyria, and Babylonia* (London: British Museum, 1861). For a modern edition of the Assyrian text, see A. Kirk Grayson, *Royal Inscriptions of Mesopotamia, Assyrian Periods*, vol. 2 (Toronto: Univ. of Toronto Press, 1991).
17. Edwin Norris to Talbot, Jan. 15, 1857, Doc. No. 7353, Correspondence (original: TC, BL). See also Norris to Talbot, Dec. 18, 1856, Doc. No. 7333, Correspondence (original: TC, BL); and Norris to Talbot, Dec. 29, 1856, Doc. No. 7341, Correspondence (original: TC, BL).
18. Notebook "Egyptian Archaeology / Inscription of Tiglath-Pileser I (back to front)," 1856–1857, Add. MS 88942/1/146, TC, BL; Notebook "The Octagon of the Tiglath Pileser I," 1857, Add. MS 88942/1/147, TC, BL; and Notebook "Tiglath," 1870s, Add. MS 88942/1/148, TC, BL.

19. Edwin Norris to Talbot, March 13, 1857, Doc. No. 7374, Correspondence (original: TC, BL). The letter is reprinted on the first page of William Henry Fox Talbot et al., "Comparative Translations," *Journal of the Royal Asiatic Society of Great Britain and Ireland* 18 (1861): 150–219.

20. Edwin Norris to Talbot, April 20, 1857, Doc. No. 7391, Correspondence (original: TC, BL); and Norris to Talbot, April 24, 1857, Doc. No. 7395, Correspondence (original: TC, BL).

21. Edwin Norris to Talbot, May 21, 1857, Doc. No. 7408, Correspondence (original: TC, BL).

22. Whewell had been Talbot's mathematics tutor at Trinity College, Cambridge, some four decades earlier. Brusius, "Preserving the Forgotten," 36.

23. Talbot et al., "Comparative Translations," 150–219; and Rawlinson and Norris, *Cuneiform Inscriptions*, vol. 1, pls. 9–16.

24. Horace Hayman Wilson, "Introductory Note," in Talbot et al., "Comparative Translations," 151–52.

25. See, for example, David Diringer, *The Alphabet: A Key to the History of Mankind* (London: Hutchinson, 1968), 26; C. B. F. Walker, *Cuneiform* (London: British Museum Press, 1987), 62; and Peter T. Daniels, "The Decipherment of Ancient Near Eastern Scripts," in *Civilizations of the Ancient Near East*, ed. Jack M. Sasson (New York: Scribner's, 1995), 85–87.

26. "The gates of cuneiform were thus, as it were, officially declared open." Pope, *Story of Decipherment*, 117. "From then on, the decipherment of Babylonian cuneiform was a matter of refining a generally accepted system." Andrew Robinson, *The Story of Writing*, 2nd ed. (London: Thames & Hudson, 2007), 79.

27. Milman and Grote wrote, "Mr Fox Talbot, who was later in the field, though, on the whole, mostly arriving at the same conclusions [as Rawlinson and Hincks], was less positive and precise" (Talbot et al., "Comparative Translations," 153).

28. E. A. Wallis Budge, *The Rise and Progress of Assyriology* (London: Hopkinson, 1925), 92.

29. See the essay in this volume by Mirjam Brusius; and Brusius, "Preserving the Forgotten," 153, pl. 5.2.

30. Rawlinson and Norris, *Cuneiform Inscriptions*, vol. 1.

31. Henry Creswicke Rawlinson and Edwin Norris, *Cuneiform Inscriptions of Western Asia*, vol. 2, *A Selection from the Miscellaneous Inscriptions of Assyria* (London: British Museum, 1866).

32. Henry Creswicke Rawlinson and George Smith, *Cuneiform Inscriptions of Western Asia*, vols. 3–4, *A Selection from the Miscellaneous Inscriptions of Assyria* (London: British Museum, 1870, 1875).

33. Brusius, "Preserving the Forgotten," pl. 2.7. I have not yet been able to identify the tablets concerned.

34. "By his liberality the Council were enabled to print valuable papers, to publish which, at that time, but for his generous assistance, the income of the Society would have been inadequate." Richard Cull, "A Biographical Notice of the Late William Henry Fox Talbot, FRS," *Transactions of the Society of Biblical Archaeology* 6 (1878): 555.

35. "Introduction," *Transactions of the Society of Biblical Archaeology* 1 (1872): ii.

36. Elegantly recounted in Damrosch, *Buried Book*, 9–80. George Smith, *The Chaldean Account of Genesis: Containing the Description of the Creation, the Fall of Man, the Deluge, the Tower of Babel, the Times of the Patriarchs, and Nimrod; Babylonian Fables, and Legends of the Gods; from the Cuneiform Inscriptions* (London: Low, Marston, Searle, and Rivington, 1876).

37. William Henry Fox Talbot, "The Revolt in Heaven, from a Chaldean Tablet," *Transactions of the Society of Biblical Archaeology* 4 (1876): 349–62.

38. The expected second and third parts are mentioned in Talbot, "Revolt in Heaven," 349. "The Fall of Man" was never published, but drafts survive in Talbot's notebooks. Talbot read the tablet from which these passages are taken from Friedrich Delitzsch, *Assyrische Lesestücke* (Leipzig, 1876), pls. 42–43, which he entitled *Weltschöpfungs und Sündenfallenzählung* (Creation of the World and Account of the Fall), and Smith interpreted them in much the same way. Smith, *Chaldean Account*, 91. But they are now known to be extracts from *Enūma Eliš*, Tablet VII, the concluding chapter of the composition. There is no Fall; rather, the fifty names of the god Marduk are enumerated and elaborated on, to celebrate his destruction of Tiamat, his creation of the world, and his consequent elevation to supreme deity. A short epilogue exhorts readers to ensure that future generations continue to tell the tale.

39. Talbot, "Revolt in Heaven," 349. Similarly Smith wrote, "It is quite clear that the dragon . . . is connected with the Fall like the serpent in the book of Genesis, and in fact is the equivalent of the serpent." Smith, *Chaldean Account*, 90.

40. Talbot, "Bel and the Dragon," 2. By comparison Smith, in *Chaldean Account* (95–96), has:

> 1. . . . . . he fixed . . . . .
> 2. . . . . . to his right hand he distributed
> 3. . . . . . and quiver his hand hurled,
> 4. the lightning he sent before him,
> 5. fierceness filled his body.
> 6. He made the sword to silence the dragon of the sea,
> 7. the seven winds he fixed not to come out of her wound.
> 8. On the South, the North, the East and the West,

41. Talbot, "Bel and the Dragon," 1.

42. Translated by Frances Reynolds in 2008 for the Geography of Knowledge in Assyria and Babylonia (GKAB) Project, funded by the United Kingdom Arts and Humanities Research Council (AHRC), 2007–12, http://oracc.org/cams/gkab/P338319/html, released under Creative Commons Attribution-ShareAlike license 3.0.

43. K 3437 has since been joined to another fragment, Rm 641 (found by Rassam). It is one of sixteen known sources now for Tablet IV of *Enūma Eliš*, and doubtless more will be discovered in years to come. See Talon, *Standard Babylonian Creation Myth*, xvi.

44. On Thompson's sending Talbot photos of the "Deluge" tablet in December 1872, see Brusius, "Preserving the Forgotten," 166. On Talbot's continued fraught correspondence with Birch over field photography, see Brusius, "Preserving the Forgotten," 169–70.

45. Talbot to Samuel Birch, Feb. 27, 1873, Doc. No. 9956, Correspondence (original: Department of the Middle East, BM). See also Brusius, "Preserving the Forgotten," 168.

46. It went through five editions over the subsequent decades, each a substantial improvement and enlargement of its predecessors. This first edition lacked a grammar or glossary—though there were several useful grammars available by now—but it did provide a convenient sign list, which gave syllabic values of more than 220 signs and nearly 180 logograms.

47. George Smith, "On Some Fragments of the Chaldean Account of the Creation," *Transactions of the Society of Biblical Archaeology* 4 (1876): 363–64, pl. 6; Talbot, "Bel and the Dragon," 4.

48. Delitzsch, *Assyrische Lesestücke*, [vi].

49. The first is Notebook "The Fight between Bel and the Dragon," 1876, Add. MS 88942/1/121, TC, BL, which also contains Talbot's preliminary editorial work on "The Fall of Man." Text references hereafter are to this notebook. The second is Notebook "Bel & Dragon," 1876, Add. MS 88942/1/122, TC, BL, which is otherwise blank; the third is Notebook "Notes and Observations," Add. MS 88942/1/123, TC, BL, which also contains Talbot's preliminary "Notes and Observations" to his edition of "Ishtar and Izdubar" (now known as the *Epic of Gilgamesh*, Tablet VI), published as William Henry Fox Talbot, "Ishtar and Izdubar, being the Sixth Tablet of the Izdubar Series," *Transactions of the Society of Biblical Archaeology* 5 (1877): 97–121.

50. William Henry Fox Talbot, "Contributions towards a Glossary of the Assyrian Language," *Journal of the Royal Asiatic Society of Great Britain and Ireland*, n.s., 3 (1867): 1–64; 4 (1869): 1–80; 6 (1873): 1–80; and Edwin Norris, *Assyrian Dictionary: Intended to Further the Study of the Inscriptions of Assyria and Babylonia*, 3 vols. (London: Williams and Norgate, 1868–72).

51. Talbot, "Bel and the Dragon," 19.

52. Valentin Schindler, *Schindleri Lexicon Pentaglotton, Hebraicum, Chaldaicum, Syriacum, Talmudico-Rabbinicum, et Arabicum* (London: William Jones, 1634).

53. Wilhelm Gesenius, *Hebräisches und Chaldäisches Handwörterbuch über das Alte Testament* (1812).

54. Samuel Prideaux Tregelles, trans., *Gesenius's Hebrew and Chaldee Lexicon to the Old Testament Scriptures* (London: Bagster, 1853).

55. In modern alphabetic transliteration, logograms (whole-word signs) are shown in capitals, determinatives (semantic markers) in superscript lowercase, so that ^{giš}PAN is a determinative-logographic writing for the Assyrian word *qaštu*, a wooden bow.
56. Similarly Smith wrote, "The *saparu*, or sickle-shaped sword, is always represented both in the sculptures and inscriptions as a weapon of Bel in this war." Smith, *Chaldean Account*, 95.
57. Smith, *Chaldean Account*, 96.
58. Likewise Smith wrote, "the dragon Tiamat . . . closely corresponds in all essential points to the dragon conquered by Michael." Smith, *Chaldean Account*, 100.
59. 2 R 19 = K 38 in Jerrold S. Cooper, *The Return of Ninurta to Nippur* (Rome: Pontifical Biblical Institute, 1978), 80, referring to l. 138 of source e.
60. See J. G. Brooke, "Additions to Daniel," in *The Oxford Bible Commentary: Oxford Biblical Studies Online*, http://www.oxfordbiblicalstudies.com/article/book/obso-9780198755005/obso-9780198755005-div1-295. In reality Astyages (r. ca. 585–550 BCE) was king of Media in present-day northwestern Iran and an ally of Nebuchadnezzar II of Babylon; he did not rule Babylonia himself; see http://www.livius.org/as-at/astyages/astyages.htm.
61. Jeremy Black and Anthony Green, *Gods, Demons, and Symbols of Ancient Mesopotamia* (London: British Museum Press, 1992), 166.
62. Larsen, *Conquest of Assyria*, 258–60.
63. As first argued by William Hayes Ward, "Bel and the Dragon," *American Journal of Semitic Languages and Literatures* 14 (1898): 94–105. For a more recent overview, see Black and Green, *Gods, Demons, and Symbols*, 107–8.
64. Smith, *Chaldean Account*, 122. The standard modern edition is Amar Annus, *The Standard Babylonian Epic of Anzu* (Helsinki: Neo-Assyrian Text Corpus Project, 2001); the best modern English translation is Foster, *Before the Muses*, 555–78.
65. Smith, *Chaldean Account*, 90; note the erroneous appeal to Hebrew etymology here too.
66. See the essay by Anne Secord in this volume.
67. In particular, Talbot has been written out of the editorial history of *Enūma Eliš*: "The Babylonian myth of Creation has been known to Assyriologists for more than a century through the works of G. Smith, L. W. King, and St. Langdon." Talon, *Standard Babylonian Creation Myth*, ix. On his role in decipherment, see Austen Henry Layard, *Nineveh and Babylon* (London: John Murray, 1867), xlvi–xlvii; and a few secondary sources such as Gordon, *Forgotten Scripts*; and Christopher Walker, "The Kouyunjik Collection of Cuneiform Texts: Formation, Problems, and Prospects," in *Austen Henry Layard Tra l'Oriente e Venezia: International Symposium; Selected Papers*, ed. Frederick M. Fales and Bernard. J. Hickey, vol. 2 (Rome: "L'Erma" di Bretschneider, 1987), 183–93, esp. 184. Walker acknowledged Talbot's part in these early attempts at decipherment and mentioned that Talbot had offered Layard the chance to take a camera to Nineveh. See also Peter T. Daniels, "Edward Hincks's Decipherment of Mesopotamian Cuneiform," in *The Edward Hincks Bicentenary Lectures*, ed. Kevin J. Cathcart (Dublin: University College Dublin, 1994), 30–57, esp. 49; Larsen, *Conquest of Assyria*, 115–24, 177–88, 215–27, 293–305, 336–37, 358; and Frederick Bohrer, *Orientalism and Visual Culture: Imagining Mesopotamia in Nineteenth-Century Europe* (Cambridge and New York: Cambridge Univ. Press, 2003), 127. In addition, Nele Diekmann is currently preparing a PhD thesis at Freie Universität Berlin entitled "Talbot's Tools: Scientific Notebooks as a Laboratory of Victorian Scholarship," which seeks to trace Talbot's scholarship in Assyriology in his notebooks and in the wider context of Victorian science.

From Photographic Science to Scientific Photography: Talbot and Decipherment at the British Museum around 1850

Mirjam Brusius

On May 19, 1854, Edward Hawkins, keeper of the Department of Antiquities at the British Museum, received a letter containing the following lines:

> As I take a great interest in the Assyrian inscriptions, and have made considerable progress in the study . . . and as the Museum is now photographing a series of the clay tablets &c. for distribution among persons engaged in the study of this branch of antiquarian research; I should be much gratified if you would give me a copy of the work at the time of its completion.[1]

The author of this letter was William Henry Fox Talbot. The letter suggests that Talbot's interest in the pictures arose not because they were photographs, but because the photographs he mentioned should be distributed for the purpose of antiquarian research. Numerous letters (more than ten are known) were to follow over a period of nine years, some of them insistent and even displeased, in which Talbot asked for lithographs as well as photographs of the clay tablets. Most of these letters, usually sent to the British Museum librarian Antonio Panizzi, remained unanswered.[2]

It took nearly ten years before Talbot received a parcel containing the photographs he sought (fig. 89). Examination of the examples in the Talbot Collection in the British Library reveals peculiar handwritten notes on the surface (fig. 90), indicating that the photographs had experienced a vital transformation at Lacock Abbey. The two shown here are merely examples of each of these two stages, and they depict different tablets, but their comparison is revealing.

In figure 90 we can see a triangular, slightly damaged tablet that, like the one in figure 89, bears some ancient inscriptions. The tablet is either on the floor or on a desk and is leaning against a wall. Some white bricks on the left support the broken corner of the tablet. In the bottom left-hand corner are two numbers, one on a brick, the other on a piece of paper. The numbers suggest that the picture was intended as a visual aid in classifying the object. The light comes from the right,

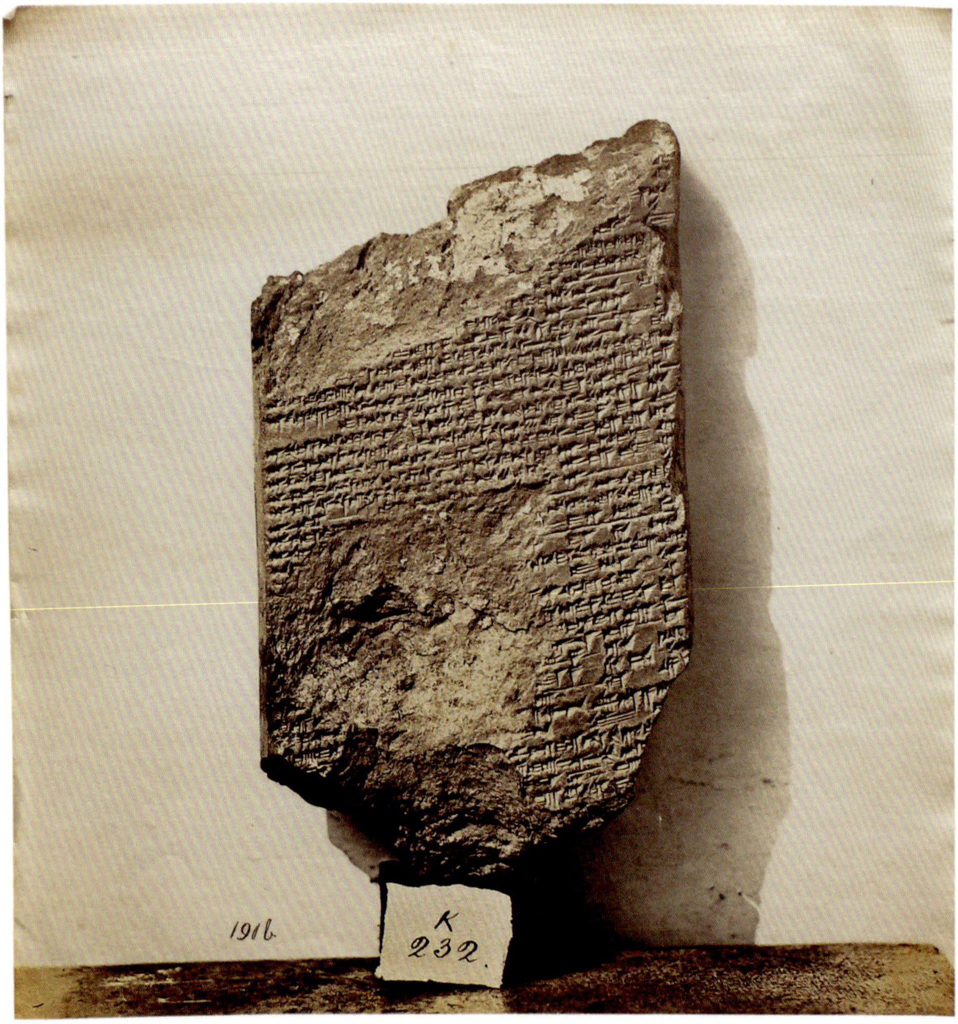

89. Roger Fenton, *Cuneiform Tablet*, 1854–56, salted paper print, 8¼ x 7½ in. (21 x 19.2 cm). Photo 22 (244), Talbot Collection, British Library.

so the script on the tablet seems legible—irrespective of the fact that no one at that time could read the entire script, as it had yet to be fully deciphered. But there is more script here, and this is the main difference from figure 89. Unlike the latter, figure 90 shows words written in Latin script and in pencil. At first sight it is unclear where exactly the penciled notes—in fact, made on the surface of the photograph—are "located": in the background behind the tablet, or on the photograph. It is also hard to tell whether we are looking at an object shown life-size or a close-up of something tiny.

The pictures are photographs of ancient items, clay tablets with cuneiform script, connecting photography and antiquity. Although several scholars have

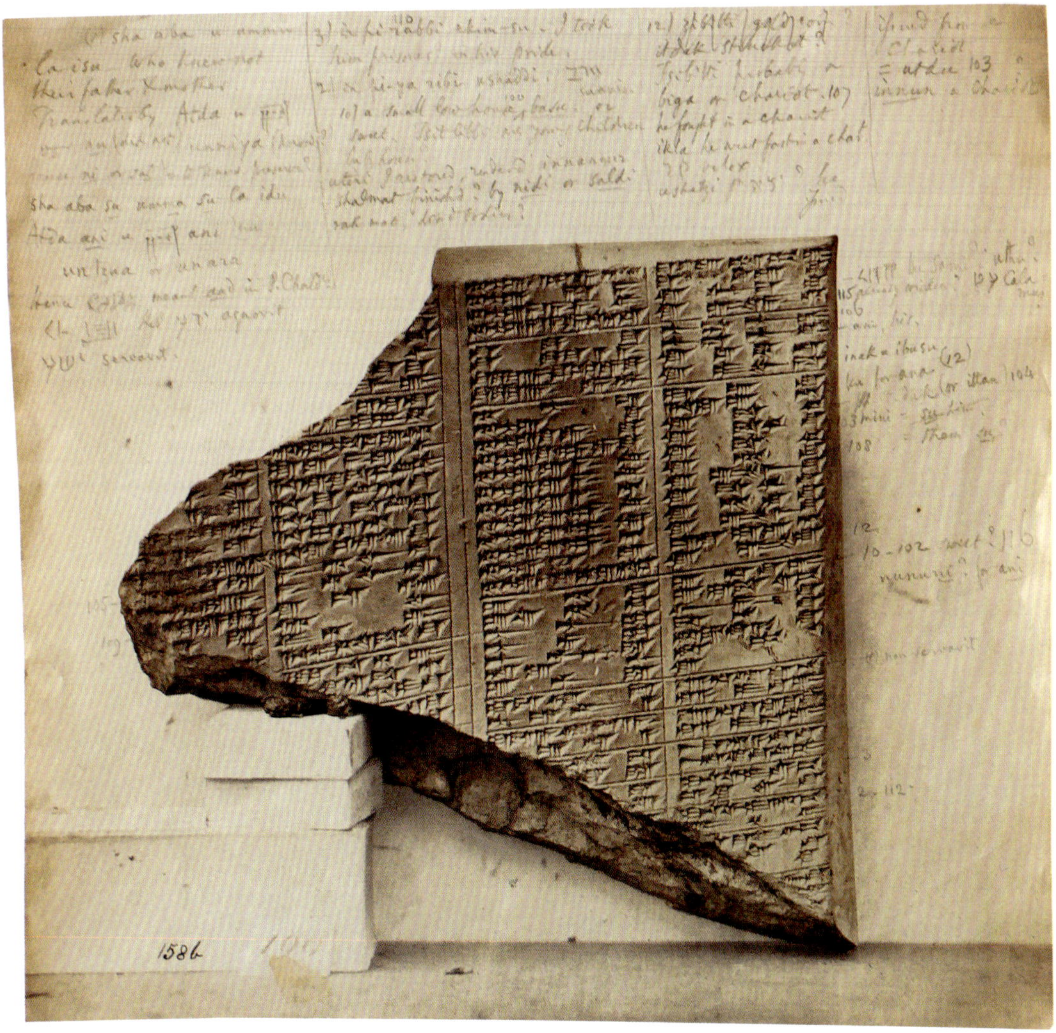

90. Roger Fenton, *Cuneiform Tablet*, 1854–56, salted paper print, 7⅝ x 7⅝ in. (19.5 x 19.5 cm). Photo 22 (201), Talbot Collection, British Library.

attributed the series to Talbot, Roger Fenton, who was employed by the British Museum between 1853 and 1858, can be firmly identified as the photographer.[3]

The photographs were taken in the early to mid-1850s. According to his notebooks Talbot had then moved on to intellectual pursuits beyond photography, and—though he continued to advocate applications of photography and his new invention, photogravure, as well as to pursue his patent rights—he would have declined had someone asked him to take photographs.[4] Nevertheless, the history of these photographs is strongly connected to Talbot.

The pictures mark a shift in his interests from the invention of photography toward its application in the sciences; as Kelley Wilder aptly put it, they are evidence for a transition from "photographic science" to "scientific photography," defined as "the use of photography to represent scientific phenomena, and photographic science is the investigation of any part of the photographic emulsion in the attempt to find out just what is happening when a photograph is made (after exposure, development or fixing)."[5]

Ten years earlier, in about 1840, a key concern in photography was "photographic science," the demonstration of its operability.[6] The earliest photographs were mainly *about* photography, rather than about the content of the photographs. Toward the midpoint of the century, however, scholars became seriously concerned with the question of how one could practically implement photography as a scientific tool. Talbot, driven by a personal interest, was actively involved in this intricate process, though he had already theoretically and practically demonstrated the possible uses of scientific photography years earlier.

THE BRITISH MUSEUM AROUND 1850

In 1843, only four years after Talbot had announced his invention, he suggested to the British Museum that photography could be used to study the antique, for example by taking a photographic apparatus on the expedition of the British archaeologist Charles Fellows to Lycia (in present-day Turkey). Talbot offered to instruct Fellows in the art and visited the British Museum. But Fellows's impending departure may have pushed Talbot to rush his experiment, which led to an unsatisfying result. This in turn may have caused the museum to decide not to use the Talbotype.

Even though Talbot kept trying, Fellows did not change his mind: "I have not the slightest doubt that I could already make use of the Calotype apparatus, . . . —but considering the great imperfection in the manufacture of paper suited to the purpose, the nicety required in using the chemicals, and the extreme cleanliness and exclusion of light in the process, I fear the science is not yet ripe enough for the use of the rough traveller."[7] We do not know of any photographic specimens deriving from Talbot's experiment in 1843; it is likely, however, that the outcome depicted something along the lines of an early photograph depicting the vague outlines of the British Museum's facade (fig. 91).

The museum thus learned about the photographic process early on, but it was another ten years before photography was taken up there. The notion of photography as a tool, both in the field and in the museum, had initially produced very different reactions among scholars. Their mixture of enthusiasm and distrust was rarely expressed as explicit criticism in these early years; however, it can be traced

91. Artist unknown, earliest known photograph of the British Museum (Montagu House), 1843, in Marjorie Caygill, *The Story of the British Museum* (London: British Museum Press, 2002), 29.

in the exchange of many letters describing complex logistical and funding questions, slow decision making, and major delays in application. Although initially quick to broach the issue of the new invention, the British Museum did not act on it speedily. Photography there not only had a mixed story at its beginning but also came to an abrupt temporary end, due to financial and administrative difficulties, after the medium had been applied for only a few years.

The advent of photography coincided with a particularly important moment of transformation in the history of the British Museum that had a general impact on the role of museums in knowledge production. The institution was facing major administrative, curatorial, and architectural challenges in about 1850.[8] Distinguished scientists had been elected trustees in the 1840s. This was one attempt to transform the British Museum from a collection of items on display into a major research institution in various fields, including the study of antique objects.[9] The museum increasingly came to be regarded as the center of British scholarly life, scientific knowledge, and education, not just academically but geographically, with

Oxford, Cambridge, and other universities establishing research facilities in the vicinity.[10] It took time for these ideals finally to gain ground, however, as uncertainty regarding the reorganization of collections prevailed.

This uncertainty applied especially to some of the most important new arrivals around 1850, the as-yet-undeciphered Assyrian cuneiform tablets. In theory photographing cuneiform tablets would unify two superlatives, combining a very recent technology with objects from one of the oldest civilizations of mankind. In practical terms these two elements were no easy match. The rest of this essay explores the role that photography played in the scholarly upheaval of the British Museum. It reveals the gap between ideals about new technologies, brought forward by scholars like Talbot, and the actual applications of photography. Indeed, the link between the application of photography and the transformations in the museum has formerly led to the plausible statement that "photography never came to be utilized and commercially exploited as thoroughly as one might have expected."[11] The trustees' lack of technical knowledge of photography, along with their different priorities regarding improvement of the museum as an institution, may have made new suggestions by amateurs such as Talbot appear second rate and thus not worthy of attention.

THE MUSEUM AND THE FIELD

Two components formed the museum in the nineteenth century: the institution itself and the field where specimens and artifacts were collected. In 1807 the natural philosopher Georges Cuvier stated that, despite having the pleasure of investigating within the surrounding vivid environment of the field, the field scientist, imprisoned in instantaneous time and space, was often deprived of books and comparisons. By contrast, the sedentary scientist may not have access to the field but could "compare them [the products of nature] with each other as often as it [is] necessary to reach reliable conclusions. He chooses and defines his own problems; he can examine them at his leisure."[12] According to Cuvier, true knowledge lay in the study of objects detached from their original surroundings, not in the physical experience that a researcher could gain in the field. Cuvier's sentiment was held by many naturalists and archaeologists alike. If, however, the sedentary scientist has access to the research tools that a museum can offer, such as comparison, why take photographs in the museum at all? For the trustees the answer to this question initially was not very obvious.

The case of the British Museum is by no means singular. The hesitation that accompanied the use of photography in the newly emerging antique sciences can also be observed in the simultaneous discussions taking place in other institutions and scientific fields. Lack of technical knowledge, logistical failures, and financial difficulties would often serve as arguments against the use of photography, as did objections to the epistemological utility of the medium. These objections would not necessarily have been discussed in public, but rather are made evident, for example, by the persistence of drawings and other types of media that were perceived by the trustees and scholars as having obvious advantages. Fenton's photographs of cuneiform tablets are a first result of the final, brief application of photography at the British Museum.

For this essay the importance of the years around 1850 is twofold: this brief time period marked a turning point in Talbot's life and also a change in the appreciation of photography. Both are interwoven. While he stopped being an active photographer, his personal scholarly curiosity expedited the use of photography as a tool for the sciences. Neither shift, however, appeared without difficulties.

Given Talbot's interest in cuneiform, it is unsurprising that the handwriting on the photograph (see fig. 90) is his, though the photographs are not by him. The picture annotated by Talbot was part of a photographic series of Assyrian cuneiform tablets of the Kouyunjik Collection, the most important group of cuneiform tablets in the British Museum. The tablets were excavated by Austen Henry Layard and others at Nineveh and brought back to the museum.[13] Talbot must have received the photographs from colleagues at the British Museum for the purpose of decipherment.[14] When asking for the photographs he described himself as a person "engaged in the study of this branch of antiquarian research."[15] Indeed, with the huge number of specimens and art objects being carried to Britain from around the world, systems of documentation and classification were beginning to be challenged. Furthermore, the demand for copies of certain items forced museums to think about possibilities of reproduction.[16] Interested in both photography and antiquarianism himself, Talbot advocated the use of photography in the antique sciences from its inception. Nevertheless, the scholars around him were still not convinced.

LORD ROSSE AS A MEDIATOR

It was not that the trustees did not know about photography, but rather that they knew about it and chose not to use it. In short, they ignored it. At the beginning of 1839, shortly before Talbot announced his process at the Royal Society, some of his photogenic drawings had also been exhibited at its library. In the same year the astronomer John Frederick William Herschel, who would later become a trustee of the British Museum, presented to the Royal Society a photographic process similar to Talbot's. Scholars who would later become highly influential in the institutionalized museum and art world were thus directly informed about the new technology. Photography was now a topic in the major art journals, and one of its first applications was copying works of art. Photographic studios emerged in London around 1850. Furthermore, Talbot's publication *The Pencil of Nature* (1844–46), which was surely known to the trustees, included many examples of the possible application of photography to the fine arts, architecture, and antiquities.

The possibilities of using the process to record statues, paintings, and drawings were thus noted but not carried out immediately. The most obvious contact of the British Museum trustees with photography, however, was Talbot's visit to the museum in 1843, a landmark in the idiosyncratic relationship between photography and that institution.

Many previous occasions had given the trustees the possibility of engaging with the new medium, but between 1843 and 1854 there is no clear evidence of any systematic photographic activities at the museum. The trustees seemed to have little interest in the application of photography in almost all fields.[17] It was only with a recommendation in 1852 by William Parsons, third Earl of Rosse, astronomer and president of the Royal Society, that they were convinced photography could be considered a useful tool for recording, reproducing, and circulating antique objects. From Talbot's perspective, as he pointed out in 1852 in a letter to Lord Rosse, the Great Exhibition, at the Crystal Palace, London, the year before, had also played a crucial role:

> Ever since the Great Exhibition I have felt that a new Era has commenced for Photography as it has for so many other useful arts and inventions. Thousands of persons have now become acquainted with the art. . . . A variety of new applications of it have been imagined and doubtless many more remain to be discovered. I am unable myself to pursue all these numerous branches of the invention in a manner that can even attempt to do justice to them.[18]

In this letter Talbot complied with Rosse's request to offer the patent gratis to the public, with the single exception of portrait photography. The patent remained an issue in the correspondence between the two men and was subtly interwoven with references to spreading the use of photography in the sciences.[19] It was in the same year, 1852, that Talbot felt it was time to pass his invention on to other people. He himself then moved on to other fields. That is, at the very moment when Talbot stopped taking photographs himself and opened the process to the public, photography slowly started to take off. A year or two later, he began filling various notebooks on the decipherment of cuneiform. These accumulated greatly in the 1860s and especially toward his final years, shortly before his death in 1877 (see fig. 5).

Rosse played a major role in the eventual use of photography at the British Museum, mediating the vexed relationship between Talbot and the institution. The earl had a strong personal interest in photography. He was especially acquainted with problems regarding the visualization of the moon and the nebulae, ephemeral objects that were by no means easy to grasp. Having hopes for Talbot's new instantaneous photographic process, he asked for some advice.[20] No lunar photographs by Rosse are known to have been made, so it is unknown if he followed what Talbot told him. Even though problems of credibility and trust were inevitable issues, drawings were still much in use in the earl's circle in the 1860s.[21] Drawing, sketching, and tracing were constitutional epistemological elements in a process through which scientists were trying to become familiar with the object they observed.[22] The parallel use of various media dominated the methodology of scholars in both archaeology and astronomy.[23] It seems ironic, therefore, that the camera was praised for lacking the subjective and psychological aspects of these other techniques. In both cases, the photographing of cuneiform tablets and astronomical photography, almost a decade passed without any traces of photographic experiments. Another practical problem was that photographs could not be reproduced without the aid of further printing techniques until about 1880, and even then they often had to be manipulated in order to be easily reproduced. One conclusion that can be drawn when looking at this story is that the need for photographs was simply not pressing at this time.

It was when Rosse took an interest in lunar photography in 1852, however, that he suggested to the trustees at the British Museum—assuming that his idea would be more feasible than astronomical photography—that an attempt should be made to photograph some of the cuneiform inscriptions in order to distribute them among Assyriologists:

> I feel very conscious that some attempt should be made to photograph the Inscriptions in as to place them conveniently within the reach of persons who have a taste for that line of research. . . . The continually varying indication of the meteorological and magnetic instruments at Greenwich are recorded by Photography . . . and to photograph inscriptions would be a much easier operation. Besides by photography numerous inscriptions can be taken off, using the original photograph.[24]

Rosse significantly underestimated the difficulty with which cuneiform tablets could be photographed; they later turned out to be intractable, almost visually "perverse" objects.[25] Unlike sculptures, for which a photograph allowed different ways of seeing and understanding the object, a photograph of a cuneiform tablet needed to show legible inscriptions. Legibility was relative, however, for as of that year nobody could yet read the script—or even be certain if every marking on a tablet indicated a symbol. Neither did anyone know how photographic recording operated in this respect.

On the other hand, the photographs were meant to place the original objects conveniently within reach, which is precisely what could not be done with the clay tablets themselves. That is, even though the tablets were in the museum, they were not mobile enough to be examined fully by scholars—least of all on the Continent, where interest in cuneiform expanded simultaneously. Photography was thus meant to serve as a reproducible proxy. The trustees finally welcomed the suggestion, and Rosse invited Talbot back to the museum.[26] This time the trustees' interest in the demonstration was clearly more along the lines of scientific photography than of photographic science.

IN A GLASS HOUSE ON THE ROOF

Thanks to the persistence of Rosse and Hawkins (the keeper of the antiquities department), certain advances were made. The trustees met in the summer of 1853 to discuss the establishment of a photographic studio on the roof of the museum. Shortly afterward, Fenton offered his services as a photographer.[27] Fenton was employed in the spring of 1854 but initially paid for only two months. He was supposed to dedicate his time chiefly to the clay tablets, which he had to photograph as "experimental trials"; after that the trustees would think about possible further employment. Meanwhile news from the excavations in Nimrud, an ancient capital of Assyria, constantly reached the British Museum. The use of photography on-site was discussed at the same time. First attempts to send out photographers had been made, but no suitable specimens had returned.

Fenton's experimental trial followed up on the first experiment in 1843, when Talbot had intended to demonstrate the advantages and efficiency of the technology. Talbot's was more of an experiment in photochemistry than a trial regarding the object to be rendered. The shift to scientific photography was possible because "the more controlled and predictable the emulsions became, the more reliable their evidence."[28] A decade had passed since then, and photography was more established and easier to handle. Unlike the initial experiment in photographic science, the subject of the one in 1854 was clearly the iconic content of the picture itself. It now mattered whether the "translation skills" of the medium were adequate according to the purposes of decipherment. When we consider, however, that the trustees were to decide on the success or failure of photography based on the results of the experiment, a problem emerges: what exactly was the content of the cuneiform photographs supposed to be, and how would success be determined?

Like reliefs, cuneiform tablets sat somewhere between flat text and contoured object when used for research: they were three-dimensional tablets from which two-dimensional texts were to be extracted. Sometimes the imprints in the clay were very deep, sometimes shallow. They therefore required complex lighting techniques. Legibility was the criterion by which a researcher would evaluate the usefulness of reproductions of ancient scripts. But for undeciphered tablets, many of the symbols to be read were as yet undefined. It could be unclear which markings were part of a sign and which were incidental marks on the surface of the tablet, subsequent damage to the clay, or traces of the photographic process itself. Thus the desired outcome of the photograph was not clearly defined. How was the legibility of a photograph of an undeciphered tablet to be judged, if the ciphers themselves had yet to be determined? What specific information, or type of information, was required in the photograph? What were the conditions for the closure of the experiments? If the end is not defined, can this be called a useful experiment at all? Or was it simply, as in the almost miraculous experiment of 1843, an experiment in a science "without any clear consensus about what can be counted as an experimental failure or success," so that the experiment could also be considered successful even if it failed?[29] In other words, was it successful if it showed a tablet, no matter what it looked like?

It was therefore necessary to make assumptions regarding the performance of the instrument and to regard the camera as unproblematic.[30] The camera could work as a tool only if the instrument itself remained transparent and invisible; like windows in a house, the camera worked best if it could not be seen. Otherwise the procedure would result in an experiment within an experiment. Whether a photographic experiment was successful or failed, no matter whether it was of the first or second "type" of experiment described here, depended on the success or failure of its method, which lay "in its transparency, its ability to be repeated by other scientists and replicability of its results."[31] Hence, in order to ensure that the photographs provided evidence, the transparency of the medium and the exertion of control over chemicals were essential, as was belief in the camera as an instrument.[32] As time had passed, Fenton had better conditions technically than Talbot did in 1843. Furthermore, Fenton was a photographer who took special care, having a great interest in making his own prints to achieve "a good photographic picture."[33]

Fenton's chamber was planned in detail.[34] In March 1854 Hawkins recommended "that twenty impressions should be taken of each of the Assyrian Tablets and five of ordinary new acquisitions. The Assyrian inscriptions were to be transcribed and printed, and photographic impressions to be sent to such Scholars as are known to be engaged in their translation, and interpretation for the verification of the printed copies."[35] This is where Talbot reenters the story. In May he asked the museum whether he could obtain these copies. Many letters were to follow in the coming years, in which Talbot also asked for lithographs, but no one immediately sent the copies that he requested. It seems ironic that the person who fought with such enthusiasm for the application of photography at the museum found it so difficult to take personal advantage of it. Talbot now wanted to *use* his invention as a scholar, but he was treated simply as the inventor of photography, not as a future expert in Assyriology.

DECONTEXTUALIZATION

Photography was believed to record everything reliably, whether important or not. It might have been precisely the fact that it remained unclear just how the important information on clay tablets needed to be defined that made the choice to photograph these clay tablets especially appealing. As Talbot had pointed out in the text accompanying the photograph entitled *Entrance Gateway, Queen's College, Oxford* (fig. 92) in *The Pencil of Nature* (1844), in which he referred to the clock in the back and some placards in the foreground, sometimes a photograph revealed something that was not noted before the picture was taken, such as inscriptions or the hour of the day "at which the view was taken."[36] Not coincidentally, his argument refers precisely to inscriptions, dates, printed placards, and dial plates being unconsciously recorded.

Nevertheless, the disadvantages of a photograph of an inscription were no less obvious. Just as drawing nebulae allowed scientists to become familiar with the objects, drawing a cuneiform tablet was part of the process of decipherment. Here the "thinking hand" was of major importance, for in the act of tracing, the mind and the hand got to know and understand the possible form of the characters.[37] They made choices, discarded possibilities, and thereby took a first step toward decipherment. When tablets were merely photographed, this step was absent. The cuneiform tablet photographs with Talbot's penciled inscriptions on top are an outstanding example of this problem. The photographs were meant to be proxies, but they were proxies with an unstable connection to the object depicted. The scholarly work, decipherment, still had to be done—which is why they were, in a

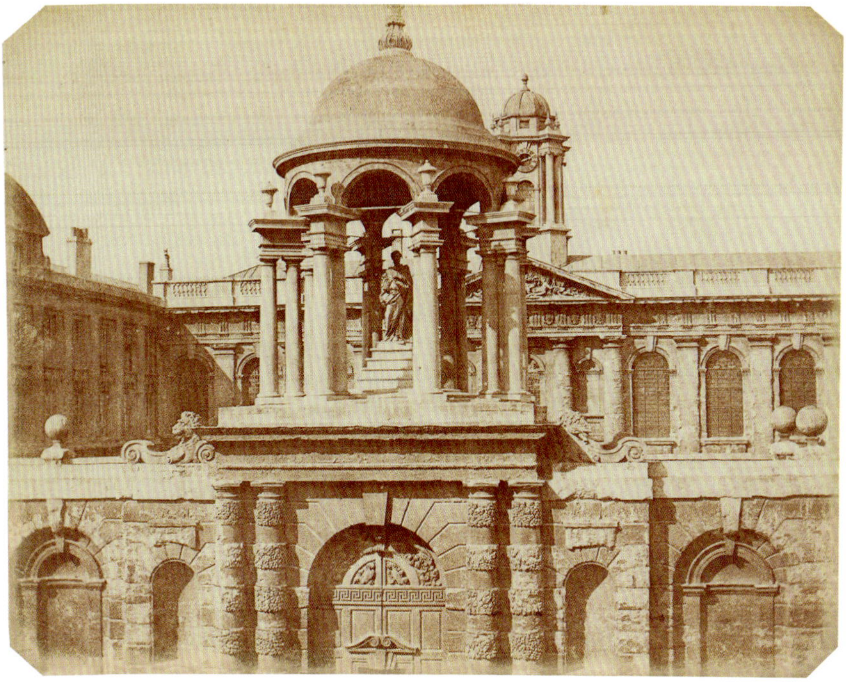

92. William Henry Fox Talbot, *Entrance Gateway, Queen's College, Oxford*, 1843, salted paper print from paper negative, in Talbot, *The Pencil of Nature* (London: Longman, Brown, Green and Longmans, 1844), plate XIII. The Metropolitan Museum of Art, New York.

sense, illegible. Their initial function as a tool was primarily restricted to their reproducibility and their mobility.

In order to determine the function of the photographs, it is worth looking at the way Fenton rendered the tablets (see figs. 89 and 90). The procedure for photographing the tablets involved their systematically being taken to one place, leaned against the wall, provided with a tag, and photographed. With their consistent and neutral form the series of cuneiform tablet photographs anticipates systematic police photography.[38] Like Fenton's later skeleton photographs, made for a publication by Charles Darwin's opponent Richard Owen (fig. 93), they differ from the context-based photographs that Fenton partly took in the museum's environment (figs. 94 and 95).

While the items Fenton photographed in the exhibition space appear in the prints as part of an arranged collection of items, the sculptures he photographed on pedestals and blocks (fig. 96), as well as the cuneiform tablets with their more neutral standardized depictions, wiping out everything that might be distracting,

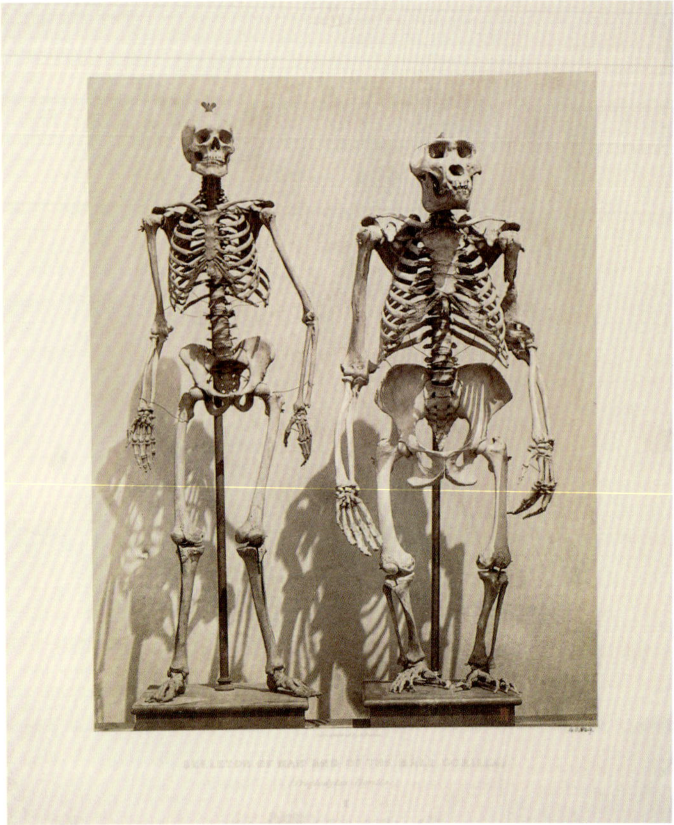

93. Roger Fenton, *Skeleton of Man and of the Male Gorilla (Troglodytes Gorilla)*, ca. 1855, albumen print mounted on printed card, 14⅜ x 11 in. (36.5 x 28 cm). Victoria and Albert Museum, London.

appear as subjects of an attempt to inventory. This seeming drive toward objectivity is undermined by the displacement and detachment of the objects from their original contexts.

Nevertheless, placed on small pedestals in front of a neutral background with a sheet, the tablets "share the same fate" as every single one in the series.[39] The background in each photograph seems temporary and improvised. The viewer is meant to focus on the object. That object appears on its own in the picture and yet is part of a series. It is therefore depicted as a common (in both senses) and replaceable object. It is a mise-en-scène without an actual mise-en-scène, fading out everything arbitrary. The tablets were meant to be preserved in this way as permanent inscriptions that not only could testify to and save what was readable on them, but also could present the tablets—once and for all—as artifacts detached

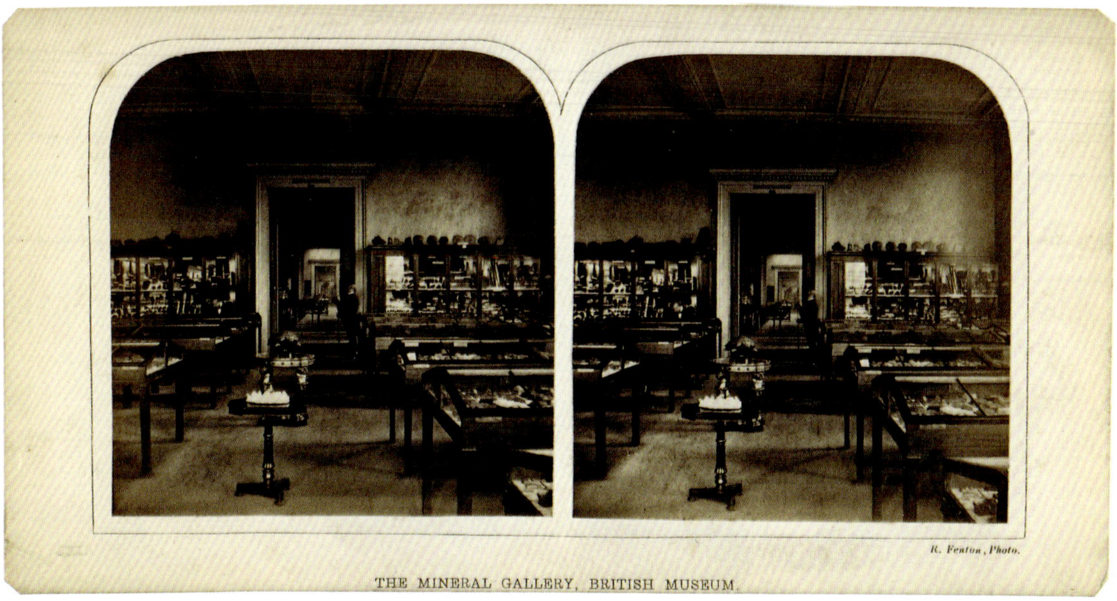

94. Roger Fenton, *The Mineral Gallery, British Museum*, 1857, stereoscopic albumen prints, each 3 x 2¾ in. (7.6 x 7.1 cm). British Museum.

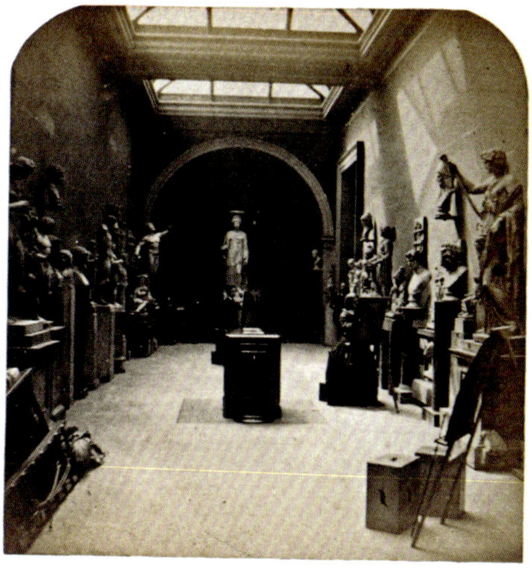

95. Roger Fenton, *The Third Graeco-Roman Saloon on Artists' Day at the British Museum*, 1857, albumen print, 3 x 2¾ in. (7.6 x 7.1 cm). Victoria and Albert Museum, London. This is probably one of a pair of stereoscopic photographs.

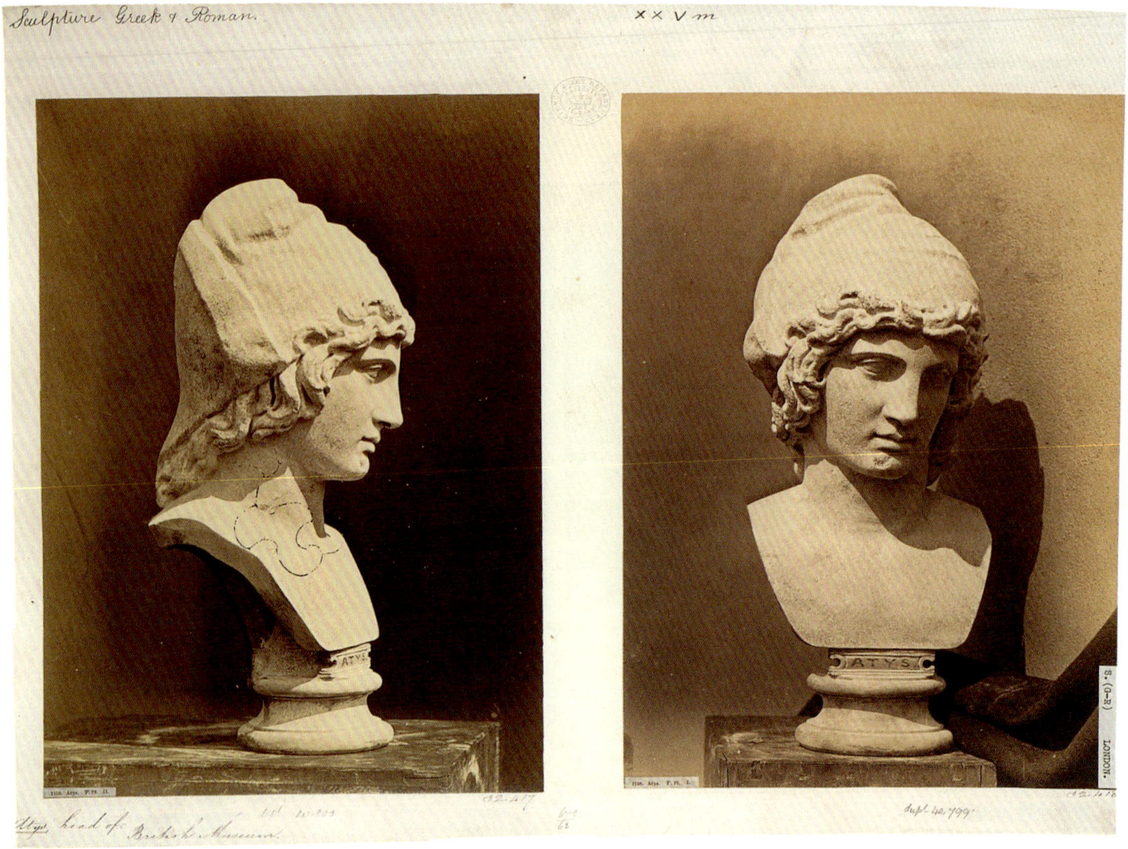

96. Roger Fenton, *View of a Bust of Aty*, ca. 1850, albumen prints, sheet: 14⅜ x 18⅞ in. (36.5 x 48 cm). Victoria and Albert Museum, London.

from their cultural origin. They were now independent research objects, awaiting classification and study by scholars possessed by a nation, not by an extinct culture. With Fenton's photographs of the tablets, the inscriptions are all the more detached from their original context. The photographs turned the clay tablets into flat objects that fit into files, folders, and briefcases. Unlike drawings, however, they seemed to convey the authentic appearance of the tablets. They served as mobile proxies that could be distributed among scholars and also used as tools for the archive and the inventory. As part of a large series they are products of a production line. Fenton's arrangement anticipated the fact that most of these clay tablets would never be chosen for public display but would instead be locked away in drawers to be consulted by scholars.

CIRCULATION

Most of the decisions the trustees made regarding the use of photography in the British Museum were connected to the dissemination and circulation of their objects in order to make scholarly activities easier. Photography became part of the effort to turn the museum into a research institution.

Circulation of the photographs, however, was a serious problem. Lithography was still the preferred method of printing reproductions in books. To publish the photographs themselves without transforming them into another medium, which would result in the loss of the "photographic," the only solution was to paste photographic prints into the publication.

Fenton's cuneiform photographs were now cut out and mounted on a board, so that the photographed object appeared like a proxy of the tablet (fig. 97). The way the tablet floated within the frame as a tangible, separate object against a white background isolated it all the more from its ancient surrounding. As much as this made comparison and annotation easier, it completely denied perception of scale.[40] Several scholars throughout Europe had heard of the photographed tablets and asked for copies. Talbot was among the first to ask, but most of them received their copies before him. The recipients were primarily the more established scholars holding prestigious positions at academic institutions and societies.[41]

At the end of two months, in May 1854, Fenton's contract was renewed for another identical period.[42] Then, after his return from an eighteen-month-long expedition in the Crimea, Fenton started to make photographs of other objects in the museum (see figs. 94–96) while continuing to photograph a great number of the clay tablets, which were to be mounted on paper and annotated. Having taken into consideration the expenditure of photographing the Assyrian tablets, however, the trustees decided in 1856 "that no more Photographs of these Tablets should be made at present."[43]

LETTERS TO THE TRUSTEES

Talbot sent an enormous number of letters to the trustees requesting lithographs and photographs of the tablets. In 1856 there was another "flood of letters."[44] The letters reflect issues related to access and the authorship of the translations. Many drafts of these contain an unusually high number of crossed-out sections, revealing his anxious attitude regarding the matter.

Talbot was constantly put off and eventually told that there "appeared to be some misunderstanding between him and Henry Creswicke Rawlinson on this

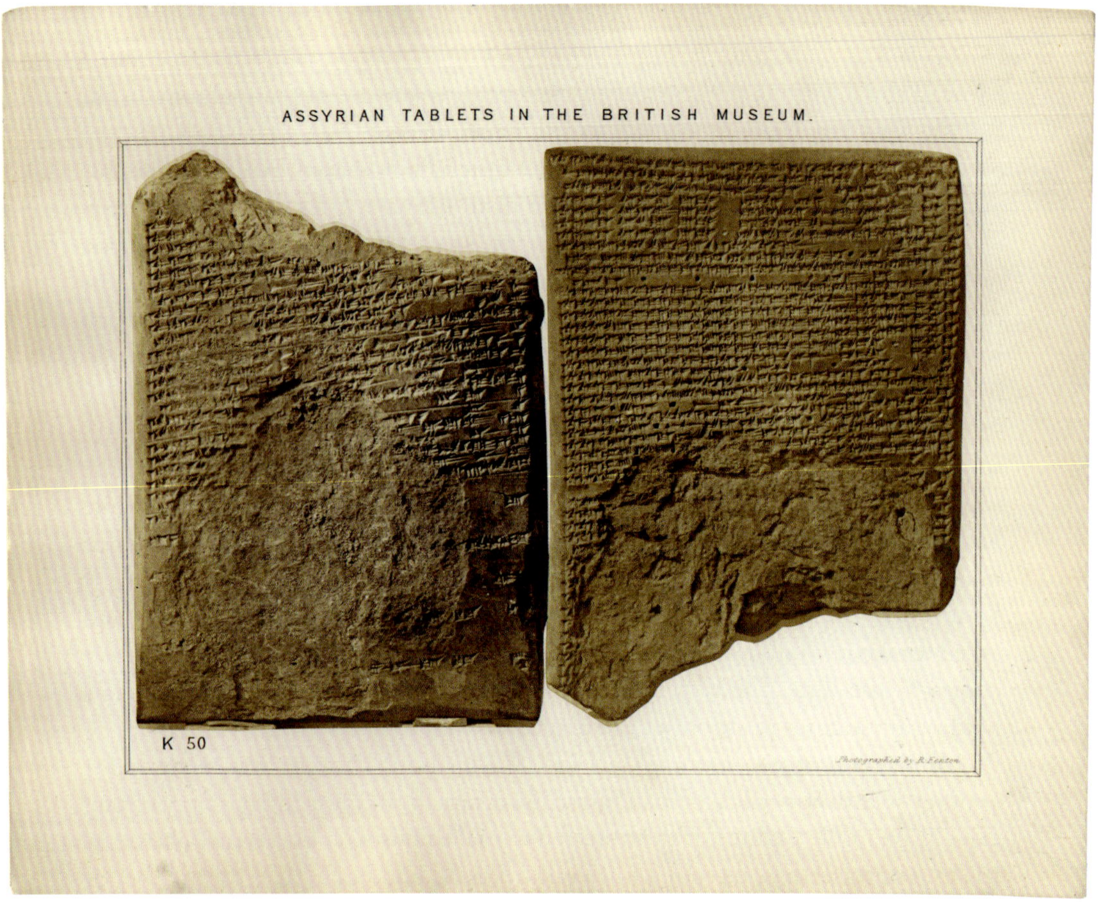

97. Roger Fenton, *Cuneiform Tablet*, 1854–56, salted paper print mounted on paper, 10⅞ x 13 in. (27.5 x 33 cm). British Library (uncatalogued).

subject which ought to be cleared up before the lithographs could be forwarded to him."[45] Rawlinson was an ambitious and leading decipherer at the British Museum at the time, and he must have considered Talbot a threat. He seems not to have supported any attempts to decipher inscriptions that he had not already deciphered.[46] Rawlinson planned to publish transliterations and translations of inscriptions Talbot was asking for and wanted to avoid any possibility of being preempted.[47] But Talbot insisted on improving his decipherment skills by taking part in their translation.[48] In another meeting consideration of his long letter was deferred, probably because the trustees were worn down by the dispute.[49] The whole story was fatiguing. Talbot's letters stop being mentioned in detail in the minutes of the trustees' meetings. Finally, Rawlinson suggested the terms on which this request might be granted: Talbot should not publish any translations.[50] Talbot agreed to that condition. Furthermore, it was agreed that two or more independent translations of the same text should be made and compared in public. Rawlinson picked the translations, among them the recently excavated clay cylinder from the reign of the Assyrian king Tiglath-pileser, which Rawlinson was preparing for publication, and Talbot offered to deposit his translation under seal until Rawlinson's translation was published. Not being aware that a parcel was then sent off to him, Talbot mailed yet another request to the museum on the same day.[51] He seemed both restless and relentless.

In the summer of 1858 the trustees, considering Fenton's employment a costly luxury, decided that no further objects at the museum should be photographed.[52] After a dispute over payment and the property rights to his negatives, Fenton's time at the British Museum came to an abrupt end.

Talbot, however, remained involved for the rest of his life. For him the game had only just begun. In 1860 Rawlinson gave him permission to examine and study the inscriptions "without interfering with the ordinary business of the establishment."[53] Lithographs of the copies Talbot had requested of the newly published cuneiform inscriptions were now sent to him. This was nowhere near enough for Talbot; he still had not received Fenton's photographs of the challenging clay tablets from Nineveh that yet had to be deciphered. In a last attempt, in November 1863—almost ten years after his first letter—Talbot asked for a copy "of the Collection of Photographs of clay tablets bearing cuneiform inscription, which were made some years ago by direction of the Trustees. This series of documents would be of great use to me in the prosecution of my researches into the Assyrian languages."[54]

Talbot must have been relieved when a month later he finally received the first proper answer to the letter he had sent nine years earlier: "I have the pleasure to acquaint you that the Trustees have complied with your request for a copy of the collection of Photographs of Clay Tablets bearing Cuneiform Inscriptions. Will you be so good as to let me know in what way you would like to have them sent to you?"[55]

CONCLUSION

Taking Talbot as a starting point, this essay has drawn a new picture of the problems of access, the status of the Victorian gentleman scientist, the role of amateurs and professionals, and above all the early use of photography in Victorian Britain around 1850. It has shown, first, that there was no obvious answer to the question that some people were no doubt asking: "What is photography good for?"[56] The reasons cannot merely be found in the nature of photography. Rather, they are grounded in the process of transition at the British Museum itself. That is to say, photography did not arrive as a self-driven new invention, but instead was one of many features that emerged as a result of a grander project that encompassed the integration, access, and therefore mobility of new collection items that had yet to be made sense of and defined. In the context of the museum's unstable and often disordered enterprises, photography and its hesitant application were thus, first and foremost, determined by the museum's cultural, social, and economic structures. The restructuring and attempts to turn it into a scholarly institution simultaneously aided and hindered the application of photography.

Another crucial reason for the new medium's comparative lack of use was the ill-defined status of the newly arriving objects that were to be visually depicted—and, in fact, of photography itself. This status was principally negotiated at the museum rather than in the field, and even in London it never took place without semantic difficulties. Naturally it was here that photographic experiments were carried out rather than in the unstable environment of the field.[57] The study of objects detached from their original distant surroundings, which Cuvier had described as an advantage for the sedentary scientist, was enforced by the photographic camera as a sedentary tool. It decontextualized objects not only from their surroundings but also from the museum space in which they arrived.

The practical side of things was different. The trustees seemed content with the use of drawings and lithographs, and photography was also nowhere near feasible or inexpensive enough to be widely applied.[58] Fading prints, artifacts suddenly appearing on the photographic surface, color-changing chemicals, and broken glass would later provide evidence that photography was by no means a permanent record.[59] In considering the problem from this point of view, instead of asking why the museum did not employ a photographer more immediately, one could turn the question around to ask why the institution employed a photographer at all. Knowledge of objects, traces, and materials depends on and rests with human interests and concerns rather than with isolated technical systems. Decisions about photography's fruitfulness as a visual device are thus directed by specific and often complex judgments regarding the appropriate combination of photographic techniques with the materials to be registered.

Second, this essay has identified several shifts around 1850, for which Talbot serves as a case study. The British Museum institutionalized and funded scholarship but also controlled it. Talbot, a sedentary scientist and example par excellence of the domestication of gentlemanly science, asked for photographs because objects in the museum were not necessarily mobile, tangible, or comparable there either. His story, as told here, shows that not all scholars counted as equal actors in scientific networks: although the line between amateurs and professionals was still fluid in the period around 1850, it had started to take shape. Consequently, access to intellectual networks was controlled more starkly than before: the distribution of information was hierarchical, with access being initially restricted to scholars holding positions at prestigious institutions, even though access for the public was being increasingly improved for all social classes.[60]

After Talbot decided to become an Assyriologist, he also pushed forward the implementation of photography in the antique sciences. This implied a shift from his role as the inventor of a medium to that of its user in an increasingly professionalized context. But neither his attempt to become more "professional" in these emerging areas of study nor the application of photography itself was self-evidently likely to succeed. Talbot was surely recognized as the inventor of photography, but this certainly did not entail the right of access to photographs taken at the British Museum. This essay thus has highlighted the discrepancy between innovation in and practical usage of (visual) technologies on the one hand, and between expert knowledge and amateur application on the other. Talbot was an expert in photography, but he was certainly not so in the fields he targeted as potential platforms for its application. His hitherto stable status as genteel amateur was at stake due to the increasing but uneven professionalization in the second half of the nineteenth century. He was an inventor, a scholar, and a funding agent, but these manifold functions often troubled his status in the scholarly network of Victorian London, rather than easily stabilizing it. That is, Talbot's own invention proved useful for his studies but took on a dynamic of its own.

Nevertheless, in the story told here, Talbot appears more like an Assyriologist who happens to have invented photography as well. Roger Fenton's photographs with Talbot's handwriting on the surface are the embodiment of a productive fusion between the antique, photography, and note-taking, and thus between photography and Talbot's mind-set.

EPILOGUE: TEN YEARS LATER

A permanent force of photographers was not set up at the British Museum until as late as 1927.[61] Before that, photography's marginal status at the museum continued throughout the 1860s, the 1870s, and the 1880s; so did the disorder regarding the holdings, especially the Kouyunjik Collection, within the museum's walls.

Talbot's final receipt of Fenton's photographs by no means represents a resolution of the story of photography's use. The historiography of photography in the later decades of the nineteenth century has generally assumed that, due to its

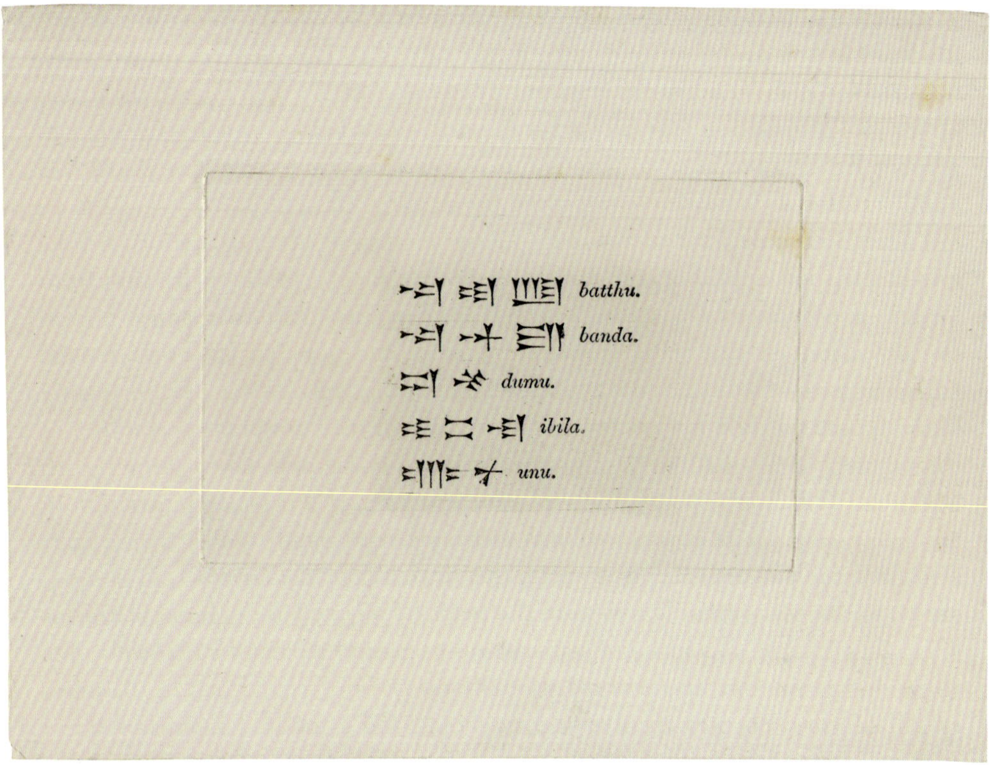

98. William Henry Fox Talbot, *Assyrian Cuneiform Writing*, ca. 1874, photoglyphic engraving, plate: 2½ x 3⅞ in. (6.5 x 9.7 cm), paper 5 x 6⅜ in. (12.6 x 16.1 cm). Houghton Library, Harvard University.

apparent authenticity, photography was thoroughly in use and more appreciated than in earlier decades. There were of course numerous occasions on which the use of the photographic camera seemed to have gained ground. Yet many archival sources tell—once again—a different story. At the British Museum Talbot continued fighting for the application of his invention until his death in 1877. During the course of the promotion of his invention, it almost seemed as if he had partly lost trust in the medium itself, perhaps because others could not see its potential, and he himself had experienced difficulties in applying the medium in particular fields. In 1874 Talbot finally used a second important visual technique he had developed, photogravure, to reproduce an Assyrian cuneiform inscription, possibly to contrast the shortcomings of his photographic invention with the clarity of a new visual device (fig. 98).[62] But it was too late for this move to gain support. In the years before Talbot's death attitudes toward photography were still highly ambiguous and contested.

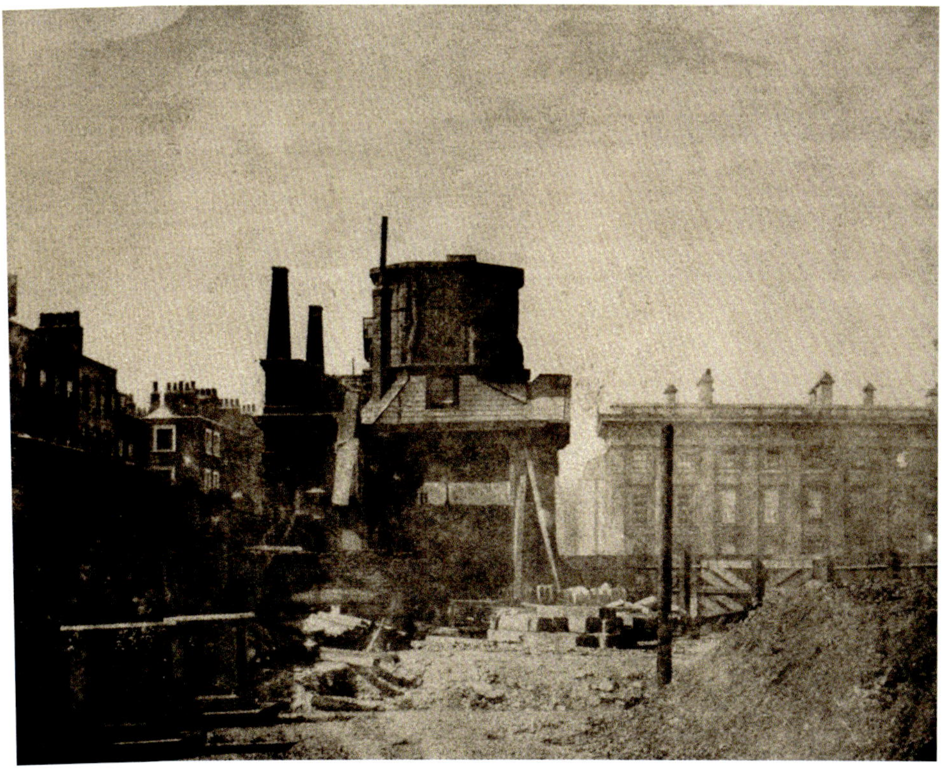

99. Artist unknown, *Demolition of Montagu House Gateway,* 1847, in Marjorie Caygill, *The Story of the British Museum* (London: British Museum Press, 2002), 34.

It would be wrong simply to replace the historical assumption of photography's effortless deployment with an entirely revisionist story of its startling failure. The aim here is to abandon a history of self-evidence (when describing the technique's adoption) and surprise (when reacting to its absence), and instead adopt a more focused attention to the specific, local, and often uneven circumstances in which photography was reckoned worthy, or ignored, or contingently put to use.

A photograph taken on the site of the British Museum—possibly one of the very few at the time that depict the dusty demolition of the former museum space at Montagu House gateway—was made in 1847 (fig. 99). Shortly afterward, the first Mesopotamian finds arrived at the museum. The photograph depicts the museum as a construction site in the most literal sense, by the very medium that itself became a difficult case in the museum's institutional reconstruction. The building site as yet lacked any clear definition—just as did the newly arriving objects, photography, and perhaps even Talbot himself.

I would like to express my gratitude to Simon Schaffer, Eleanor Robson, Katrina Dean, Chitra Ramalingam, Lorraine Daston, and the curators at the British Library and the British Museum.

1. William Henry Fox Talbot to Edward Hawkins, May 19, 1854, Doc. No. 6965, The Correspondence of William Henry Fox Talbot, http://foxtalbot.dmu.ac.uk (accessed Jan. 31, 2012), hereafter "Correspondence" (original: Department of the Middle East, British Museum, London, hereafter "BM").
2. See, for example, Talbot to Antonio Panizzi, Nov. 19, 1863, Doc. No. 8728, Correspondence (original: Talbot Collection, hereafter "TC," British Library, hereafter "BL").
3. For the attribution of the photographs, see Aug. 12, 1854, April 1847–July 1855 (C6), Minutes of the Committee Meetings (hereafter "MCM"), BM. The attribution of the cuneiform tablet photographs to Roger Fenton is still the subject of controversy. Talbot seems never to have had a strong role in Fenton's photographic education; the only records of their contact are four rather fierce letters about a dispute over patent rights from the 1850s. Several scholars have dealt with the history of photography at the British Museum, some of them in connection with Fenton. See *Roger Fenton: Photographer of the 1850s* (London: South Bank Board, Hayward Gallery, 1988). See also John Hannavy, "Roger Fenton and the British Museum," *History of Photography* 12 (1988): 193–204; Christopher Date, "Photographer on the Roof," *British Museum Society Bulletin* 61 (1989): 10–12; Christopher Date and Anthony Hamber, "The Origins of Photography at the British Museum, 1839–1860," *History of Photography* 14 (1990): 309–25; Anthony Hamber, *"A Higher Branch of the Art": Photographing the Fine Arts in England, 1839–1880* (Amsterdam: Gordon and Breach, 1996), 377; and Frauke Kreutler, "Sonnenlicht und Druckerschwärze: Paul Pretsch (1808–1873) und sein Beitrag zu den fotografischen Reproduktionsverfahren," *Fotogeschichte* 24 (2004): 17. Some secondary sources mention the photographs of the cuneiform tablets but do not examine their history in more detail. The connection between these prints and Talbot's role at the British Museum has not been explored.
4. See the essay by Larry Schaaf in this volume.
5. Kelley Wilder, *Photography and Science* (London: Reaktion Books, 2009), 13.
6. Peter Geimer, "Photographie und was sie nicht gewesen ist: Photogenic Drawings, 1834–1844," in *Wahrnehmung der Natur, Natur der Wahrnehmung*, ed. Gabriele Dürbeck et al. (Dresden: Verlag der Kunst, 2001), 135–49, esp. 137. See also Wilder, *Photography and Science*.
7. Charles Fellows to Talbot, Aug. 1, 1843, Doc. No. 6436, Correspondence (original: BM Archives, London). See also Mirjam Brusius, "Inscriptions in a Double Sense: An Early Scientific Photograph of Script," *Nuncius: Journal of the History of Science* 24 (2009): 367–92.
8. Date and Hamber, "Origins of Photography," 309.
9. See John V. Pickstone, "Museological Science? The Place of the Analytical/Comparative in Nineteenth-Century Science, Technology and Medicine," *History of Science* 32 (1994): 119–20.
10. See James Arthur Bennett, "Museums and the Establishment of the History of Science at Oxford and Cambridge," *British Journal for the History of Science* 30 (1997): 29–46. For the function of museums around 1850, see Sophie Forgan, "Building the Museum: Knowledge, Conflict, and the Power of Place," *Isis* 96 (2005): 572–85.
11. Date and Hamber, "Origins of Photography," 309.
12. Quoted in Dorinda Outram, "New Spaces in Natural History," in *Cultures of Natural History*, ed. Nick Jardine, James A. Secord, and Emma C. Spary (Cambridge: Cambridge Univ. Press, 1996), 260–61.
13. The photographic series, which is part of the Talbot Collection at the British Library, consists of the remains of a complete set (K. 50–278) of photographs of Assyrian cuneiform tablets (both front and back). The K number on each tablet refers to the Kouyunjik Collection.
14. For Talbot as an Assyriologist, see the essay by Eleanor Robson in this volume. Talbot's notes do not seem to be related to a particular publication but might have been studies for his glossary of Assyrian, which he published in the late 1860s. See William Henry Fox Talbot, "Contributions towards a Glossary of the Assyrian Language," *Journal of the Royal Asiatic Society of Great Britain and Ireland*, n.s., 3 (1867): 1–64; 4 (1869): 1–80; 6 (1873) 1–80. With thanks to Eleanor Robson.

15. Talbot to Edward Hawkins, May 19, 1854, Doc. No. 6965, Correspondence (original: Department of the Middle East, BM).
16. *Roger Fenton: Photographer of the 1850s*, 10.
17. Date and Hamber, "Origins of Photography," 315. Here a few commissions, donations, and purchases are listed. The detailed and pedantic character of the minutes suggests that any such incident would have been recorded. Research on the scale of photographic holdings in the different departments of the British Museum has yet to be carried out.
18. Talbot to William Parsons (Lord Rosse), July 30, 1852, Doc. No. 6668, Correspondence (original: Birr Scientific and Heritage Foundation / Earl of Rosse, Ireland). Various scholars also mention the Great Exhibition of 1851 and the display of photographic objects as an inducement to apply photography at the British Museum.
19. See the thirteen letters between Talbot and Lord Rosse, Correspondence. For Talbot and patents, see Keith Adamson, "Early British Patents in Photography," *History of Photography* 15 (1991): 313–23.
20. William Parsons to Talbot, Feb. 2, 1854, Doc. No. 6909, Correspondence (original: TC, BL). For Talbot and instantaneous photography, see the essay by Chitra Ramalingam in this volume.
21. See Simon Schaffer, "On Astronomical Drawing," in *Picturing Science, Producing Art*, ed. Caroline A. Jones and Peter Galison (New York: Routledge, 1998), 455.
22. See Omar Nasim, "Observation, Working Images and Procedure: The 'Great Spiral' in Lord Rosse's Astronomical Record Books and Beyond," *British Journal for the History of Science* 43 (2010): 353–89.
23. See Alex Soojung-Kim Pang, "Victorian Observing Practices, Printing Technologies, and Representations of the Solar Corona (I): The 1860s and 1870s," *Journal for the History of Astronomy* 25 (1994): 270–72; and Stefanie Klamm, "Bilder im Wandel: Der Berliner Archäologie Reinhard Kekulé von Stradonitz und die Konkurrenz von Zeichnung und Fotografie," *Jahrbuch der Berliner Museen* 49 (2007): 115–26.
24. May 15, 1852, April 1847–July 1855 (C6), MCM, BM. Also quoted in Date and Hamber, "Origins of Photography," 316.
25. For the resistance of cuneiform tablets to photography, see the essay by Eleanor Robson in this volume.
26. William Parsons to Talbot, June 13, 1852, Doc. No. 6644, Correspondence (original: TC, BL).
27. Letterbook 10, Oct. 1853–Jan. 1856, MCM, BM; and *Roger Fenton: Photographer of the 1850s*, 10.
28. Wilder, *Photography and Science*, 78.
29. James A. Secord, "Extraordinary Experiment: Electricity and the Creation of Life in Victorian England," in *The Uses of Experiment: Studies in the Natural Sciences*, ed. David Gooding, Trevor Pinch, and Simon Schaffer (Cambridge: Cambridge Univ. Press, 1989), 338.
30. This is a version of what Harry Collins has called "the experimenter's regress." Harry Collins, *Changing Order: Replication and Induction in Scientific Practice* (London: Sage, 1985) 79–111. See also Simon Schaffer, "Where Experiments End: Tabletop Trials in Victorian Astronomy," in *Scientific Practice: Theories and Stories of Doing Physics*, ed. Jed Z. Buchwald (Chicago: Univ. of Chicago Press, 1985), 263–67, 269; Peter Galison, *How Experiments End* (Chicago: Univ. of Chicago Press, 1987), 277; and Schaffer, "Glass Works: Newton's Prisms and the Uses of Experiment," in Gooding, Pinch, and Schaffer, *Uses of Experiment*.
31. Wilder, *Photography and Science*, 64.
32. Wilder, *Photography and Science*, 64.
33. Fenton to South Kensington Museum (1859), quoted in Hannavy, "Roger Fenton and the British Museum," 202.
34. See March 11, 1854, April 1847–July 1855 (C6), MCM, BM; and Robert Smirke to Henry Ellis, March 10, 1854, quoted in Oct. 1853–May 1854, MCM, BM. Smirke was an architect; Ellis was principal librarian of the British Museum. See also John Hannavy, *Roger Fenton of Crimble Hall* (London: Gordon Fraser Gallery, 1975), 34.
35. See Hannavy, *Roger Fenton of Crimble Hall*, 34; and March 11, 1854, April 1847–July 1855 (C6), MCM, BM.
36. William Henry Fox Talbot, *The Pencil of Nature* (London: Longman, Brown, Green and Longmans, 1844), plate XIII.

37. Horst Bredekamp, *Galilei der Künstler: Der Mond, die Sonne, die Hand* (Berlin: Akademie Verlag, 2007), 337.
38. See Hans-Jürgen Lechtreck, "Evolution vor der Kamera: Roger Fenton und Richard Owen im British Museum 1856–1858," *Fotogeschichte* 28 (2008): 38–56.
39. Lechtreck, "Evolution vor der Kamera," 49.
40. More sets (up to five) of fifty cuneiform photographs were cut out and mounted on paper boards. The British Museum (Department of the Middle East) holds several copies of Fenton's photographs. These "leftovers" were probably the copies that were supposed to be sent to scholars. One further set is bound in a leather volume.
41. On October 10, 1854, Friedrich Rietschl applied for copies of photographs to be sent to Bonn. It was decided to send copies to him four days later. See Original Papers, June–Dec. 1854, and the note in the Committee Meeting Minutes, April 1847–July 1855 (C6), MCM, BM. Julius Oppert also ordered copies, which were sent to him. See Aug. 26, 1855, July 1855–July 1862 (C7), MCM, BM. See also "Inscriptions cunéiformes. Photographies envoyées le 27 mai 1857, par Jules Oppert," MS 5677, Institut de France.
42. May 13, 1854, April 1847–July 1855 (C6), MCM, BM.
43. April 26, 1856, July 1855–July 1862 (C7), MCM, BM.
44. Talbot to Antonio Panizzi, Dec. 7, 1856, Doc. No. 5007, Correspondence (original: TC, BL), among many others.
45. Nov. 22, 1856, July 1855–July 1862 (C7), MCM, BM.
46. See Mogens Larsen, *The Conquest of Assyria: Excavations in an Antique Land, 1840–1860* (London: Routledge, 1996), 50, 203, 208, 215–27.
47. Antonio Panizzi to Talbot, Jan. 12, 1857, Doc. No. 92, Correspondence (original: TC, BL).
48. There are numerous letters and several notebooks related to this incident. For a critical view of the historiography, see the essay by Eleanor Robson in this volume.
49. Dec. 13, 1856, July 1855–July 1862 (C7), MCM, BM.
50. Jan. 10, 1857, July 1855–July 1862 (C7), MCM, BM.
51. Talbot to Antonio Panizzi, Jan. 13, 1857, Doc. No. 7352, Correspondence (original: TC, BL).
52. See Marjorie Caygill, *The Story of the British Museum* (London: British Museum Press, 2002), 40.
53. Antonio Panizzi to Talbot, Oct. 31, 1860, Doc. No. 8233, Correspondence (original: TC, BL).
54. Talbot to Antonio Panizzi, Nov. 19, 1863, Doc. No. 8756, Correspondence (original: BM Archives, London).
55. Antonio Panizzi to Talbot, Dec. 23, 1863, Doc. No. 8776, Correspondence (original: BM Archives, London).
56. As also posed by Wilder, *Photography and Science*, 83.
57. See Forgan, "Building the Museum," 140.
58. Christopher Date and Anthony Hamber also made the museum responsible for this missed opportunity, rather than restricting their argument to photography itself. Date and Hamber, "Origins of Photography," 323.
59. See Wilder, *Photography and Science*, 84.
60. See Aileen Fyfe, "Reading Natural History at the British Museum and the Pictorial Museum," in *Science in the Marketplace: Nineteenth-Century Sites and Experiences*, ed. Aileen Fyfe and Bernhard Lightman (Chicago: Univ. of Chicago Press, 2007), 197.
61. See Caygill, *Story of the British Museum*, 40. For an overview of photography between the late 1850s and the early 1870s at the British Museum, see Hamber, *"A Higher Branch of the Art,"* 384–87. For an account of the time up to 1877, see Mirjam Brusius, "Preserving the Forgotten: William Henry Fox Talbot, Photography and the Antique" (PhD diss., Univ. of Cambridge, 2011; Akademie Verlag, forthcoming, and English edition in preparation), 162–77.
62. This photogravure consists of five lines of "typeset" cuneiform signs and appears in the article "Four New Syllabaries and a Bilingual Tablet: Edited, with notes and remarks, by H. F. Talbot, F.R.S., &c. Read 3rd November, 1874," *Transactions of the Society of Biblical Archaeology* 3 (1874): 508. (Personal correspondence between Larry J. Schaaf and John Huehnergard, with thanks to the authors.) See also the essay by Larry Schaaf in this volume.

"The Most Transitory of Things": Talbot and the Science of Instantaneous Vision

Chitra Ramalingam

THE FIRST INSTANTANEOUS PHOTOGRAPH?

AMONG the many photographic firsts attributed to William Henry Fox Talbot is the first successful photograph of a rapidly moving object, frozen in the position it held for only a single instant. Working alone in the basement laboratory of the experimental philosopher Michael Faraday at the Royal Institution in London, Talbot fastened a piece of paper printed with text onto a disk attached to a mechanism that could spin it rapidly about its center.[1] He installed a camera containing one of his new and highly sensitive "Amphitype" plates in a suitable position for capturing its image. He then set the wheel in motion, increasing the speed until the disk was visible only as a dull gray blur. Closing the shutters, he plunged the room into darkness and opened the camera before discharging a carefully positioned Leyden jar, a device for storing and discharging static electricity. The brilliant, sudden light of the spark very briefly illuminated the rotating disk. The printed letters in the resulting photograph, Talbot wrote to Faraday the following day, were "just as sharp as if the disk had been motionless."[2]

The year was 1851, and photography was far from instantaneous, although the language of capturing the ephemeral instant permeated early writings on the new medium, including Talbot's.[3] In a famous passage from his first official description of his invention in 1839, Talbot had marveled that "the most transitory of things, a shadow, the proverbial emblem of all that is fleeting and momentary, may be fettered by the spells of our 'natural magic,' and may be fixed for ever in the position which it seemed only destined for a single instant to occupy."[4] More than a decade later, when exposure times still severely limited the range of photographic subjects, Talbot's experiment was intended to redefine the very notion of photographic time and its relation to ocular vision. It is tempting to see the experiment purely within the developing discourse of photographic instantaneity: as the fulfillment of his original photographic fantasy of capturing "the most transitory of things." Although the image he produced has not survived, this event is now

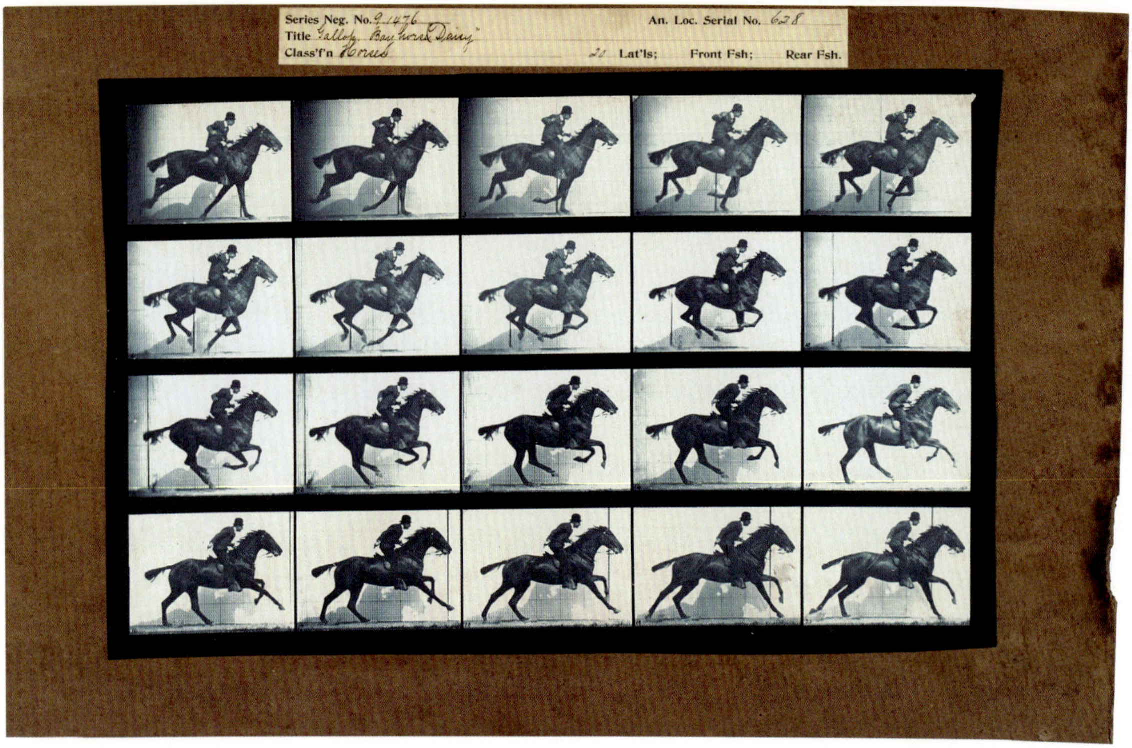

100. Eadward Muybridge, *Gallop, Bay Horse Daisy*, 1884–86, cyanotype print, mounted on card, printed ca. 1887, 12½ x 19 in. (31.8 x 48.4 cm), full-plate proof of plate 628 from the series *Animal Locomotion*. Photographic History Collection, National Museum of American History, Smithsonian Institution, Washington, DC.

remembered as the first in which a photograph rendered a crisp image of a rapidly moving object. Decades later, more famous photographic interventions into instantaneity, such as Eadweard Muybridge's running horses or Étienne-Jules Marey's birds on the wing, would become iconic representations of photography's power to go beyond the powers of the eye to bring into view individual instants that were otherwise unperceivable (fig. 100).

Talbot's experiment in 1851 has long been seen as an originating moment in this development. Yet in the historiography of photography it occupies a puzzling position, simultaneously central and yet utterly marginal to the history of the medium. On the one hand, as Helmut and Alison Gernsheim wrote, it seems to have "laid the foundation of an entire branch of scientific photography," promising to slice individual instants mechanically from the continuous flow of time.[5] On the other hand it was a historically isolated episode, remembered and referenced by none of the other pioneers of radically instantaneous photography. It seems to bear little connection to their later, more well-known and visually stunning interventions into the photography of the instantaneous.

What might we find if instead we consider Talbot's spark photography trial as part of the story of visual experimentation in the natural sciences? Mechanically revolving disks and spark devices were an important category of image machine in this period. This was particularly true at the Royal Institution, the site where Talbot chose to carry out the experiment. There Faraday and his friend and collaborator Charles Wheatstone had been using these instruments for years to investigate the optical illusion of visual persistence, which was understood to affect how the eye perceives rapid movement. These machines were incorporated into both investigative experiments and public lectures at the Royal Institution, where they were used to generate new visual experiences of movement otherwise inaccessible to the eye. This included the production of "instantaneous" views of rapidly moving objects, apparently frozen in time. But these ocular instantaneous views were only one of several new kinds of ephemeral image summoned by these scientists' manipulation of the visual experience of time. A variety of "pictures of time" generated by experimentation on human perception proliferated well before photography.[6]

Talbot's own fascination with the temporality of human perception and his pursuit of instantaneous vision began in the 1820s. This early interest was rekindled by a series of encounters and exchanges with Faraday and Wheatstone in London in 1833, shortly before Talbot began the research that would lead to his invention of photography. Eventually his spark photography experiment of 1851 brought together these two strands of interest, combining the photographic surface with elements of their visual experiments using spinning disks, rotating mirrors, and electric sparks, in order to create a permanent, photographic "instantaneous image." This essay traces Talbot's pursuit of instantaneity (both ocular and photographic) through a series of encounters both enabled and constrained by the intricate social geography of early Victorian London science. Faraday and Wheatstone developed their techniques for manipulating vision within the city's demonstration-centered culture of scientific performance, a culture of showmanship simultaneously useful and distasteful for men of Talbot's class. His encounters with their work, his translation of their public demonstration techniques into a privately made photographic trial, and the corresponding challenges he faced in gaining public attention for his experiment all indicate the social complexity of this scientific culture and its networks. They also suggest a richer understanding of what it might mean to say that experimental science was a "context" for the development of photography.

PERFORMING VISUAL PERSISTENCE AT THE ROYAL INSTITUTION

The location Talbot chose to conduct his experiment with photographic instantaneity, the laboratory of the Royal Institution, is a useful starting point.[7] His persona and photographic oeuvre were, throughout his life and after, strongly identified with Lacock and the aesthetics and ideology of landed society.[8] During his frequent stays in London, however, he moved through far more socially diverse scientific circles in which images and image machines were situated in a clamorous display culture with a very different aesthetics indeed. His experiment of 1851 was predicated on the experimental resources of the Royal Institution and its well-developed performance culture, centered on the illusion of visual persistence. But as we will see, his genteel distaste for (or lack of skill in) the showmanship that characterized visual experimentation there prevented him from fully realizing his ambitions for instantaneous photography.

In Regency and Victorian London, optical illusions and innovative image machines were incorporated first into entertaining spectacles such as magic lantern shows and phantasmagoria, and later into popular domestic toys like the stereoscope and phenakistoscope. These sites and devices for public encounters with visual deceptions were descendants of the natural magic tradition, a centuries-old mode of interacting with and usefully manipulating the natural world.[9] But the extraordinary proliferation of occasions for experimentation on human vision in popular and scientific culture in the early nineteenth century has led many scholars to identify this period with a fundamental and far-reaching transformation in the character of visual experience: a "reorganization" or "modernization" of vision.[10]

The Royal Institution was a prominent site where vision was scrutinized and manipulated, both behind the scenes in Faraday's basement laboratory and before audiences in the public lecture theater. Wheatstone and Faraday shared a long-standing fascination with the illusion of visual persistence. It had been known since antiquity that a burning coal whirled swiftly in a circle would be seen as a brilliant ring of light. Writers on vision in the 1820s and 1830s referred to this as "the duration of visual impressions" and tended to attribute it to the sluggishness of the retinal surface itself, which was understood to retain the succession of visual impressions of the glowing point in each of its positions as it raced around its circular path.[11]

Faraday and Wheatstone were the first to carry out a systematic investigation of this visual illusion. In the early 1830s they collaborated on an extended series of evening lectures at the Royal Institution that frequently featured demonstrations of visual persistence. In the city's clamorous marketplace of ideas, lecturing was a

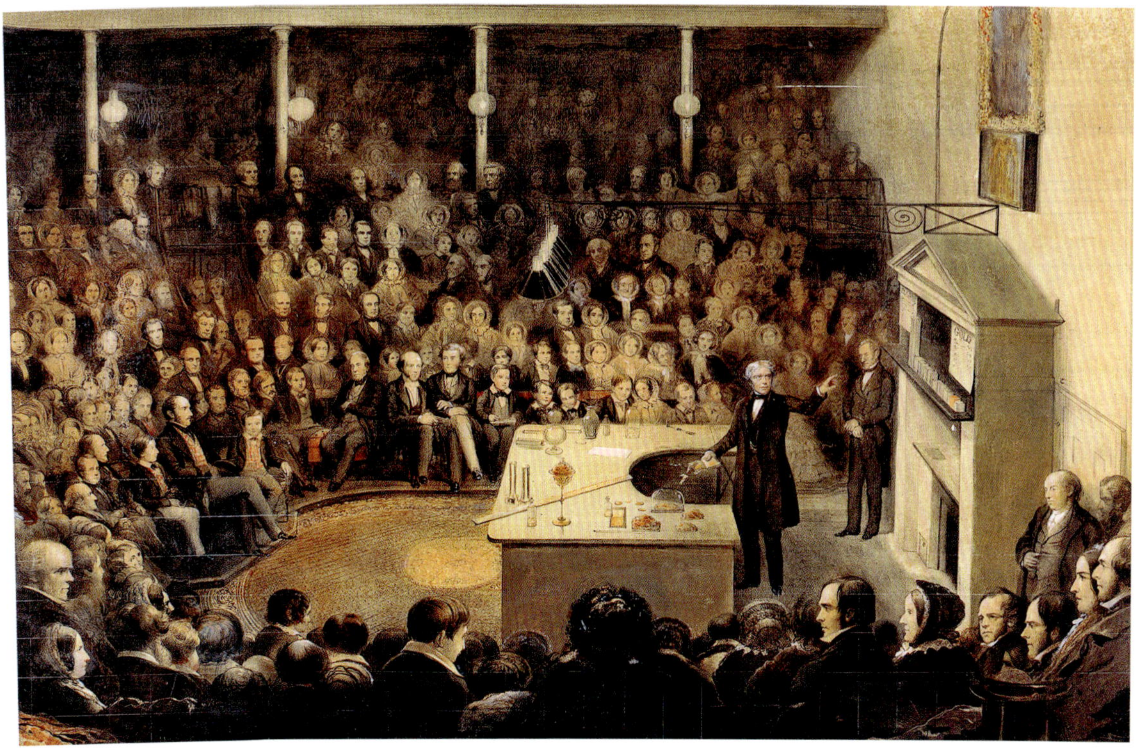

101. Alexander Blaikley, *Faraday Lecturing at the Royal Institution in 1856 to Prince Albert and the Prince of Wales*, 1856, oil over lithograph on paper, 19⅛ x 28½ in. (48.5 x 72.5 cm). Hunterian Art Gallery, University of Glasgow.

key means to promote and communicate scientific knowledge as well as to consolidate a scientific career.[12] This was certainly the case for both Faraday and Wheatstone, who had worked their way up from humble origins through lecturing.[13] Faraday had built a reputation as the city's premier experimental scientist through his popular series of "Friday Evening Discourses," in which the latest scientific discoveries were presented to a genteel audience (fig. 101). For such individuals experimental research and experimental performance were completely interdependent. Their investigation of visual persistence was motivated by the pragmatic need to devise visually diverting public performances, as well as by their growing fascination with techniques for observing extremely rapid transient events in the laboratory. Using rotating mirrors and mechanically rotating disks as optical instruments, they manipulated not only their own eyes in their private laboratories but also the eyes of their audiences. In the process they generated novel visual experiences of movement, temporality, and rest.[14]

Faraday, for example, lectured in 1831 on the optical illusions seen in spinning wheels, all of which he explained through visual persistence. He was inspired by the strange illusions he had seen in spinning carriage wheels and coordinated factory cogwheels.[15] For the performance he built machinery that could spin pairs of cardboard disks at exactly determined relative speeds (fig. 102). Manipulating the wheels' shape and their relative and absolute velocities, or visually superimposing a wheel with its own shadow, its reflection in a mirror, or its projection from a magic lantern, he could summon a bewildering variety of shadowy, illusory wheels before his audience. These apparitions might stand stock-still, whirl in unexpected directions, or sprout additional cogs or mysteriously curved spokes. Each demonstrated the eye's power of retaining and fusing successive images of position over time into a single composite image. Faraday's wheel was an important inspiration for the popular optical toy eventually known as the phenakistoscope (fig. 103). This was a spinning disk with regularly placed slits, decorated on one side with a series of sequential images. Spinning the disk before a mirror and watching its reflection through the slits, the observer saw a circle of enchanting animated figures. The succession of images seen through the slits was fused by the eye into a visual experience of movement, just as in the ghostly cogwheels in Faraday's device.[16] In the following decades he incorporated the lessons from his careful study of visual persistence into his experiments on the observation of electric discharges.[17]

Meanwhile, Wheatstone was also working on the phantom optics of spinning wheels. He designed a photometer—a device for measuring the intensity of light—based on a set of interlocking wheels that spun a reflective bead in intricate patterns, seen through visual persistence as a glowing figure in the air.[18] He then designed a circular mirror that rotated about its axis, and he began studying the distorted reflections seen when moving objects were observed by reflection in its rapidly spinning surface (fig. 104). The reflections seen in the rotating mirror, generated by visual persistence, could be used to analyze the trajectories of high-speed objects. In early 1833 Wheatstone attempted to use the mirror to "spread out" the image of an electric spark and thus analyze its temporality: how fast it moved, and in what direction. Although the experiment failed it brought Wheatstone fame for the unprecedented precision with which the whirling mirror allowed the measurement of minute periods of time.[19]

In February of that year Talbot arrived in London from Lacock, as he frequently did at this time of year, for he usually spent much of the London "season" in town. He had just been elected to Parliament as a reformer in the massive Whig victory following the passage of the Reform Act of 1832, and the House of Commons was scheduled to sit from February through August 1833. Talbot found the tumultuous politics of the day exciting and engrossing. But he also found time to

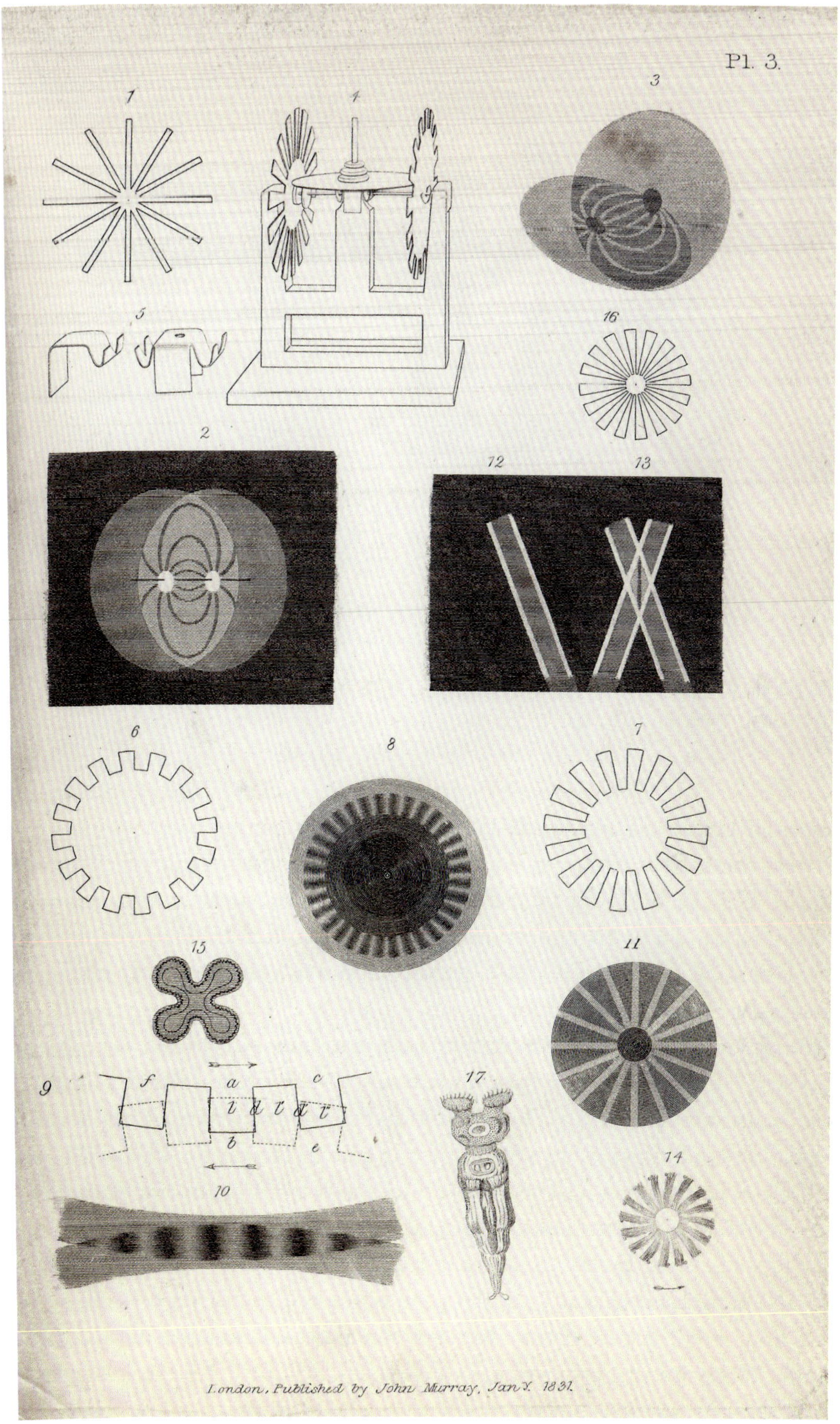

102. Michael Faraday, "On a Peculiar Class of Optical Deceptions," *Journal of the Royal Institution* 1, no. 2 (1831): plate 3. British Library. The device that came to be known as "Faraday's wheel" is at top, marked 4; around it appear depictions of the various illusions it produced.

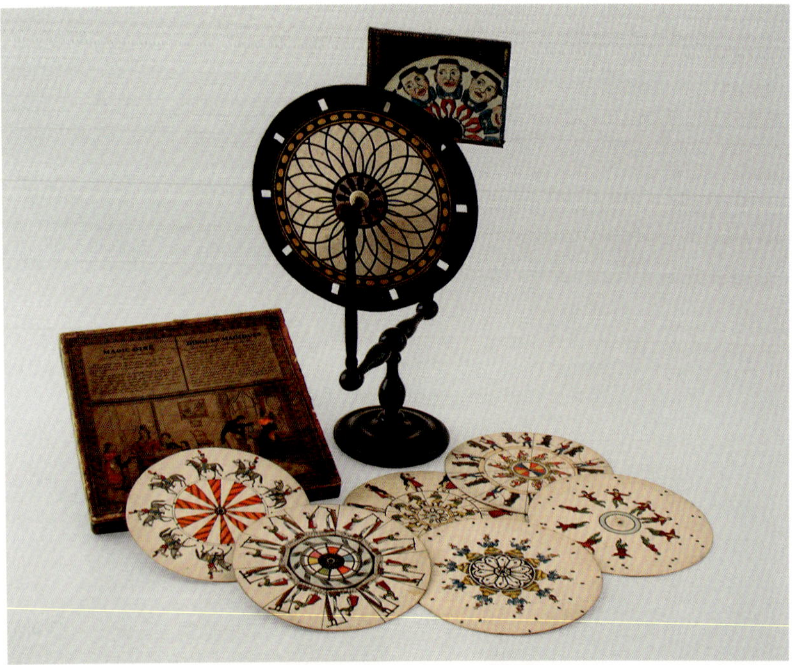

103. Phenakistoscopes, ca. 1840. Bill Douglas Centre for the History of Cinema and Popular Culture, University of Exeter.

engage in his philosophical interests by attending teas at the Athenaeum, browsing objects at the British Museum, viewing panoramas, visiting the recently opened National Gallery of Practical Science, Blending Instruction and Amusement (commonly known as the Adelaide Gallery of Practical Science, referring to its location in Adelaide Street, off the Strand), and witnessing an early performance of the oxyhydrogen microscope.[20]

On the first two Fridays in March, Talbot attended a pair of evening lectures at the Royal Institution along with more than two hundred of the city's most well-heeled consumers of science. In the first Faraday presented an early account of Wheatstone's ongoing experiments to use a rotating mirror to analyze the temporality of the electric spark. The second, delivered by Wheatstone himself, was a survey of the many instruments he had recently devised to exploit and explore the persistence of vision, including rotating mirrors and a spinning disk as a photometer.[21] Talbot took in the lecture with both excitement and dismay. He wrote to his friend John Frederick William Herschel a few days later that the lectures had "brought to my mind a train of experiments which I formerly executed, attended with interesting results, one of which astonished me much when I first beheld it. . . . I don't know why I have suffered these experiments to slumber in my portfolio for seven years."[22]

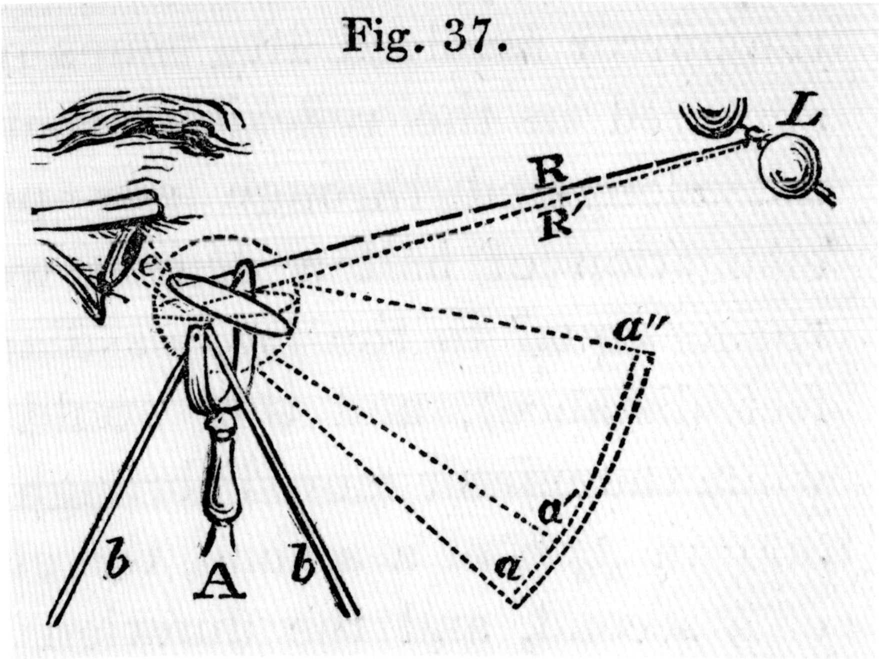

104. Charles Wheatstone's revolving mirror, 1833, in Charles Tomlinson, *Introduction to the Study of Natural Philosophy*, 3rd ed. (London: John Weale, 1853), 149. Yale University Library.

TALBOT ON VISION

Talbot's agitation stemmed from the fact that for more than a decade he had been intermittently investigating the use of rotating mirrors and rotating disks as optical devices, exploiting visual persistence. His earliest research notebooks display great familiarity with the illusions seen when the eye was presented with a rapid and regular movement. In a series of undated notebook entries in 1822, for example, he had mulled over the idea that visual persistence might bring quantitative rigor to the science of photometry. He wondered about the quantitative laws governing the difference in brightness between a luminous point and the continuous circle of light that was seen when it was whirled around, or the difference in color observed between a rapidly spinning white disk with one sector painted black and a disk with two black sectors; he experimented with how the image of the sun was converted by a rapidly rotating mirror into a circular band of light that lingered on the eye due to visual persistence; and he suggested placing a luminous point on a vibrating object to see the persisting figures it traced (fig. 105).[23] A few years later he recorded a proposal for another photometric device involving a "circle revolving *against the light*, with a sector cut away, whose angle measures the apparent light (fig. 106)."[24]

After the lectures at the Royal Institution, Talbot threw himself back into experiments on the optical effects of revolving disks and mirrors.[25] He resolved

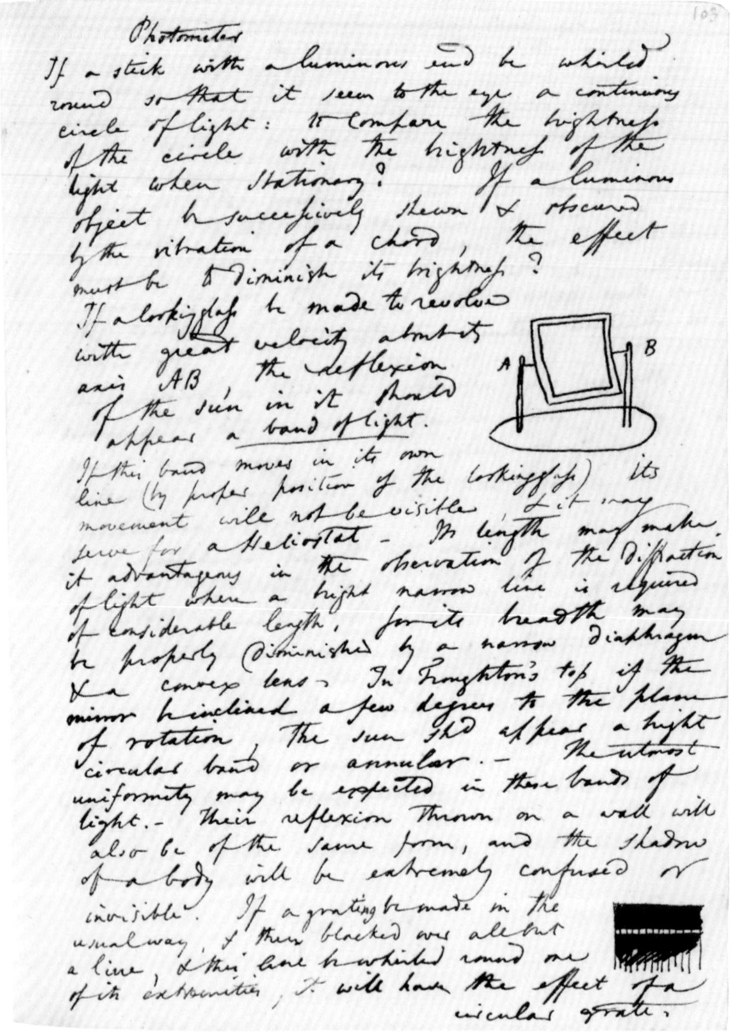

Photometer
If a stick with a luminous end be whirled round so that it seem to the eye a continuous circle of light: to compare the brightness of the circle with the brightness of the light when stationary? If a luminous object be successively shewn & obscured by the vibration of a chord, the effect must be to diminish its brightness?
If a lookingglass be made to revolve with great velocity about its axis AB the reflexion of the sun in it should appear a band of light.
A B
If this band moves in its own line (by proper position of the lookingglass) its movement will not be visible & it may serve for a Heliostat - Its length may make it advantageous in the observation of the diffraction of light when a bright narrow line is required of considerable length, for its breadth may be properly diminished by a narrow diaphragm & a convex lens - In Troughton's top if the mirror be inclined a few degrees to the plane of rotation the sun shd appear a bright circular band or annular. - The utmost uniformity may be expected in these bands of light. - Their reflexion thrown on a wall will also be of the same form, and the shadow of a body will be extremely confused or invisible. If a grating be made in the usual way, & then blacked over all but a line, & this line be whirled round one of its extremities, it will have the effect of a circular grating.

105. William Henry Fox Talbot, early notes on a revolving mirror, in Notebook B, 1822. Add. MS 88942/1/184, Talbot Collection, British Library.

immediately to make public his own private studies of manipulating the ocular experience of light, movement, and time.[26] He spent the next few months predominantly in London in a whirlwind of scientific and political activity, racing to develop and publish his own proposals for experimenting on the eye with whirling disks and mirrors, while participating in parliamentary politics before his scheduled departure for the Continent.

Within a month he had sent off to the *Philosophical Magazine* a proposed modification to Wheatstone's rotating mirror experiment on the velocity of electricity.[27] Purchasing a "revolving apparatus" with multiple paper-covered wheels from an inventor on Oxford Street, Talbot also set about investigating the queries

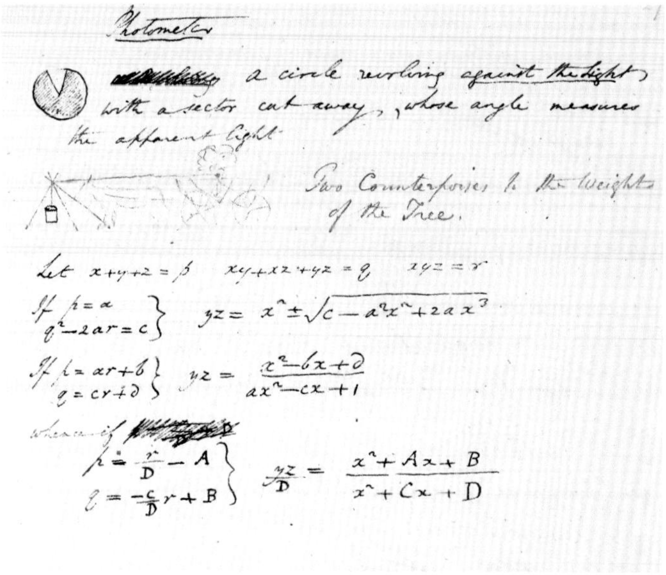

106. William Henry Fox Talbot, early proposal for a revolving disk photometer, in Notebook E, 1827–1828. Add. MS 88942/1/187, Talbot Collection, British Library. Talbot's proposal reads: "Photometer. A circle revolving *against the light*, with a sector cut away, whose angle measures the apparent light."

he had recorded years before, using the spinning wheels to manipulate light intensities and color mixtures.[28] When a disk with sectors cut out of it was rotated rapidly in front of a light source at high speeds, for example, visual persistence prevented the observer from seeing the periodic interruptions in the light beam. Instead the light appeared continuous but slightly dimmer than when viewed directly, and Talbot argued that its intensity diminished by a factor that could be calculated from the number of cut sectors. When disks with alternately colored sectors were mounted onto the rotation apparatus, the colors mixed. A spiral-shaped figure in black mounted onto a white disk and spun rapidly produced "a gray surface varying gradually and regularly, in concentric shades, from perfect blackness in the centre to perfect whiteness in the circumference" (fig. 107).[29] The distance from the center could be used as a quantitative, linear measure of the degree of color mixture at any point on the disk. By fusing visual impressions over time, spinning disks could generate objective scales for measuring color or light; thus "*time* may be employed to measure the intensity of *light*."[30] He presented a paper before the Royal Society on this method of photometry; it was eventually published in the *Philosophical Magazine* (fig. 108).[31]

These techniques, like those of Faraday and Wheatstone, depended on the eye's tendency to retain repetitive sensory stimuli over time, to combine separate "impressions" of the duration of a complex movement into composite still images. But Talbot also explored a very different way of generating static images

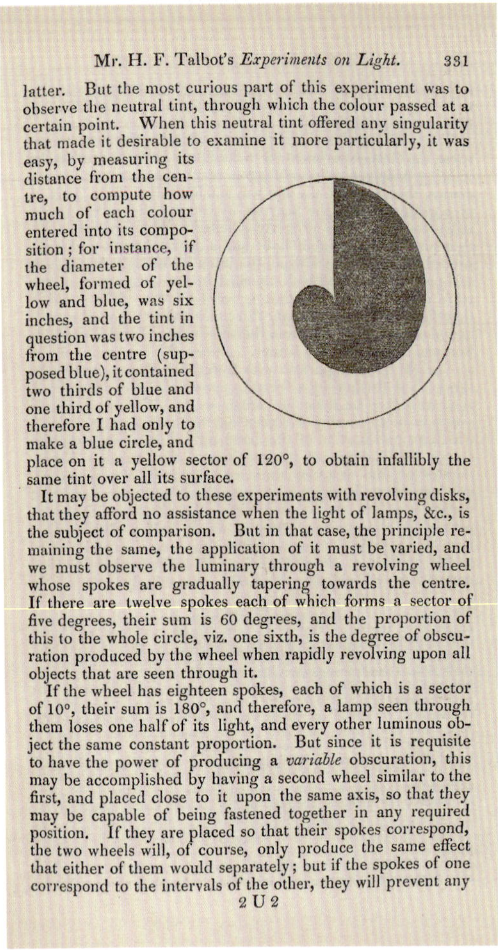

Mr. H. F. Talbot's *Experiments on Light.* 331

latter. But the most curious part of this experiment was to observe the neutral tint, through which the colour passed at a certain point. When this neutral tint offered any singularity that made it desirable to examine it more particularly, it was easy, by measuring its distance from the centre, to compute how much of each colour entered into its composition; for instance, if the diameter of the wheel, formed of yellow and blue, was six inches, and the tint in question was two inches from the centre (supposed blue), it contained two thirds of blue and one third of yellow, and therefore I had only to make a blue circle, and place on it a yellow sector of 120°, to obtain infallibly the same tint over all its surface.

It may be objected to these experiments with revolving disks, that they afford no assistance when the light of lamps, &c., is the subject of comparison. But in that case, the principle remaining the same, the application of it must be varied, and we must observe the luminary through a revolving wheel whose spokes are gradually tapering towards the centre. If there are twelve spokes each of which forms a sector of five degrees, their sum is 60 degrees, and the proportion of this to the whole circle, viz. one sixth, is the degree of obscuration produced by the wheel when rapidly revolving upon all objects that are seen through it.

If the wheel has eighteen spokes, each of which is a sector of 10°, their sum is 180°, and therefore, a lamp seen through them loses one half of its light, and every other luminous object the same constant proportion. But since it is requisite to have the power of producing a *variable* obscuration, this may be accomplished by having a second wheel similar to the first, and placed close to it upon the same axis, so that they may be capable of being fastened together in any required position. If they are placed so that their spokes correspond, the two wheels will, of course, only produce the same effect that either of them would separately; but if the spokes of one correspond to the intervals of the other, they will prevent any

2 U 2

107. William Henry Fox Talbot, "Experiments on Light," *Philosophical Magazine* 5 (1834): 331. Beinecke Rare Book and Manuscript Library, Yale University. This is Talbot's published account of his spinning disk photometer.

out of movement. In a letter to Herschel after the first of the two lectures at the Royal Institution, Talbot wrote that he soon planned to publish another long-slumbering idea of his, "a method of rendering the image of a body which is in the most rapid motion entirely fixed, so that you can see what happens to it . . . with as much ease as if it were at rest."[32] In a paper for the *Philosophical Magazine*, he described how a repetitive movement, such as a rotating object or a vibrating rod, could be seen as frozen in place in an otherwise unseeable momentary position, when viewed under a bright light source made intermittent through reflection in a rotating mirror.[33] The eye was presented with a series of individual, instantaneous images of the rod, repeating so rapidly that they were fused by visual persistence into a single, apparently permanent image.

Such a way of producing an image of stillness from a swift mechanical movement could be more widely applied, Talbot argued, whenever "it might be desirable to see what passes in the interior or immediate vicinity of a body during the time of its being subjected to a swift motion."[34] Privately he experimented over

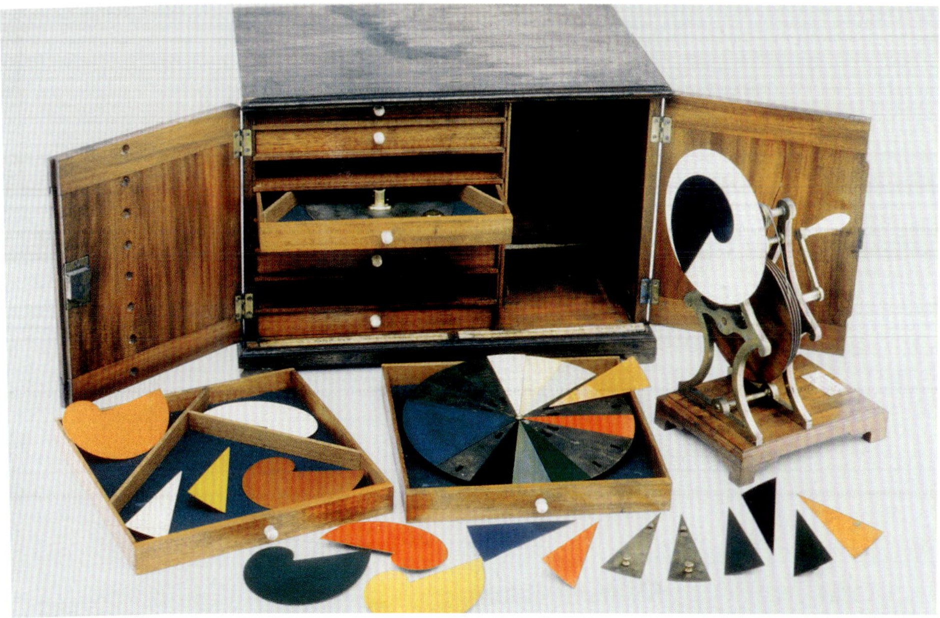

108. Optical photometer for the measurement of the intensity of light and color, to the design of William Henry Fox Talbot, constructed by W. & S. Jones, London, ca. 1830. National Museums Scotland. This is Talbot's spinning disk photometer.

the next few years with other ways to stabilize the image of a swiftly and periodically moving object through an ingenious assemblage of mirrors. A few years later Talbot experimented with another method for reducing motion to rest: observing an object in motion under the intermittent light produced by hiding a constant light source behind a spinning slotted disk (fig. 109). In his notebook he recorded a succession of static, frozen images that would emerge from the dark: "'The immovable drop' or 'selfsuspended liquid.' . . . The 'motionless tetotum.' A swinging pendulum which would have the appearance of 2, 4, or more distinct pendulums. The motionless wheel wch nevertheless will not allow anything to be passed between its spokes."[35]

WHEATSTONE: A PICTURE OF STILL LIFE

Even Talbot's list of stroboscopically illuminated revelations echoed Wheatstone's own experiments on visually stopping time. Wheatstone had had his first accidental taste of instantaneous vision during his attempt in 1833 to observe a measurable velocity for an electric spark through air. Sitting in his pitch-black laboratory watching hundreds of electric sparks pierce the gloom, he found that their frustrating instantaneity led to further revelations. The light they produced was so brief

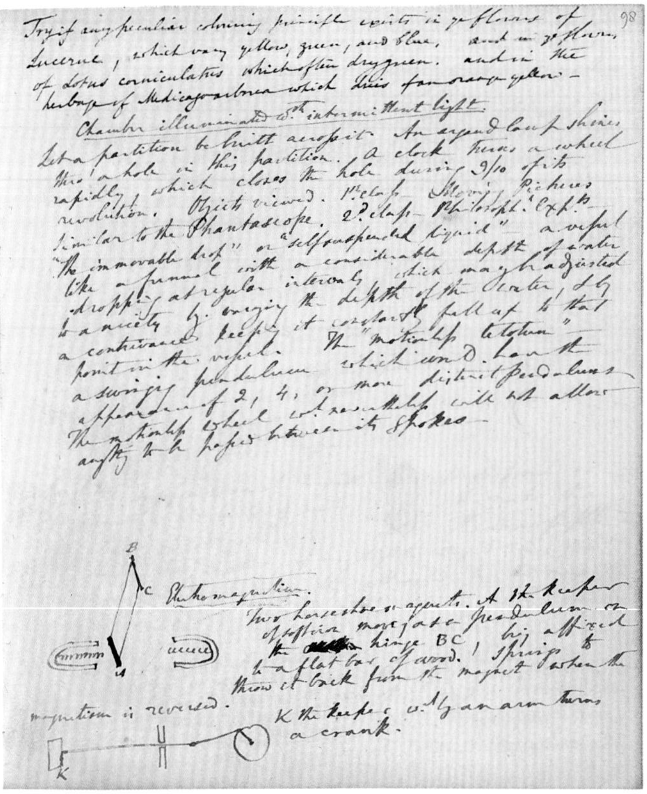

109. William Henry Fox Talbot, notes on stroboscopic vision, in Notebook M, 1834–1835. Add. MS 88942/1/193, Talbot Collection, British Library.

in duration that any moving object in the room with him flashed into view as a static image, seemingly frozen in time. Characteristically, Wheatstone immediately turned this into a demonstration experiment. He gave his first performance before a small group of prominent men of science in Cambridge:

> Mr. Wheatstone showed a very ingenious experiment. . . . [The rotating circle] appeared grayish, on account of the duration of the impression of light upon the retina. This determined, if the circle is put in motion in a perfectly dark chamber, which is suddenly illuminated by an electric spark . . . we shall see, very distinctly, the black and white sectors as if the circle were in a state of perfect rest. . . . The circle has been enlightened only during an instant infinitely short. Yet the image on the retina is so vivid and continuous as to render the image of the circle very distinct.[36]

Watching things by spark light, Wheatstone proposed in a paper on the velocity of electricity, would allow the possibility of "observing rapidly changing phenomena during a single instant of their continued action . . . when their successive positions follow each other too quickly to be seen."[37] He cited a few examples of what he had seen:

> A rapidly moving wheel, or a revolving disc on which any object is painted, seems perfectly stationary. . . . Insects on the wing appear, by the same means, fixed in the air. Vibrating strings are seen at rest in their deflected positions. A rapid succession of drops of water, appearing to the eye a continuous stream, is seen to be what it really is, not what it ordinarily appears to be.[38]

Wheatstone suggested that a spark could serve as a kind of imaging device, isolating static images of a single instant in time before the eye and producing illuminated but ephemeral revelations of things as they really were. In an earlier draft of the paper, he had connected this artificially instantaneous vision to the more familiar visual experience of lightning, observing that "a flash of lightning during a sea storm at night renders all the surrounding objects visible in a state of unnatural tranquility. . . . This transforms as it were a stormy sea with its tumultuous objects into a picture of still life."[39] He followed up on this vision of turning maritime tumult to rest with a fantasy about shattering the anonymity and isolation of rail travel by a single lightning flash of recognition:

> The rapidity with which two trains of carriages impelled by steam power, pass each other in opposite direction on a railway, renders the carriages of one train and the passengers which travel in them perfectly indistinguishable by a passenger in the other train; but should the transit happen in a dark night at the instant a flash of lightning occurs, all the objects would be seen by him perfectly still, and he might even distinguish the features of an acquaintance, should his eye chance to be directly towards him.[40]

Wheatstone's drafts on visual persistence and instantaneous vision were never published, but they were written in 1833 and 1834, long before any comparable fantasies of photographic instantaneity had taken shape.[41] Meanwhile, the vivid experiment of the stopped wheel took on a life of its own. When the experiment was done in perfect darkness, as it had been in Cambridge, the image of the frozen wheel was most vivid. But the effect was even more astonishing in a dimly lit room where the viewer could just see the blurred surface of the whirling disk. Then, when the Leyden jar discharged, the instantaneous "fixed figure" appeared to "start into existence [as if] by the touch of a magic wand."[42] He invited friends and colleagues to his rooms in King's College London to witness the marvelous fixed image of the wheel flashing over its blurred surface. Talbot himself witnessed one such demonstration in 1840.[43] The writer Mary Somerville included a description of this experiment in the revised second edition of her popular book *On the Connexion of the Physical Sciences.*[44] The book went through numerous editions and gave Wheatstone's disk canonical status. By the mid-1840s popular books and articles on natural philosophy referred regularly to the "famous experiment" of the spinning disk frozen by a flash of lightning, although Wheatstone had never published an account of it beyond the passing reference to a spinning wheel in his paper on the velocity of electricity in 1834.[45] For Wheatstone instantaneity was to be apprehended through visual demonstration, not through publication. In 1841 he turned the demonstration into a device for measuring the duration of flashes of ignited gunpowder at the Royal Artillery Institute at Woolwich.[46] Eventually he arranged for Faraday to perform this vivid experiment as part of a lecture at the Royal Institution, where he showed

> the revolution of a disc divided into colours, seen altogether while revolving, and appearing white. Thus, then, are we cheated by our senses out of the true observations. And as proof, and to remove false impressions, the revolving wheel was exhibited to the audience under temporary lights; a flash of gunpowder partially resolving the white appearance into its elementary colours, and an electric spark, by its momentary existence, shewing the colours as though the wheel were at rest.[47]

Wheatstone's disk had become an iconic embodiment of visual instantaneity, of the possibility to snatch a "true" image at rest from the flickering confusion of movement.

TALBOT, PHOTOGRAPHY, AND INSTANTANEITY

During the 1830s Faraday, Wheatstone, and Talbot exploited the duration of visual impressions in order to produce a variety of new visual experiences of movement that could reveal spatial and temporal information otherwise inaccessible to the eye. These emerged from the interplay between two things: the durational quality of the human perceptual apparatus, which was understood to aggregate, accumulate, and fuse persisting impressions; and the mechanical control of exposure (by the use of spark illumination, sectored or slotted disks, or other shutter-like devices), which tended to slice, isolate, and separate.[48] This rhetorical, technical, and aesthetic interplay between durational and instantaneous vision emerged for Faraday and Wheatstone within a performance-centered culture of scientific display: ephemeral visual experiences were fixed through being witnessed and affirmed by a roomful of attentive eyes.

For Talbot, as for many early photographers, this tension between duration and instantaneity instead became a defining problem of the fixed images produced by the new medium. In his earliest attempts to capture images on light-sensitive paper, which began in early 1834, the technique required hours of exposure to full summer sunlight. Only the most immobile subjects were rendered in his early experimental prints. Yet when Talbot first made the process public, he portrayed his new method for making pictures as a means for capturing the ephemeral, for fixing "the most transitory of things."[49]

The notion of ensnaring the single, fleeting instant would become a common trope of early writings on photography, but it always carried with it the signs of its own elusiveness.[50] The many temporal references in this passage identified the profound challenges to overcome: the temporal limitations on possible photographic subjects, the time required to form the image, and the temporal disintegration of the image over time. These were precisely the experimental problems with which Talbot grappled between 1834 and 1836, his most fervent years of photochemical research.[51]

Those years also overlapped with Talbot's exploration of the eye's own power to produce instantaneous and durational images. His notes on stroboscopic vision, for example, are dated March 7, 1835, right in the middle of his most productive period of photographic research (see fig. 109).[52] It seems possible—even likely—that he would have imagined capturing these frozen images of movement in photographs, but no documentary record supporting such speculation exists. It may have been the frustrating weakness of existing photochemistry that prevented him from planning such an experiment for more than a decade after photography was made public.

On June 15, 1851, Talbot returned to Faraday's laboratory at the Royal Institution, the site of so many visual manipulations over the previous decades, to carry out the experiment with which this essay began. It should now be clear that the photograph of the spinning disk was directly modeled on the now classic demonstration experiment known as "Wheatstone's disk." The rotation device Talbot used to spin the paper disk in Faraday's laboratory was most likely the same device that Faraday had used to demonstrate Wheatstone's disk before an audience in 1846. Rather than marking some new, uniquely photographic experience of time, Talbot's experiment directly mimicked the formation of images of movement and instantaneity in the eye. It overcame the sluggishness of light-sensitive chemical emulsions by the same visual-instrumental system of spinning disks and sudden illumination that Faraday, Wheatstone, and Talbot himself had previously used to overcome and exploit the sluggishness of human visual perception.

Talbot's success convinced him that it was time to redefine the very notion of photographic instantaneity. Previous photographs might be called instantaneous, he wrote to the French scientist Jean-Baptiste Biot the next day, "but that has only been a figure of speech. Strictly speaking they are far from being so, for the objects which move with great speed have escaped photographic representation until the present time."[53] Now, he declared triumphantly in a letter to the *Athenaeum* a few weeks later, there would be a new kind of photography, one that for the first time could present "distinct images of bodies moving swiftly across the field of view. . . . We shall be able to arrest and embody the most fugitive phenomena which present themselves to the eye."[54] The *Literary Gazette* rhapsodized that now "nothing will be more easy than to take the most agile ballet dancer during her rapid movements, or to catch the image of the bird of swiftest flight during its passage instantaneously over the field of the lens of the camera obscura. . . . Here we have the means of fixing a shadow which passes as swiftly by as the flash of lightning."[55] The echo of Talbot's original reference to "fixing a shadow" was deliberate, for he was eager to represent his latest experiment as the fulfillment of the medium's original promise to capture "the most transitory of things." Although previously moving objects had eluded the photographic surface, now it would be "in our power to obtain the pictures of all moving objects, no matter in how rapid motion they may be, provided we have the means of sufficiently illuminating them with a sudden electric flash."[56]

Unfortunately, this last requirement—of a bright but sudden electric spark—was a tall order. Although Talbot was eager to conduct the experiment again before witnesses, the singular success of the "delicate experiment of the rotating disk" proved too difficult to repeat reliably.[57] In order to perfect this temperamental procedure into a generalizable technique for instantaneously capturing movement, he needed a more reliable means of obtaining single sparks bright enough and (most importantly) actinic enough to register vividly on his photographic plate, which was sensitive to a much narrower range of light than the eye was. This was beyond his expertise. He asked for help from Faraday, who offered him free use of the laboratory and all his equipment at the Royal Institution but could not spare the time to work on it himself.[58]

Despite an initial flurry of attention in the press, Talbot failed to inspire much lasting interest in his technique. Over the next few years he repeatedly tried to turn his singular and private experiment into a striking public demonstration.[59] To William Robert Grove, an expert electrical experimenter and barrister who also happened to be serving as Talbot's lawyer at the time, he fantasized about what kinds of images he would use to demonstrate his new technique: "A child spinning a tetotum would make a pretty little picture, tho' I am afraid it would be beyond present means of accomplishment — The tetotum alone taken spinning, with the figures legible would be of itself sufficiently curious."[60] With no reliable strategy for repeating the experiment on such a scale, Talbot abandoned his ambition to photograph rapidly moving objects. He soon returned to using the phrase "instantaneous photography" to refer to images requiring less than three seconds of exposure.[61]

None of the future pioneers of radical photographic instantaneity appear to have cited Talbot's experiment in 1851 as a precursor or inspiration. His largely forgotten experiment was instead rediscovered by twentieth-century art historians seeking to pinpoint the origins of photographic instantaneity.[62] Understanding Talbot's strange, lost photograph of the spinning wheel as the "first spark photograph," first in a long line of instantaneous revelations, obscures the peculiar combination of interests through which he approached exploring the representational possibilities of the photographic surface.

It also obscures Talbot's failure. His experiment was modeled on Wheatstone's disk, but this was a demonstration experiment, dependent for its vivid persuasiveness on the presence of an audience that witnessed the frozen disk magically flashing into view, experiencing its brilliant fixity against a backdrop of blurred movement. Translating this into a photographic procedure, Talbot found quickly that published reports of a crisp photograph of a spinning disk apparently standing still could not themselves engender public recognition that photography's representational possibilities had been transformed. This is clear from his concern with generating brighter sparks to make a more vivid image, and the proposed images he promised soon to make that were nonetheless far beyond his technical feasibility, like a ballet dancer fixed in mid-leap. His unfulfilled plans for a public lecture are telling, for they suggest that Talbot recognized quite clearly what had been lost in the disk experiment's transformation from performance to privately made fixed image.[63] Talbot was not a natural showman by temperament or by breeding, and his usual practice to persuade people of the value of a new photographic process was to send around and exhibit specimens of his work.[64] His failure to pursue this strategy, along with the fact that the image of the printed text on the disk has not survived, suggests that it may not have been as visually satisfactory or as sharply distinct as he claimed.

Lacking the immediacy and the drama of Wheatstone's disk, Talbot's photograph of the stilled disk could not serve as a persuasive "golden image," representing at once a specific visual revelation of such clarity that it commands acceptance and the more general promise of a new procedure for making things visible.[65] Unlike later famous radical interventions into photographic instantaneity, it would not have shocked viewers with unimagined, awkwardly arrested poses. In fact, as he had proudly proclaimed, it looked *exactly* as it might have had the disk been perfectly motionless and the photograph taken in full daylight.[66] In translating a vivid visual experience from the showmanship of a lecture demonstration experiment into a mundane fixed image, Talbot's attempt to redefine photographic instantaneity had failed, not just technically but socially. He was a regular consumer of science as public spectacle, but he was not himself comfortable with or skilled at practicing such showmanship himself. His early experiments with visual persistence had been private ones, recorded only in notebooks and letters, and he responded to seeing Faraday's and Wheatstone's spectacular demonstrations on vision not with a striking performance of his own but with a flurry of publication and further private experimentation. But visual persistence was the stuff of public science in London, and Talbot misjudged how to engage appropriately with the demonstration-based culture of experimentation on visual persistence in London. Wheatstone's disk was remembered for decades, until the more spectacular revelations of photographic instantaneity eclipsed it in the 1880s; Talbot's photograph of Wheatstone's disk was forgotten almost immediately.

I would like to thank Mirjam Brusius, Katrina Dean, Peter Galison, Sharrona Pearl, Simon Schaffer, Jennifer Tucker, and all the participants in the Talbot conference in 2010 for their incisive feedback on my various attempts over the last five years to tell the story presented in this essay. This research was funded by the National Science Foundation, the Mellon Foundation/American Council for Learned Societies, and the Paul Mellon Centre for Studies in British Art.

1. In twentieth-century retellings of this event the printed paper is sometimes described as a page from that day's *Times*, but this appealing detail is apocryphal. Many accounts also erroneously state that Talbot performed the experiment before an audience, but he conducted it in the private laboratory of the Royal Institution, not its public lecture hall. He represented himself as working alone, though he undoubtedly had one of Faraday's laboratory assistants on hand.
2. William Henry Fox Talbot to Michael Faraday, June 15, 1851, Doc. No. 6429, The Correspondence of William Henry Fox Talbot, http://foxtalbot.dmu.ac.uk (accessed Jan. 31, 2012), hereafter "Correspondence" (original: Royal Institution, London). He also made the same comparison to a motionless disk, emphasizing the text's legibility, in Talbot to Jean-Baptiste Biot, June 16, 1851, Doc. No. 6432, Correspondence (original: Talbot Collection, hereafter "TC," British Library, hereafter "BL"); and Talbot, "Note on Instantaneous Photographic Images," *Abstracts of the Papers Communicated to the Royal Society of London* 6 (June 1851): 82.
3. Phillip Prodger, *Time Stands Still: Muybridge and the Instantaneous Photography Movement* (New York: Oxford Univ. Press, 2003), 24–111; see also André Gunthert, "La conquête de l'instantané: Archéologie de l'imaginaire photographique en France, 1841–1895" (PhD diss., École des Hautes Études en Sciences Sociales, Paris, 1999).
4. William Henry Fox Talbot, "Some Account of the Art of Photogenic Drawing; or, The Process by Which Natural Objects May Be Made to Delineate Themselves without the Aid of the Artist's Pencil," in *Photography in Print*, ed. Vicki Goldberg (Albuquerque: Univ. of New Mexico Press, 1981), 41.
5. Helmut Gernsheim and Alison Gernsheim, *The History of Photography: From the Camera Obscura to the Beginning of the Modern Era* (New York: McGraw-Hill, 1969), 444–45. Other chronologies of the photography of movement that incorporate Talbot's experiment in 1851 as a starting point include Marta Braun, "The Expanded Present: The Photography of Movement," in *Beauty of Another Order: Photography in Science*, ed. Ann Thomas (New Haven: Yale Univ. Press, 1997), 151; Beaumont Newhall, *The History of Photog-*

raphy: From 1839 to the Present, 5th ed. (New York: Museum of Modern Art, 1999); and Peter Geimer, "Picturing the Black Box: On Blanks in Nineteenth Century Paintings and Photographs," *Science in Context* 17, no. 4 (2004): 467–501.

6. This argument is more fully developed in Chitra Ramalingam, "Fixing Transience: Photography and Other Images of Time in 1830s London," in *Time and Photography*, ed. Jan Baetens, Alexander Streitberger, and Hilde Van Gelder (Louvain: Leuven Univ. Press, 2010), 3–26; see also Ramalingam, "A Science of Appearances: Vision, Visualization, and Experimental Physics in Victorian England" (PhD diss., Harvard Univ., 2009). Contrast this to Marta Braun's suggestion that "photography is the first way we have had of picturing time. It is the first medium to provide a picture of the exact temporal interval in which it was made." Marta Braun, "Time and Photography," in *Tempus Fugit: Time Flies*, ed. Jan Schall (Kansas City, MO: Nelson-Atkins Museum of Art, 2000), 132. This common claim appears to owe its origins to attempts by critics and curators in the 1960s and 1970s to extend modernist formalism to photography by identifying the intrinsic capacities of the medium. See, for example, John Szarkowski, *The Photographer's Eye* (New York: Museum of Modern Art, 1966).

7. For more on Talbot in London's scientific scene, particularly the Royal Institution, see the introduction to this volume.

8. Joel Snyder, "Enabling Confusion," *History of Photography* 26, no. 2 (2002): 154–60; see also Carol Armstrong, *Scenes in a Library: Reading the Photograph in the Book, 1843–1875* (Cambridge, MA: MIT Press, 1998). A notable exception is Steve Edwards's recent work, which argues that Talbot's work must also be read in relation to the emerging industrial order and a dialogue about labor and skill in which Talbot and many of his gentlemanly friends were enthusiastic participants. Steve Edwards, *The Making of English Photography: Allegories* (University Park, PA: Pennsylvania State Univ. Press, 2006), chap. 1. Simon Schaffer's illuminating study of the country house of the Victorian industrialist in a slightly later period avoids this apparent dichotomy by showing how certain bucolic country estates were rendered important sites for scientific, industrial, and commercial innovation. Simon Schaffer, "Physics Laboratories and the Victorian Country House," in *Making Space for Science: Territorial Themes in the Shaping of Knowledge*, ed. Crosbie Smith and Jon Agar (Basingstoke, UK: Macmillan, 1998), 149–80. For more on Talbot and on Lacock as both country house and scientific site, see the essay by Katrina Dean in this volume.

9. On natural magic and demonstration, see Thomas Hankins and Robert Silverman, *Instruments and the Imagination* (Princeton, NJ: Princeton Univ. Press, 1995); and Iwan Rhys Morus, "'More the Aspect of Magic Than Anything Natural': The Philosophy of Demonstration," in *Science in the Marketplace: Nineteenth-Century Sites and Experiences*, ed. Bernard Lightman and Aileen Fyfe (Chicago: Univ. of Chicago Press, 2007), 336–70. See also Simon Schaffer, "Natural Philosophy and Public Spectacle in the Eighteenth Century," *History of Science* 21 (1983): 1–43.

10. The most influential version of this thesis is Jonathan Crary, *Techniques of the Observer: On Vision and Modernity in the 19th Century* (Cambridge, MA: MIT Press, 1992).

11. For a sketch of the history of this illusion, see Nicholas Wade, *Perception and Illusion: Historical Perspectives* (New York: Springer, 2005), 112–16.

12. Bernard Lightman, "Lecturing in the Spatial Economy of Science," in Lightman and Fyfe, *Science in the Marketplace*, 97–132; and Morus, "More the Aspect of Magic." See also J. N. Hays, "The London Lecturing Empire, 1800–50," in *Metropolis and Province: Science in British Culture, 1780–1850*, ed. Ian Inkster and Jack Morrell (London: Hutchinson, 1983), 91–119.

13. David Gooding, "'In Nature's School': Faraday as an Experimentalist," in *Faraday Rediscovered: Essays on the Life and Work of Michael Faraday, 1791–1867*, ed. David Gooding and Frank A. J. L. James (New York: Stockton, 1985), 104–35; and Iwan Rhys Morus, *Frankenstein's Children: Electricity, Exhibition, and Experiment in Early-Nineteenth-Century London* (Princeton, NJ: Princeton Univ. Press, 1998), 13–42.

14. Ramalingam, "Science of Appearances," chaps. 1–2.

15. Michael Faraday, "On a Peculiar Class of Optical Deceptions," *Journal of the Royal Institution* 1, no. 2 (1831): 206. Jonathan Crary has suggested a connection between the increased awareness of optical illusions and

the prevalence of new forms of machine movement in the early nineteenth century, though his interpretation of Faraday on optical deceptions differs from mine. Crary, *Techniques of the Observer*, 111–12. The new visual and bodily experiences generated by early industrial transportation systems are discussed in Wolfgang Schivelbusch, *The Railway Journey: The Industrialization of Time and Space in the 19th Century* (Berkeley: Univ. of California Press, 1986).

16. The phenakistoscope's inventor, the Belgian physicist Joseph Plateau, wrote to Faraday to acknowledge his debt to him. Such devices are most commonly discussed as part of the "prehistory" of the cinema, as in Laurent Mannoni, *The Great Art of Light and Shadow: Archaeology of the Cinema*, trans. Richard Crangle, Exeter Studies in Film History (Exeter: Univ. of Exeter Press, 2000); see also Michael Chanan, *The Dream That Kicks: The Prehistory and Early Years of Cinema in Britain*, 2nd ed. (London: Routledge, 1996). The phenakistoscope has become strongly associated with the notion of a peculiarly "modern" observing subject, whose visual experience was firmly rooted in the physiology of the eye and subject to mechanical deception, embodying the modernist "double effect" of being simultaneously a subject and object of vision. See Crary, *Techniques of the Observer*; and Rosalind Krauss, "The Im/pulse to See," in *Vision and Visuality*, ed. Hal Foster (Seattle: Bay Press, 1988), 51–75.
17. Ramalingam, "Science of Appearances," chap. 2.
18. "Mr. Wheatstone on the Duration of Luminous Impressions on the Organ of Vision," *Athenaeum*, no. 281 (March 16, 1833): 170–71. There are more notes on the photometer in Charles Wheatstone, "New Optico-Electrical Experiments" (unpublished draft manuscript, ca. 1834), folder 4/1, Papers of Charles Wheatstone, Archives of King's College London. A smaller version was manufactured and sold in London. See Watkins & Hill, *Descriptive Catalogue of Optical Instruments and Apparatus Constructed and Sold by Watkins and Hill, 5 Charing Cross, London* (London: Taylor and Francis, 1850), 15–16. Another, similar photometer built to Wheatstone's design is held by the Science Museum, London: William Ladd, Wheatstone-type photometer (London, ca. 1870), Object No. 1872.38, Science Museum, London.
19. Charles Wheatstone, "An Account of Some Experiments to Measure the Velocity of Electricity and the Duration of Electric Light," *Philosophical Transactions of the Royal Society of London* 124 (1834): 583–91.
20. H. J. P Arnold, *William Henry Fox Talbot: Pioneer of Photography and Man of Science* (London: Hutchinson Benham, 1977), 57–65. Detailed lists of Talbot's activities in London during this period are to be found in Talbot, diary, Jan. 1, 1833–April 30, 1833, acc. no. 24558, Add. MS 88942, TC, BL (partially catalogued). On the Adelaide Gallery and the oxyhydrogen microscope, a device for projecting views through a microscope before large audiences, see Morus, *Frankenstein's Children*; Iwan Rhys Morus, "Seeing and Believing Science," *Isis* 97 (2006): 101–10; and Morus, "More the Aspect of Magic."
21. Talbot's attendance at these two lectures are recorded in his diary as "1 March — R. Institution, Faraday on ye Electric Spark, revolving mirrors &c." and "8 March — R. Institution. Wheatstone on Light." Talbot, diary, Jan. 1, 1833–April 30, 1833, acc. no. 24558, Add. MS 88942, TC, BL (partially catalogued). The actual titles of the lectures were "On the Velocity and Nature of the Electric Discharge" and "On the Duration of Luminous Impressions on the Organ of Vision." For the content of these two lectures, see Ramalingam, "Science of Appearances," 88–90, 93–107.
22. Talbot to John Frederick William Herschel, March 4, 1833, Doc. No. 2625, Correspondence (original: Royal Society, London).
23. Notebook B, 1822, Add. MS 88942/1/184, TC, BL.
24. Notebook E, 1827–1828, Add. MS 88942/1/187, TC, BL. Emphasis in original.
25. Talbot's research notebook for the period features numerous notes written in March and April 1833 about visual persistence, photometry, revolving wheels and disks and mirrors, and electric sparks. Notebook J, 1832–1833, Add. MS 88942/1/191, TC, BL.
26. Talbot to John Frederick William Herschel, March 4, 1833, Doc. No. 2625, Correspondence (original: Royal Society, London).
27. This appeared, somewhat improperly, before Wheatstone had even written up his own results. The resulting exchange underscores the stark difference in status between the two men, as well as the ambiguities and tensions in this period between public performance and publication as acts for securing scientific priority.

William Henry Fox Talbot, "Proposed Philosophical Experiments," *London and Edinburgh Philosophical Magazine and Journal of Science* (hereafter *Philosophical Magazine*) 3 (1833): 81–82; Charles Wheatstone, "Remarks on One of Mr. Talbot's Proposed Philosophical Experiments," *Philosophical Magazine* 3 (1833): 204–5; and Talbot, "Facts Relating to Optical Science, No. 1," *Philosophical Magazine* 4 (1834): 114. On shifting norms for establishing and adjudicating discovery through print journals at about this time, see Alex Csiszar, "Broken Pieces of Fact: The Scientific Periodical and the Politics of Search in Nineteenth-Century France and Britain" (PhD diss., Harvard Univ., 2010), chap. 2.

28. Talbot obtained this apparatus from John Lane Higgins, an engineer and inventor who held a number of patents related to steamships and carriage design. Talbot's visit is recorded as "Mr. Higgins exp^{ts} on rotation, 370 Oxf^{d} S^{t}" in entry for March 15, 1833, Talbot, diary, Jan. 1, 1833–April 30, 1833, acc. no. 24558, Add. MS 88942, TC, BL (partially catalogued). A letter from Higgins confirms the delivery of the machine to Talbot the next day: John Lane Higgins to Talbot, March 16, 1833, Doc. No. 2644, Correspondence (original: TC, BL). There are copious notes on Talbot's investigation of spinning disk photometry with this apparatus in March 1833 in Notebook J, 1832–1833, Add. MS 88942/1/191, TC, BL.
29. William Henry Fox Talbot, "Experiments on Light," *Philosophical Magazine* 5 (1834): 330.
30. Talbot, "Experiments on Light," 329.
31. The photometer in figure 11.9 was sold soon afterward by a London instrument maker, but Talbot may never have had many built himself. See A. D. C. Simpson, "Talbot's Photometer, or Developments before Photography," *Studies in Photography* (1996): 8–10.
32. Talbot to John Frederick William Herschel, March 4, 1833, Doc. No. 2625, Correspondence (original: Royal Society, London).
33. This came under the heading "A Body in Rapid Motion, Yet Apparently at Rest" in Talbot, "Facts Relating to Optical Science," 113–14.
34. Talbot, "Facts Relating to Optical Science," 113.
35. Notebook M, 1834–1835, Add. MS 88942/1/193, TC, BL. A teetotum was a four-sided spinning top on a spindle with letters or figures inscribed on each sides, commonly used in children's games of chance instead of dice. It is not clear from this notebook entry whether Talbot actually carried out this experiment and saw these images himself or merely proposed it and imagined them.
36. Lambert Adolphe Jacques Quetelet, "Notes Extracted from a Tour in England, during the Months of June and July, 1833, by Mr. Quetelet, of Brussels," *American Journal of Science and Arts* 28, no. 1 (1835): 60–61.
37. Wheatstone, "Velocity of Electricity," 591.
38. Wheatstone, "Velocity of Electricity," 591. Talbot undoubtedly read this paper on his return to England, given his public correspondence with Wheatstone concerning the electrical experiments described in it; see note 27 above.
39. Wheatstone, "New Optico-Electrical Experiments."
40. Wheatstone, "New Optico-Electrical Experiments."
41. The most famous of these is in Oliver Wendell Holmes, "The Stereoscope and the Stereograph," *Atlantic Monthly* 3 (1859): 748. Revealingly, Holmes also used the example of Wheatstone's wheel, though not by name.
42. Wheatstone, "New Optico-Electrical Experiments."
43. An undated memoranda entry (between May and Sept. 4, 1840) states that Talbot saw "Wheatstone's disk." Notebook "Memoranda," 1838–1840, Add. MS 88942/1/277, TC, BL.
44. Mary Somerville, *On the Connexion of the Physical Sciences*, 2nd ed. (London: John Murray, 1835), 313.
45. Most of the references to Wheatstone's disk experiment in the 1830s and 1840s are clearly based on Somerville's account. Some examples include "Optical Illusions II," *Saturday Magazine*, June 12, 1841, 228–29; William Snow Harris, *Rudimentary Electricity*, 3rd ed. (London: John Weale, 1853), 166–67; and Charles Tomlinson, *Introduction to the Study of Natural Philosophy: For the Use of Beginners*, 6th ed. (London: John Weale, 1859), 148–51.
46. Charles Wheatstone, "Note on the Electro-magnetic Chronoscope," in *Brewster and Wheatstone on Vision*, ed. Nicholas J. Wade (London: Academic Press, 1983), 213–15.

47. "Royal Institution," *Literary Gazette; and Journal of the Belles Lettres, Arts, Sciences, &c.* (hereafter *Literary Gazette*), April 11, 1846, 340–41.
48. Ramalingam, "Fixing Transience," 15–16.
49. Talbot, "Art of Photogenic Drawing," 41.
50. Prodger, *Time Stands Still*, 24–111; and Gunthert, "La conquête de l'instantané."
51. See the valuable account of Talbot's early photochemical and photographic research in Larry J. Schaaf, *Out of the Shadows: Herschel, Talbot & the Invention of Photography* (New Haven and London: Yale Univ. Press, 1992), 39–44.
52. A few pages before in the same notebook, dated one week earlier, is the first record of Talbot's original idea of using a negative image to make multiple positive prints. See Notebook M, 1834–1835, Add. MS 88942/1/193, TC, BL.
53. Talbot to Jean-Baptiste Biot, June 16, 1851, Doc. No. 6432, Correspondence (original: TC, BL). Since 1839 Biot had been Talbot's primary advocate in the Académie des Sciences in Paris, and he served as an intermediary whenever Talbot wanted to share photographic discoveries with the French savants. See Theresa Levitt, "Biot's Paper and Arago's Plates: Photographic Practice and the Transparency of Representation," *Isis* 94 (2003): 463–66.
54. William Henry Fox Talbot, "Important Experiment at the Royal Institution," *Athenaeum*, no. 1235 (June 28, 1851): 688.
55. "Important Discovery in Photography," *Literary Gazette*, June 28, 1851, 443. It is possible that this extremely complimentary piece was written either by Talbot himself or by William Jerdan, his friend and former editor of the *Literary Gazette*. For more on Talbot's relationship with Jerdan, see the essay by Herta Wolf in this volume.
56. William Henry Fox Talbot, "On the Production of Instantaneous Photographic Images," *Athenaeum*, no. 1258 (Dec. 6, 1851): 1286–87.
57. Talbot to Jean-Baptiste Biot, Nov. 24, 1851, Doc. No. 6493, Correspondence (original: published).
58. Talbot to Michael Faraday, June 15, 1851, Doc. No. 6429, Correspondence (original: Royal Institution, London); and Faraday to Talbot, June 16, 1851, Doc. No. 6431, Correspondence (original: TC, BL).
59. He proposed a performance of instantaneous photography at the Royal Polytechnic Institution, but this lecture never took place. Talbot to William Parsons, Feb. 5, 1854, Doc. No. 6914, Correspondence (original: Birr Scientific and Heritage Foundation).
60. Talbot to William Robert Grove, Dec. 1854, Doc. No. 89, Correspondence (original: Royal Society, London). Like the printed text in the 1851 experiment, this proposal highlights the importance of textual legibility as Talbot's model for photographic signification. His fascination with linguistic signs as a model for photographic meaning has been noted in Robin Kelsey, "Photography, Chance, and *The Pencil of Nature*," in *The Meaning of Photography*, ed. Robin Kelsey and Blake Stimson (Williamstown, MA: Sterling and Francine Clark Art Institute, 2008), 24–25; and Armstrong, *Scenes in a Library*, 160 n.19. Talbot's concern with the legibility of inscriptions as an important criterion for evaluating the success of photographic images also comes up in essays by Larry J. Schaaf, Eleanor Robson, and Mirjam Brusius in this volume.
61. Talbot to Henry Petty Fitzmaurice, May 31, 1852, Doc. No. 6628, Correspondence (original: private collection).
62. See note 5 above.
63. Talbot to William Parsons, Feb. 5, 1854, Doc. No. 6914, Correspondence (original: Birr Scientific and Heritage Foundation).
64. In January 1839, for example, when news of Louis-Jacques-Mandé Daguerre's work reached him, Talbot's first public move was to arrange an exhibition of photogenic drawings in the Royal Institution Library for viewing after a Friday evening meeting. The vast number of images he sent to friends, journals, institutions, and rivals is recorded in detail in Notebook "Memoranda," 1838–1840, Add. MS 88942/1/272, TC, BL.
65. The term "golden image" is based on Peter Galison's notion of the "golden event" as captured by image devices in twentieth-century particle physics. Peter Galison, *Image and Logic: A Material Culture of Microphysics* (Chicago: Univ. of Chicago Press, 1997), 22–23.
66. See note 2 above.

Commentary

Simon Schaffer

How hard it seems to put William Henry Fox Talbot in his proper place. Early photographs stripped the world of anything that moved too rapidly for the slow exposure times then dictated by the limits of photochemical processing. Just so, Talbot's brisk dynamics have posed a challenge to the ponderous divisions of the fields of learning and technique that historians customarily employ. By the age of eight, so he calculated, he had already journeyed more than three thousand miles; Larry Schaaf has confirmed that Talbot was "born with wanderlust and cultivated into a polymath."[1] This volume argues for the significance for his career of the fluid social geography of the sciences in earlier Victorian Britain, a period when, as the editors rightly insist, most of the terms describing the work of invention, discovery, experiment, and property were in dispute.

At the time some shrewd observers somehow knew exactly where to place this well-traveled, genteel, but cash-strapped Whig savant on the map of public knowledge. In September 1839 *Bentley's Miscellany*, a widely read literary journal that Charles Dickens had just left as editor and star contributor, carried a skit on the principal scientific institutions of the metropolis. This was "the intricate social geography of early Victorian London science" described in the essay by Chitra Ramalingam in this volume, "the city's clamorous marketplace of ideas." The journal's satire reports that an eminent (if imaginary) German scholar, Baron von Dullbrainz, has been welcomed at the Geological Society, where Sledgeweak (the geologist Adam Sedgwick) compared him with a fossil beast; has met (John Frederick William) Hershell [*sic*] at the Royal Astronomical Society; has referred to the inventions of Babbleage (Charles Babbage) and Veetstone (Charles Wheatstone) in a talk at the Mechanics' Institute; and has lectured the Royal Institution on balls and the Statistical Society on lying. At the much more venerable if crisis-wracked Royal Society, readers learn, "FRS" after a member's name stands for "fellow remarkably stupid." The erudite German lectures these fellows on "photogenism": it is as old as creation, since as ancient as the sun, and was well known as "sun painting" by the ancient Persians and by the indigenous peoples of North

America. So much for the host of more modern claimants to the invention of photography. While at the Royal Society, Dullbrainz also hears of "some improvements in photogenic drawing" from "Mr Talbot Foxhound," a name already familiar enough to London audiences to rank alongside the other lions of the sciences in 1839.[2]

"Talbot Foxhound" was here being gently spoofed by the canny Scottish journalist William Jerdan, who was in fact Talbot's close ally. As detailed in the essay by Herta Wolf in this volume, Jerdan had already managed Talbot's initial announcement of photogenic drawing six months earlier in his *Literary Gazette*, used the resonant term "Pencil of Nature" to title reports on daguerreotypes, and been sent specimens of Talbot's new art in the summer of 1839, just before composing this account of Dullbrainz's Royal Society visit.[3] Jokes about propriety, priority, and property showed how many constituencies were involved in early Victorian public knowledge and art. Contemporaries knew how to link enterprises that have otherwise subverted apparently self-evident distinctions between higher and lower sciences and status. When the waspish Tory administrator and reviewer John Wilson Croker launched an unusually barbed attack on Talbot's apparently unrelated work entitled *English Etymology* in the autumn of 1847, Jerdan once again helped manage the response, arranged for the publication of the author's long reply in the *Literary Gazette*, and added editorially that Talbot's "high position in the social and literary scale, and his remarkable discoveries in science (honorable to his country) ought to have saved him from rude assault."[4]

As several contributors to this volume indicate, Talbot found positions through the changed Victorian settings of new disciplinary orders and economic, legal, and political reforms of the system of property.[5] Katrina Dean argues that he does not fit current historiography of Victorian sciences. It has become a consensus of that historiography that formal scientific disciplines first emerged in the early decades of the nineteenth century. Discipline carried a double sense, both of vigilant training and of intellectual classification. However, the consensus has problems: disciplines as systems of border maintenance and regulation of practice considerably predate this period, and their powers certainly were not exercised effortlessly during the period of Talbot's career.[6] Salient transformations in this epoch of direct relevance to that career include the formation of a national rally under the aegis of a cadre of gentlemen of science in the British Association for the Advancement of Science; the deliberate construction of new physical sciences by an uneasy alliance of scientific servicemen, Cambridge wranglers, and urban mathematical practitioners; the foundation of the first teaching laboratories for aspiring chemists, physicians, and engineers; the growth of disciplinary scientific societies and public showrooms in the metropolis; the political disruption of the Royal Society;

and the displacement of generalized surveys of the cycle of the sciences by specialists' lucrative lecture courses. Organs such as the *Literary Gazette* carried reports on each of these changes. Although these developments help make sense of how a practitioner such as Talbot could pursue his range of intellectual and commercial interests, the puzzlement about his setting within disciplined milieus remains.[7] Jerdan's joke worked because his readers would recognize the allegedly authoritative landscape of the sciences and their "remarkably stupid fellows," while simultaneously finding their claims and ambitions amusing and questionable.

Secondary literature more concerned with early photography's origins tends to start with notions of Talbot's bucolic amateurism and some evocation of the breadth of his interests, as if somehow to reconcile Wiltshire seclusion with London patent suits and Ionian botanizing. Received histories assert that Talbot was "the paradigmatic amateur" and that the Royal Society, where he announced the discovery of photogenic drawing, was an "association of amateur scientists."[8] It has been stated he was "an all-around man of science and letters," whose inquiries were underwritten by "the social ideology of private experimentation,"[9] and "a world authority in what we would now separate into botany, mathematics, philology, linguistics, chemistry and optics, as well as being an amateur poet and artist."[10] More astutely, modern scholars have studied the work that Talbot performed to make this kind of public figure. Photogenic drawing, for example, involved making an image "formulated in accordance with a sensibility appropriate to a Cambridge-educated gentleman," so "static, unchanging, timeless, durable and thus capable of being copied." Some argue that stress on Talbot's landed and aristocratic standing was his deliberate tactic in the age of political reform, agrarian violence, and fiscal crisis.[11] *The Pencil of Nature*, the focus of most Talbot scholarship, has been read as an ingenious if cryptic effort to establish his genteel status. The combination of projects has also been judged typical of a gentlemanly capitalist of the age, linking commercial ambitions with amateur dabbling in general sciences.[12]

The contributors to this new volume are not, as a whole, especially impressed by Talbot's disciplinary distinction. Faint praise abounds. It appears from Eleanor Robson's study that Talbot was not a confident Semitist even if never entirely incompetent in Assyriology. Similarly, the mathematics papers for the Royal Society in the 1830s apparently lacked innovative quality; as June Barrow-Green points out, a distinction must be made between Talbot's links with leading mathematicians and others' more engaged mathematical research. Anne Secord likewise notes that Talbot's connections with eminent botanists were commonplace in the correspondence networks of early Victorian natural history. These authors provide resources for at least two rather different strategies. One is the question, as Barrow-Green puts it, not of what Talbot did in mathematics, for example, but of what mathematics (and by extension a range of intellectual and practical resources) did for him. The complementary approach examines the extent to which Talbot's projects, unusually fully documented in a well-edited correspondence and a newly catalogued set of notebooks, can illuminate Victorian scientific and social life.

Several contributors use these considerations to trace transfers between separate enterprises. Secord convincingly indicates that Talbot's early passion for mosses, and thus for microscopic measures, predated and in some respects nourished his later photographic works: the eminent naturalist William Jackson Hooker recognized in the youthful Talbot an eye worth cultivating. Mirjam Brusius and Robson both explore what was evidently the principal intellectual passion of Talbot's career in antiquarianism and Assyriology. Robson shows how, in his remarkable analysis in 1876 of drawings of cuneiform tablets made by Friedrich Delitzsch and George Smith, Talbot used comparative etymology for some tasks (the cryptography of the "flaming sword") and mixtures of ancient Near Eastern imagery for others (the delineation of the "dragon"). She concludes, with Secord, that many of Talbot's decisive resources emerged from prior experience in amateur muscology, which demanded attention to exquisite visual detail, and in mathematics, the study of which at Cambridge encouraged heuristic approaches to constrained if underdetermined puzzles. The editors of this volume argue that the Talbot notebooks reveal concerns with script, cryptography, and intense observation, to such an extent that a purely iconographic reading of his photographs does no justice to the manuscript record. Rather, practices of scientific production and communication were salient in the many different ways in which Talbot made and exchanged methods for recording and expressing knowledge.

The question of method and its institutional expression has long been of concern for scholars of Talbot and his world. Photography historians have noted his concerns with scientific discipline and method, if not always confidently or decisively.[13] Rival epistemologies and disciplines, in fact, fomented a war of methods that erupted among the British intelligentsia from the 1830s, a fight prosecuted in the troubled institutions of public sciences, the encyclopedias, periodicals, mechanics' institutes, and the British Association for the Advancement of Science. Many of Talbot's closest colleagues—the high wranglers and gentlemen of science Herschel and William Whewell, the natural philosopher and journalist David Brewster, and the lawyer and experimenter William Robert Grove—took part in these method wars in print, correspondence, and lectures on the theme of the appropriate techniques of inquiry and thus the definition of the social and moral status to be demanded from reputable inquirers.[14] It seemed important to many commentators to emphasize painstaking systematics. The Cornish mining administrator and prolific science journalist Robert Hunt insisted that "the photographic art" was one "not in any way resulting from accident, but the mature offspring of close inductive reasoning." Yet patterns of systematic reasoning and their organization have not always seemed significant in Talbot's work. Schaaf suggests that shifts in Talbot's interests were essentially "whimsical," that idiosyncratic "opportunity or inclination" governed his dynamics. In comparison with Talbot's close ally Herschel, "his contributions were generally less influential; much of this stemmed from the fact that he rarely took on questions of the organization and role of science." Indeed, Talbot was something of an innocent outsider to these conflicts, judging by his oddly ill-informed request to Herschel for support in late 1830 in an election at the Royal Society, the leadership of which Herschel in fact loathed.[15]

Dean agrees that Talbot never produced an explicit manifesto, but instead she indicates how his archive might cast light on a novel philosophy of science. Vered Maimon suggests that between 1839 and 1844, Talbot shifted his own sense of discovery and induction from that of Herschel's *Preliminary Discourse on the Study of Natural Philosophy* to that set out in Whewell's magisterial *History of the Inductive Sciences* and *Philosophy of the Inductive Sciences*, roughly a move from a rationally recoverable process of empirical inquiry to a more genial model of an induced "first and original idea." Reminiscences of the Lake Como scene were retrospectively turned into examples of unaccountable intellectual illumination.[16] A friend of Talbot's, the long-serving Cambridge head of house Thomas Worsley, keenly recommended Whewell's *Philosophy* to Talbot on its publication in 1840. Whewell, a guest at Talbot's Lacock home and patron of several of Talbot's schemes, including his election to the Royal Society, later corresponded with him about cuneiform interpretations. As Robson mentions, Whewell took part, if cursorily, in the ritual decipherments at the Royal Asiatic Society in 1857. Although the proverbially omniscient Whewell admitted to the notoriously polymath Talbot that he had "not gone into" the subject of Assyriology, the Trinity College don had considerable experience in etymology and philology. Much impressed by the projects of erudite German scholars such as the great Berlin classicist Barthold Niebuhr and his disciple the diplomat and philologist Christian von Bunsen, Whewell and Worsley both enthusiastically joined a Cambridge group of etymologists whose work closely resembled their friend Talbot's later *English Etymology*.[17]

Etymology and decipherment then both carried considerable ideological weight. In 1833 Whewell used his Cambridge colleagues' philological magazine to launch attacks both on utilitarian models of language and on the evils of modish political economy. Similarly, in his often-reprinted *Bridgewater Treatise* about physics and natural theology, published the same year, Whewell lauded the piety of "inductive habits," adopted by those discoverers who induced a new law from previously unconnected facts: "The change is like that which takes place when we attempt to read a sentence written in difficult or imperfect characters. . . . The truth of the happy conjecture seems to flash upon us from every part of the inscription." Contrast, so Whewell suggested, the habits of deductivists, who falsely imagined the world as flowing synthetically from their own strenuous mathematical labors and thus might more easily neglect or reject the divine origin of the world. Heroic discovery and the induction of general laws from phenomena sustained faith, while mere mathematical synthesis could undermine it. And pious learning was best evinced in breaking a cryptic code. This bold argument about cryptographers' devotion was one of the more telling interventions in the war of methods, a deliberate attempt to elevate the moral character of individual discoverers in the Age of Reform. Talbot's scripturalism, historicism, and orientalism fitted this pattern.[18]

Part of this pattern was its concern with antiquarian tradition and cultivation of classical reputation and aesthetic memory. These themes were as politicized, in the epoch of Niebuhr, Samuel Taylor Coleridge, Thomas Carlyle, and Thomas Babington Macaulay, as were the kindred practices of etymology and decipherment.[19] Schaaf concludes that Talbot's long investment in the processes of photogravure, much longer than his initial project on photogenic drawing, reveals his passion for permanence and for the preservation of otherwise fugitive impressions. Dean draws our attention to the archival interests centered on Talbot's Lacock estate and involved in his projects in preservation and recording that were of family and antiquarian significance, while Brusius shows how Talbot sought, if rather in vain, to introduce photography into a range of antiquarian practices. Maimon and Graham Smith both urge the importance of historicism, which attracted Talbot to Walter Scott's Waverley program and its romantic iconography of artfully cultivated tradition. Yet a simplistic reading of Talbot's allegiance to conservative and antique culture fails. Some have argued regarding the commercial failure of his Scottish calotype publication in 1845, *Sun Pictures in Scotland*, that it "represented a premature challenge to Romantic aesthetic ideology" because he "perceived himself primarily as a pioneer of science" and "clearly felt no indebtedness to art history."[20]

Many of these essays thus pose the question of the relation between novel chemical, optical, recording, and inscription techniques, and the appropriate histories into which their development should be put. As Wolf notes, the phrase "pencil of nature," used by Jerdan in February 1839 and soon by other journalists, has been subjected to intense examination for clues to Talbot's literary, aesthetic, and scientific principles. It has been argued that Jerdan found the term in the final volumes of Edward Gibbon's *Decline and Fall of the Roman Empire* (1788). This would give the terminology of early photography an apt literary-historical ancestry. It also has roots in experimental philosophy and instrumentation: the term "pencil of nature" occurs several times in descriptions in the 1760s and 1770s of the use of the camera obscura in landscape drawing by the eminent instrument maker and natural philosopher Benjamin Martin: "specimens of Painting by the Pencil of Nature may be formed at pleasure to gratify both the eye of the body and the mind at the same time."[21]

Wolf argues that such phrases as "Nature as drawing mistress" also speak to this genteel and idiosyncratically British tradition of drawing as polite pursuit, and that early photography was certainly not judged simply as a demonstration of scientific laws. Reflecting on this problem, Whewell found it hard to know exactly where to put Talbot's demonstrations in his *History of the Inductive Sciences* about the systematic advance of the sciences to their modern and, in a few privileged cases, permanent form. His *History of the Inductive Sciences* was a crucial and in many ways collective document that helped early Victorian gentlemen of science authoritatively define legitimate discoveries and discoverers.[22] In the second edition (1847) the Cambridge master confessed that he had excluded the newfangled photographic art from his chronicle because it involved "applications of discoveries already made" and therefore embodied "deductive results of laws of nature, not inductions of such laws from observation." In 1841 Talbot may have imagined that his photochemical visions involved induction with fundamental ideas, but Whewell in 1847 judged otherwise. For Whewell physical optics was almost a perfected science, cast into permanent form; photography was an impressive but merely deductive consequence. Yet a decade later he somewhat changed his view. In the intervening period Whewell learned firsthand about automatic photographic recording apparatus in the major British observatories. Now he accepted that although there was as yet no science of photochemistry, there was at least "a most exquisite art which involves the principles of such a science." Thus the inventors of photography would eventually come to count as the leaders of the prelude to the future epoch of truly scientific photochemistry. Among them would surely figure "our own accomplished countryman Mr Fox Talbot."[23]

It has thus been tempting to see Talbot's accomplishments as oddly dated, either premature visions or living fossils. Barrow-Green cites the telling verdict of Whewell's brother-in-law, the Trinity mathematician and philologist Robert Leslie Ellis, who reported to the British Association for the Advancement of Science in 1846 that Talbot's mathematical writings would have been interesting had they appeared in the previous century, but they could not now be reckoned original. There has also been a recognizable tendency to Gothicize Lacock's culture, as if Talbot's projects resembled those of eccentric country house inquiry, such as Charles Waterton's uncannily ingenious natural historical and satirical taxidermy at Walton Hall in Yorkshire from the 1820s, Andrew Crosse's extraordinary trials on the electrical generation of insects from minerals on his Somerset estate at Fyne

Court in the mid-1830s, and John Lee's ambitious combination of Egyptology and astronomy in the program run from his house at Hartwell in Buckinghamshire from 1831 on.[24] Secord mentions other examples of this kind of enterprise, notably the "model of emulation" that Talbot found in his Welsh neighbors and relatives Lewis Weston Dillwyn and his son, John Dillwyn Llewelyn, who pursued their wide-ranging scientific interests in chemistry, botany, photography, and astronomy in their rural estates and workshops. Dean emphasizes here how country houses could well aid innovation precisely because of their markedly preservative and established repute.

The terminology of magic somehow aided the task of fixing Talbot as this kind of secluded and romantic sage. In his essay Smith supposes that Talbot's collection entitled *Legendary Tales* (1830) may have been prompted by his admiration for Walter Scott and embodies their joint interest in romantic antiquities. One of the verses included in the collection, "The Magic Mirror," has been read as a meditation on transience and a means of linking Talbot's literary imagination and photochemical inquiries with those of the circle around Coleridge and Thomas Wedgwood.[25] The method wars of the 1830s about deceit and realism depended in part on protagonists' varying versions of past natural philosophers' models of projection and showmanship; the term "natural magic" was an important topic in these disputes. The Calvinist evangelical David Brewster's *Letters on Natural Magic* (1834), his demystification of optical and other illusions in the name of purified religion, was a best-selling collection of studies dedicated to Scott, who had published *Letters on Demonology and Witchcraft* in 1830. Maimon notes that Brewster may have offered one precedent for Talbot's celebrated use of the term "natural magic" in his initial explication of photogenic drawing in 1839. Brewster loathed alchemy and magic as distortions of true natural philosophy. He subjected illusion to experimental vigilance: "The progress of science and the diffusion of knowledge are alone capable of dispelling the fears which her wonders must necessarily excite even in enlightened minds." As Ramalingam points out, Talbot's own projects on stroboscopic illumination, launched in collaboration with Wheatstone in London in the 1830s, similarly were designed to turn illusion into a topic of controlled experiment.[26]

Yet the pervasive vocabulary of magic and of alchemy has often been taken as a sign of debts to some older version of natural philosophy. Speakers at a bicentenary conference on Talbot's work reckoned that his style was typified by "a confluence of mysticism and rigorous scientific experimentation." It was conjectured that his unusual enthusiasm for the negative-positive technique, his impressive account of what he called the "sciagraphic process" in early 1835, might somehow be understood as a consequence of a sensibility akin to alchemical melancholia.[27] A common strategy thus finds the discovery of photography in the remote past. This was already familiar in the 1830s and 1840s. Hence, for example, the point of Dullbrainz's story of "sun painting" among the most ancient civilizations of the Old and New Worlds. Hence, too, a comparable passage in *Chambers' Edinburgh Journal* in 1848: "In the spirit of some laborious investigators, we might go back to remotest antiquity for the origin of photography, and find it in the knowledge of the action of light possessed by the Egyptians, or detect it again in the magic mirrors and similar jugglеries of the middle ages."[28] A puzzle of Talbot's projects has been the establishment of his work's proper place in history and, simultaneously, a better account of the kind of history in which such work might find a place.

The question of whether these projects are to count as surprisingly early or distressingly late depends on tacit assumptions about the cultural and technical conditions of his various enterprises, thus presumptions about the status of discovery and invention in that epoch. This status was exactly at issue in the method wars. One of the more notorious examples is the alleged mystery that once silver salts' photosensitivity was established early in the eighteenth century, photography was not invented much earlier than it was. That mystery has been dismissed as a pseudo-problem. So has its supposition of the prior availability of all the materials and techniques used in the 1830s (such as Talbot's decisive recognition that silver chloride's light sensitivity depended on relative silver nitrate concentration in its preparation). Dubious, too, are the consequent notions that some further idealistic transformation was both necessary and sufficient to move at last from feasibility to realization, and that the implementation of photography amounted to a singular event in this history.[29]

One lesson of this volume is that such experimental practices depended on their promoters' local interests, so that their meanings were improvised, contingent, and subject to ingenious revision by subsequent commentators and historians. Compare the very similar problem of spectrochemistry in this period. As with the alleged mystery of photography's late realization, it has often been taken as puzzling that spectrochemistry was not effectively formulated until 1859–60, by the Heidelberg scientists Gustav Kirchhoff and Robert Bunsen. It has seemed as if available techniques for spectral analysis in chemistry long predated their work. Talbot has been understood as an early proponent of these techniques in his studies of the 1820s and 1830s, his failure to deliver attributed to such causes as his characteristically wide interests.[30] But just as the mystery of photography's apparently late appearance has been dismissed, in this case Talbot's projects were very different from those of the later Heidelberg professors. His work emerged from specific physical interests of natural philosophical colleagues, especially the optical disputes that formed an important component of the method wars. In the 1820s several protagonists of these disputes sought monochromatic light sources—Herschel for crystallography and Brewster for microscopy—since truly monochromatic light would avoid aberration. This drew their attention to line spectra. In 1826 Talbot published an essay in Brewster's journal on such emission spectra, which he guessed might aid analysis, but he offered no chemical account of why different substances would show different spectra. As the editors of this volume indicate, this was when the young mathematician also used the resources of Michael Faraday's Royal Institution to study absorption spectra, dark lines in the spectra of hot gases.[31]

In the early 1830s Brewster notoriously used absorption spectra in his campaign against dogmatic wave optics backed by Herschel and Whewell. Talbot then began to develop a physical model of spectrum production by vibrating atoms, thus a possible undulatory explanation of absorption lines. This vibrational model was also linked with his contemporary inquiries into phosphorescence and the darkening of silver salts. But it was never accepted by his contemporary natural philosophers.[32] Nevertheless, when, in the immediate wake of Kirchhoff and Bunsen's achievements, chemists and journalists such as William Crookes sought to vindicate a distinctively British ancestry for spectrochemistry, they reached back to Talbot's papers of the 1820s and 1830s, deliberately republished in Crookes's *Chemical News* in 1861. Talbot reinforced this ancestry in an Edinburgh publication in 1872 on the "early history of spectrum analysis." The retrospect amounted to a eulogy for his old friend Brewster and an insistence that spectrochemical technique was at least forty years old; Kirchhoff and Bunsen were ignored. The story peddled by Crookes and Talbot became part of orthodoxy, generating the notion of spectrochemistry's oddly late implementation.[33]

The historiographies of photography and of spectrochemistry show how techniques' meanings and status are fixed after the fact by appropriate disciplinary communities. Discoveries and discoverers are identified through socially sanctioned retrospective attributions.[34] It makes little sense to demand from a technique's practitioners a vivid sense of all the interests and implications later and elsewhere attributed to it. The mistake is to suppose that technological innovation is of much greater significance, worthier of historical attention, than mundane usages; hence emerge traditional obsessions with strangely delayed adoption or allegedly recalcitrant prejudice.[35] Many contributors to this volume admirably reject this anachronism and teleology, avoiding the notion that all subsequent uses are already and unproblematically present in a specific device or practice. They therefore insist on understanding Talbot's aims in the setting of his own resources and allies, and they show the work required to exploit and establish these projects. Robson argues that the collaborative event at the Royal Asiatic Society in 1857 has wrongly been seen as somehow terminating the heroic work of script decipherment, after which mere routine allegedly took over; in fact, this event did not much aid Assyriologists' practice. Ramalingam challenges art historians' favored story of Talbot's spark experiments in 1851 as playing a similar role for chronophotography. She spells out the technological and social resources required to make such images reproducible and demonstrable. In so doing she sets Talbot's program alongside Wheatstone's interests in measuring the speed of spark discharges, interests closely connected with the study of spark spectra by the new professor at King's College London.[36] This study indicates the specific interests at work in this communal experimental enterprise and usefully reminds us of the failure of Talbot's explicit aims in reorganizing and visualizing photographic instantaneity.

Just as the materiality of tools emerges when they break, the failures of projects are fine ways of mapping borders in the social and intellectual world. No object is inherently the property of an antiquarian, naturalist, or chemist. The editors of this volume point out in their introduction that authoritative interests of twenty-first-century art historians, curators, and collectors make Talbot's output count principally, if not exclusively, as an object of historically informed connoisseurship and a source of vulnerable, valuable property. But, as this volume seeks to demonstrate, the work can probe the tenuous mobile communities of nineteenth-century knowledge. Impressed by the evident failure of photography to work as botanical illustration, Maimon appeals to what she sees as philosophical contrasts between taxonomic (botanical) and dynamic (chemical) scientific disciplines.[37] Schaaf urges rather that the history of photography requires more attention to the photomechanical processes of print and thus to a connection with the intricacies of publication. The argument works well in the case of the natural sciences. Secord nicely recalls that no nineteenth-century picture of a plant was necessarily a botanical image. Talbot's photogenic drawings of plants could not simplistically be attributed to botanical interests, while naturalists were not inevitably impressed by photography's powers. Local natural historical discipline featured largely in the complex career of these images and their uses.

Talbot's political interests in institutional natural history were evident in 1838–40, when he joined an aristocratic Whig lobby—led by John Russell, sixth Duke of Bedford (until the duke's death in late 1839)—in campaigning avidly for the establishment of the Royal Botanic Gardens at Kew with William Jackson Hooker as first director.[38] This interest in the public institution of collective botany mattered. Reliably multiplied images bound the collective together. Secord suggests that Talbot's ambition for botanical photography was driven in part by his dissatisfaction with specimen books. When he sent photogenic drawings to Hooker in the summer of 1839, the naturalist at once passed them to the admiring Duke of Bedford but told Talbot they might best be used to replace etchings as copies of botanical images.[39] That summer Hooker's son Joseph joined the magnetic crusade of the British Association for the Advancement of Science, Whewell's proposal for James Clark Ross's Antarctic survey, a key enterprise in the formation of state-sponsored sciences in Victorian Britain. While Herschel lobbied through the Royal Society for Ross to carry daguerreotype equipment, Talbot offered to train the younger Hooker and his fellow surgeon naturalist Robert McCormick in photogenic drawing. Joseph Hooker later recalled the presence of "apparatus for daguerreotyping and talbotyping" on board. McCormick (who, Hooker soon decided, was a hopeless naturalist) told Talbot that this would be "an art of incalculable value," especially for the delineation of the "evanescent forms" of marine fauna. But by August 1839 Talbot was already grumbling to Ross about the frustration of his plans, and no photography was carried out.[40]

The techniques that Joseph Hooker learned on this voyage defined his future botanical practice when he succeeded his father at Kew: systematic coordination, precise delineation, and standardization. "It is a new field which none but an artist can prosecute at sea," he explained to his father regarding the indispensability of artful draftsmanship. As Secord confirms, therefore, it is not surprising that the younger Hooker's expedition to India in 1847 also generated no photographic records. William Hooker told Talbot, "I must confess that I should much prefer photographic landscapes to the kind of [botanical] representations you allude to." The uneven distribution of scientific photography tracked the distribution of socially prestigious skills in these new field sciences.[41]

Other contributors to this volume bring out the comparative redundancy of new kinds of photographic visualization under the specific institutional and technical conditions of nineteenth-century sciences. This was in part a practical challenge. Wolf recalls such judgments as that of Alexander von Humboldt on Talbot's photogenic drawings in the spring of 1839: "mirthless chlorides." In ways comparable to the problems of dramatizing stroboscopic photography, shadowy cuneiform, as Robson indicates, was barely legible in what might seem to be perfect conditions for photography: stability, steady illumination, an object devoid of defects. But communities' intellectual and institutional interests were also at stake. Brusius amplifies the discussion on photography and Assyriology with a subtle analysis of the local purposes and organization of the British Museum. Instead of surprise at the belated or uneven adoption of photography, the complex and long-drawn-out negotiations about its uses are understood as part of the construction of this scholarly and public institution, alongside recognition of its role as the site where the meanings of curios, artifacts, and relics were to be defined. It is telling that one of Talbot's principal successes in his antiquarian work was aiding the establishment of important social groups such as the Society of Biblical Archaeology, institutions indispensable for the collective work that was needed retrospectively to discipline and stipulate genuine discoveries and competent practitioners.

This retrospective work of attribution was of course a major concern for Talbot and his contemporaries, notably in the establishment of patents. Important aspects of the venomous disputes about calotyping or electromagnetic engineering in which he became embroiled, with Brewster, Grove, and Herschel among his colleagues, have been studied by some historians but deserve a more general analysis. It is regrettable that no essay in this volume addresses the issue. Similarly, no recent overview of Victorian patent disputes analyzes Talbot's experiences.[42] They matter not least because they show the explicit entanglement of discovery, property, and social status in nineteenth-century sciences. Just as the Age of Reform focused on widespread and sometimes violent conflicts about the basis of social property and the political and economic rights of workers, rentiers, and capitalists, the epoch of photogenic drawing and the calotype was marked by peculiarly intense controversies about intellectual property in invention and discovery. Talbot's political and public life placed him in these struggles: as a protagonist of defense against the Swing riots during the rural revolt of November 1830; as a Whig member of Parliament in the wake of the first Reform Act (1832); in fear of Chartist disruption of the annual meeting of the British Association in Birmingham in 1839; and as a patron of photographic works and the attempts to use them to reorganize the skill of image production.[43]

New institutions were also set up to exploit the crisis of intellectual property: London showrooms such as the Adelaide Gallery of Practical Science advertised themselves as privileged sites for patentees; in 1829 Parliament launched its first patent inquiry; the British Association was set up in 1831 in part for further reform of the patent laws; and new journals such as the *Mechanics' Magazine* lobbied for inventors and artisans. Strong analogies were made between literary and scientific invention, and the heroic inventor was frequently represented as martyr and victim.[44] During the same period, using similar jurisprudential techniques, gentlemen of science became embroiled in a series of public priority fights in which histories, archives, notebooks, and journalism all figured: during the 1830s and 1840s geologists and naturalists contested the discovery of dinosaurs; natural philosophers and optical experimenters, the discovery of the relation between magnetism and light waves; astronomers and mathematicians, the discovery of Neptune; and chemists, engineers, and natural philosophers, the discovery of the composition of water. Their eventual resolutions also involved authoritative establishment of institutional practices in paleontology, electromagnetism, celestial mechanics, and analytical chemistry.[45]

Talbot was at the British Association's Birmingham meeting in the summer of 1839 for his first large exhibition of photogenic drawings. He addressed the rival claims of his own originality and those of Louis-Jacques-Mandé Daguerre's principal publicist, François Arago, secretary of the French Académie des Sciences. Wolf argues that in the public debates about daguerreotyping and photogenic drawing, the relation between inexpert chemistry and optical physics was high on the disciplinary agenda. The problem of the status of the artisan mattered to the outcomes of these fights. Talbot's criticism of Arago's discovery story was doubtless aided by the furious denunciation of the Frenchman in Birmingham by the British Association's president, Vernon Harcourt. Harcourt's attack directly addressed the problem of assigning credit for a chemical discovery. He sought officially to demolish Arago's claim that James Watt had first discovered the compound composition of water. The genteel Harcourt held that credit should rather be granted to Henry Cavendish, preeminent Whig aristocrat, chemist, and natural philosopher. The Whig landlord and natural philosopher Talbot, perhaps understandably, agreed. This was not solely a set of disputes about priority: it was also a continuing conflict about the rights of artisans and natural philosophers, of chemists and gentlemen.[46]

Arago had clearly set out the politico-economic issue: "Men of genius, and the manufacturers of ideas, it seemed, ought to remain strangers to anything like material enjoyments, and their history, forsooth, should continue to resemble the legends of the martyrs." In contrast *Chambers' Edinburgh Journal* offered its readers "popular information on commercial economy," stating that "there can be no claim of creation, but only of finding out what has all along existed in the great laboratory of nature." Some held that the system of patents and rewards should be abolished since simultaneous discoveries showed that they depended on the spirit of the age, not on that of the hero; others asserted that the system should be overhauled since heroes were so often martyred by pirates and lawsuits.[47] Schaaf aptly cites the celebrated and fanciful proclamation, in *Blackwood's Edinburgh Magazine* in 1839, of the "sun's patent" that would license all nature to inscribe itself. Some of Talbot's closest colleagues, such as Brewster and Grove, were possessed by these issues. In his manifesto for scientific and economic reform, Brewster demanded that patent laws be overhauled and an official commission of experts in "the past and present state of the useful arts" decide on priority and invention. His writings on photogenic drawing repeated the attack. Yet in early 1839 he told Talbot to patent the process: "I do not see why a gentleman with an independent fortune should scruple to accept of any benefit that he has derived from his own genius."[48]

Grove was an expert experimenter on electromagnetism and on electrochemical etching of photographs, a successful science lecturer, and one of Talbot's patent lawyers. In October 1840 Talbot patented a design for a motor based on a sequence of electromagnets. When a Glasgow instrument maker exhibited such an engine at a Piccadilly showroom in 1843, Talbot insisted he make no others without explicit license, though since the Scotsman was a "very ingenious man" he could continue the exhibition. Meanwhile Talbot sought to persuade Grove to show his own machines at Royal Institution lectures. The two men well understood how fraught were these issues of publicity and control. A few months after the Scottish affair, Grove argued in a widely read article in *Blackwood's Edinburgh Magazine* that it was humiliating for "the dignity of philosophy" to defend patent rights and that "an entire erasure from the statute books of patent provision would be of service to science and perhaps to the community," because innovation would then be encouraged. Then he generalized the principle of publicity and social authority to propose that the fellowship of the Royal Society be completely overhauled as "oligarchs of science" who would regulate public science and its conduct in Britain. Remarkably Grove took this opportunity explicitly to denounce Talbot's defense of his photographic patents. Citing Brewster's discussion of the award of a Royal Society medal to Talbot for calotyping, Grove opined that no such scientific reward was justified to an entrepreneur who refused to publish his discovery in the scientific press.[49]

Despite attacking the ignominy of patenting and the evils of Talbot's own restrictive practices, Grove acted for the photographic inventor in the notorious calotype trials of the 1850s. Similarly, despite Brewster's lifelong campaign to change the patent laws, he stood by Talbot's rights as owner and thus controller of the discovery. The Great Exhibition of 1851 was a vast display of large-scale industrial production and the technologies of automatic copying. As Schaaf demonstrates, ingenious technologies of reproduction remained at the center of Talbot's interests in these later years. New patent disputes erupted just as his 1841 calotype patent lapsed and he launched his failed defense against men he saw as thieves. In these patent disputes the Smilesian literary genre that celebrated inventors as heroes, discoverers of genius, was developed precisely to defend the rights of patentees against all those skeptical of legal rewards for invention.[50] In exchanges with the president of the Board of Trade, Talbot resuscitated Brewster's proposal for an official commission of experts "having knowledge of trade and of the arts and sciences." The next year a new Patent Office was set up, backed by an ideology that would soon institutionalize just the kind of history of science and technology that makes figures like the inventive and genial Talbot into the protagonists of social and economic progress and the regrettably neglected visionaries of modernity.[51]

Talbot's proper place may remain ambiguous and mobile. The achievement of this collection of essays is to offer resources that help explain that mobility and use it to explore a wide range of Victorian forms of knowledge and skill. It also makes clear that Talbot's contemporaries invested much reflection on where to put him and his projects. At the conclusion of Brusius's tale of museological cuneiform and the sufferings of photo-registration, there were major distinctions between legitimate and seemingly marginal scientific and scholarly lives. The very act of geographic location can sometimes function as praise, otherwise as denunciation, depending on assumptions of the status and values of the site in question. This is the force of the remarks by Ramalingam and Brusius on the respective roles of the Royal Institution and the British Museum. Ethics and politics suggested how to locate Talbot's work. In 1839 Dullbrainz had found him among the fellows of the Royal Society in Somerset House. Others viciously parochialized him. In the midst of the patent disputes of the 1850s, the *Art-Journal* carried a notorious denunciation of Talbot's career as discoverer, patentee, and gentleman of science. This attack on Talbot's repute was anonymous, though Brewster told Talbot he reckoned that the author was Robert Hunt. That was the journalist who in 1850 had celebrated photography as a legitimate result of "close inductive reasoning" and then appeared as expert witness against Talbot in the calotype patent case four years later. The *Art-Journal* reiterated Hunt's view that photography was due "to the researches of science," so Talbot's attempt to "secure to himself a complete monopoly of the sunshine" was an outrage. Lacock's magic was transmuted into a Gothic site of reverie and error, Talbot into a mere empiric blindly stumbling upon the novelties of photogenic drawing and calotyping, plagiarizing his contemporaries and predecessors. As the anonymous author of the *Art-Journal* article put it, "It is progress by a system of accidents without a rule." According to this critic,

> We have heard of dreamy walks around the lake of Como—of a poet's meditations on the beauties of that enchanting scene awakening a desire to secure the charms of nature in permanence. Of contemplations in the green or crimson drawing-room of Lacock—we forget which—leading to the discovery of photography; and of the philosopher at home working and thinking out one of those great truths which are as stars of light to his brethren. In some future history of the inductive sciences how charmingly all this will tell. But we have to do with the reality.[52]

The question of this reality remains, as in the mid-nineteenth century, a rich theme in histories of vision and geographies of property.

1. Larry J. Schaaf, *Out of the Shadows: Herschel, Talbot & the Invention of Photography* (New Haven and London: Yale Univ. Press, 1992), 9.
2. William Jerdan, "Baron von Dullbrainz," *Bentley's Miscellany* 4 (1839): 316–22.
3. Susan Matoff, *Conflicted Life: William Jerdan, 1782–1869* (Eastbourne, UK: Sussex Univ. Press, 2011), 358; and Schaaf, *Out of the Shadows*, 138.
4. William Jerdan to William Henry Fox Talbot, Dec. 2, 1847, Doc. No. 2258, The Correspondence of William Henry Fox Talbot, http://foxtalbot.dmu.ac.uk (accessed Jan. 31, 2012), hereafter "Correspondence" (original: Talbot Collection, hereafter "TC," British Library, hereafter "BL"); Talbot to Jerdan, Jan. 1, 1848, Doc. No. 6078, Correspondence (original: published); and H. J. P. Arnold, *William Henry Fox Talbot: Pioneer of Photography and Man of Science* (London: Hutchinson Benham, 1977), 232–33. Talbot did not at once identify the review's author; Talbot to Thomas Hewitt Key, Jan. 1, 1848, Doc. No. 6086, Correspondence (original: TC, BL).
5. See also Jack Morrell, "Individualism and the Structure of British Science in 1830," *Historical Studies in Physical Science* 3 (1971): 183–204; Morris Berman, "Hegemony and the Amateur Tradition in British Science," *Journal of Social History* 7 (1975): 30–50; S. S. Schweber, "Scientists as Intellectuals: The Early Victorians," in *Victorian Science and Victorian Values*, ed. James Paradis and Thomas Postlewait (New Brunswick, NJ: Rutgers Univ. Press, 1985), 1–37; and William J. Ashworth, "The Calculating Eye: Herschel, Babbage and the Business of Astronomy," *British Journal for the History of Science* 27 (1994): 409–41.
6. Robert Kohler, "Discipline History," in *Dictionary of the History of Science*, ed. W. F. Bynum, E. J. Browne, and Roy Porter (London: Macmillan, 1981), 104; Notker Hammerstein, "The Modern World, Sciences, Medicine and Universities," *History of Universities* 8 (1989): 173; Rudolf Stichweh, "The Sociology of Scientific Disciplines: On the Genesis and Stability of the Disciplinary Structure of Modern Science," *Science in Context* 5 (1992): 4; and Jan Golinski, *Making Natural Knowledge: Constructivism and the History of Science* (Cambridge: Cambridge Univ. Press, 1998), 67.
7. Jack Morrell and Arnold Thackray, *Gentlemen of Science: Early Years of the British Association for the Advancement of Science* (Oxford: Clarendon Press, 1981), 12–29; J. N. Hays, "The London Lecturing Empire, 1800–50," in *Metropolis and Province: Science in British Culture, 1780–1850*, ed. Ian Inkster and Jack Morrell (London: Hutchinson, 1983), 91–119; David Phillip Miller, "The Revival of Physical Sciences in Britain, 1815–1840," *Osiris* 2 (1986): 107–34; and Iwan Rhys Morus, "Manufacturing Nature: Science, Technology and Victorian Consumer Culture," *British Journal for the History of Science* 29 (1996): 403–34, esp. 413–22.
8. Grace Seiberling with Carolyn Bloore, *Amateurs, Photography and the Mid-Victorian Imagination* (Chicago: Univ. of Chicago Press, 1986), 4.
9. Carol Armstrong, *Scenes in a Library: Reading the Photograph in the Book, 1843–1875* (Cambridge, MA: MIT Press, 1998), 108–9.
10. Geoffrey Batchen, *Burning with Desire: The Conception of Photography* (Cambridge, MA: MIT Press, 1997), 57–58.
11. For the quotations, see Joel Snyder, "Enabling Confusion," *History of Photography* 26, no. 2 (2002): 159. Compare Mike Weaver, "Henry Fox Talbot: Conversation Pieces," in *British Photography in the Nineteenth Century: The Fine Art Tradition*, ed. Mike Weaver (Cambridge: Cambridge Univ. Press, 1989), 13–14.
12. Armstrong, *Scenes in a Library*, 144, 168; and Steve Edwards, *The Making of English Photography: Allegories* (University Park, PA: Pennsylvania State Univ. Press, 2006), 33–34.
13. Armstrong, *Scenes in a Library*, 44; and Edwards, *Making of English Photography*, 303 n. 27.
14. For the method wars, see Susan Faye Cannon, *Science in Culture: The Early Victorian Period* (New York: Dawson and Science History Publications, 1978), 118–34; Morrell and Thackray, *Gentlemen of Science*, 267–75; Iwan Rhys Morus, "Correlation and Control: William Robert Grove and the Construction of a New Philosophy of Scientific Reform," *Studies in History and Philosophy of Science* 22 (1991): 589–621; Richard R. Yeo, *Defining Science: William Whewell, Natural Knowledge, and Public Debate in Early Victorian Britain* (Cambridge: Cambridge Univ. Press, 1993), 32–48; and Laura Snyder, *Reforming Philosophy: A Victorian Debate on Science and Society* (Chicago: Univ. of Chicago Press, 2006), 7–26.

15. [Robert Hunt], "The Pencil of the Sun," *Tait's Edinburgh Magazine* 17 (Oct. 1850): 625. Schaaf, *Out of the Shadows*, 20–21. For Talbot as an outsider to these conflicts, see Arnold, *William Henry Fox Talbot*, 79–80. For the presidential election at the Royal Society, see John Frederick William Herschel to Talbot, Dec. 28, 1830, Doc. No. 2106, Correspondence (original: Royal Society, London). See also Roy MacLeod, "Whigs and Savants: Reflections on the Reform Movement in the Royal Society, 1830–48," in Inkster and Morrell, *Metropolis and Province*, 55–90.

16. Vered Maimon, "On the Singularity of Early Photography: William Henry Fox Talbot's Botanical Images," *Art History* 34, no. 5 (2011): 964.

17. For the recommendation of the *Philosophy*, see Thomas Worsley to Talbot, Jan. 29, 1840, Doc. No. 4009, Correspondence (original: TC, BL). Whewell confessed about Assyriology in William Whewell to Talbot, Nov. 20, 1856, Doc. No. 6560, Correspondence (original: TC, BL). See also William Whewell, "An Account of the Late Cambridge Etymological Society," *Transactions of the Philological Society* 5 (1852): 133–42; John Burrow, "The Uses of Philology in Victorian England," in *Ideas and Institutions of Victorian Britain*, ed. Robert Robson (London: G. Bell, 1967), 180–204; Hans Aarsleff, *The Study of Language in England, 1780–1860* (Minneapolis: Univ. of Minnesota Press, 1983), 217–19; and Snyder, *Reforming Philosophy*, 289–90.

18. For Whewell's intervention in philology, see Yeo, *Defining Science*, 120–29. For Whewell's attack on political economy, see Whewell, "On the Uses of Definitions," *Philological Museum* 2 (1833): 263–72. His statements on inductive habits and cryptography are quoted in William Whewell, *Astronomy and General Physics Considered with Reference to Natural Theology*, 2nd ed. (London: Pickering, 1834): 305–6, 334. For the exchanges between Talbot and Whewell, see Talbot to William Whewell, Nov. 11, 1856, Doc. No. 7322, Correspondence (original: TC, BL); and Whewell to Talbot, Nov. 20, 1856, Doc. No. 6560, Correspondence (original: TC, BL).

19. Olive Anderson, "The Political Uses of History in Mid-Nineteenth-Century England," *Past and Present* 36 (1967): 87–105; Philippa Levine, *The Amateur and the Professional: Antiquarians, Historians and Archaeologists in Victorian England* (Cambridge: Cambridge Univ. Press, 1986), 7–39; and Richard Brent, *Liberal Anglican Politics: Whiggery, Religion and Reform, 1830–41* (Oxford: Oxford Univ. Press, 1987), 163–68.

20. For invention and the historical tradition, see Vered Maimon, "Displaced 'Origins': William Henry Fox Talbot's *The Pencil of Nature*," *History of Photography* 32, no. 4 (2008): 314–25; and Edwards, *Making of English Photography*, 61–66. The statements on the failure and anti-Romanticism of *Sun Pictures in Scotland* are quoted in Gillen D'Arcy Wood, *The Shock of the Real: Romanticism and Visual Culture, 1760–1860* (New York and Basingstoke, UK: Palgrave, 2001), 192–94.

21. For Jerdan's source in Gibbon, see Schaaf, *Out of the Shadows*, 138. For Martin's quotation, see Benjamin Martin, *Microscopium Pantometricum* (London: printed for the author, 1775), 18. Martin also linked the camera obscura with the "pencil of nature," in Martin, *A New and Comprehensive System of Mathematical Institutions*, 2 vols. (London: Owen, 1764), 191. For Martin's showmanship, see also Don Slater, "Photography and Modern Vision: The Spectacle of 'Natural Magic,'" in *Visual Culture*, ed. Chris Jenks (London: Routledge, 1995), 218–37. For other period uses of "pencil of nature," see Alan Trachtenberg, "Photography: The Emergence of a Keyword," in *Photography in Nineteenth-Century America*, ed. Martha Sandweiss (Fort Worth: Amon Carter Museum, 1991), 30; and Heinz Henisch and Bridget Ann Henisch, *The Photographic Experience, 1839–1914: Images and Attitudes* (University Park, PA: Pennsylvania State Univ. Press, 1994), 198–99.

22. Yeo, *Defining Science*, 147–48; and David Phillip Miller, *Discovering Water: James Watt, Henry Cavendish and the Nineteenth-Century "Water Controversy"* (Aldershot, UK: Ashgate, 2004), 25.

23. William Whewell, *History of the Inductive Sciences*, 2nd ed., 3 vols. (London: Parker, 1847), 1:vi; and Whewell, *History of the Inductive Sciences*, 3rd ed., 3 vols. (London: Parker, 1857), 2:478–79. For Whewell on technology as deductive application, see Joost Mertens, "From Tubal Cain to Faraday: William Whewell as a Philosopher of Technology," *History of Science* 38 (2000): 335–37. For Whewell and photorecording, see Charles Brooke, "On the Automatic Registration of Magnetometers and Other Meteorological Instruments by Photography," *Philosophical Transactions of the Royal Society of London* (1847): 59–77; and Charles Brooke to William Whewell, Aug. 19, 1847, Add. MS a/201/110, Trinity College, Cambridge.

24. For Waterton, see Christina Grasseni, "Taxidermy as Rhetoric of Self-Making: Charles Waterton, Wandering Naturalist," *Studies in History and Philosophy of Biological and Biomedical Sciences* 29 (1998): 269–94. For Crosse, see James A. Secord, "Extraordinary Experiment: Electricity and the Creation of Life in Victorian England," in *The Uses of Experiment: Studies in the Natural Sciences*, ed. David Gooding, Trevor Pinch, and Simon Schaffer (Cambridge: Cambridge Univ. Press, 1989), 337–84. For Lee, see Alan Chapman, *The Victorian Amateur Astronomer: Independent Astronomical Research in Britain, 1820–1920* (Chichester, UK: John Wiley, 1998), 75–91.

25. Slater, "Photography and Modern Vision," 223; and Batchen, *Burning with Desire*, 85–91.

26. David Brewster, *Letters on Natural Magic* (London: John Murray, 1832), 147. For Brewster as Talbot's "closest scientific friend," see Arnold, *William Henry Fox Talbot*, 81. For Brewster and optical illusion, see A. D. Morrison-Low, "Brewster and Scientific Instruments," in *Martyr of Science: Sir David Brewster*, ed. A. D. Morrison-Low and J. R. R. Christie (Edinburgh: Royal Scottish Museum, 1984), 59–66; Iwan Rhys Morus, "Seeing and Believing Science," *Isis* 97 (2006): 101–10; and Morus, "Worlds of Wonder: Sensation and the Victorian Scientific Performance," *Isis* 101 (2010): 806–16.

27. Anthony Burnett-Brown, Russell Roberts, and Mark Haworth-Booth, "Specimens and Marvels: William Henry Fox Talbot and the Invention of Photography," special issue, *Aperture* 161 (Winter 2000): 1, 5. Similar views are expressed in Mike Weaver, "Diogenes with a Camera," in *Henry Fox Talbot: Selected Texts and Bibliography*, ed. Mike Weaver (Oxford: Clio, 1992), 13. For the sciagraphic process, see Schaaf, *Out of the Shadows*, 41–42. For Talbot as alchemical melancholic, see Mike Ware, "Luminescence and the Invention of Photography," *History of Photography* 26 (2002): 9.

28. Jerdan, "Baron von Dullbrainz," 317; and "The Photographic Art," *Chambers' Edinburgh Journal*, n.s., 9 (June 1848): 402.

29. For the "mystery," see Helmut Gernsheim and Alison Gernsheim, "The Photographic Arts: Photography," in *A History of Technology: The Late Nineteenth Century*, ed. Trevor I. Williams (Oxford: Clarendon Press, 1958), 716; Gernsheim and Gernsheim, *The History of Photography: From the Camera Obscura to the Beginning of the Modern Era* (New York: McGraw-Hill, 1969), 13; Schaaf, *Out of the Shadows*, xi; Batchen, *Burning with Desire*, 24–25, 254–55; and Ware, "Luminescence," 6, 11. For criticism, see Joel Snyder, "Batchen, *Burning with Desire*," *Art Bulletin* 81 (1999): 540–42; Snyder, "Enabling Confusion," 154–55, 159; and Edwards, *Making of English Photography*, 301. In an example of the reification of discovery, Carol Armstrong misidentified Talbot's announcement of photogenic drawing (1839) with that of the calotype (1841); Armstrong, *Scenes in a Library*, 108.

30. William McGucken, *Nineteenth Century Spectroscopy: Development of the Understanding of Spectra, 1802–97* (Baltimore: Johns Hopkins Univ. Press, 1969), 6–8; M. A. Sutton, "Spectroscopy and the Chemists: A Neglected Opportunity?," *Ambix* 23 (1976): 16–26; and Arnold, *William Henry Fox Talbot*, 71.

31. Frank A. J. L. James, "The Creation of a Victorian Myth: The Historiography of Spectroscopy," *History of Science* 23 (1985): 2–3; and James, "The Discovery of Line Spectra," *Ambix* 32 (1985): 61–63, citing William Henry Fox Talbot, "Some Experiments on Coloured Flames," *Edinburgh Journal of Science* 5 (1826): 81. James mentions Talbot's Royal Institution work with Faraday and Herschel on dark spectral lines in 1827; see Michael Faraday to Talbot, May 9, 1827, Doc. No. 309, Correspondence (original: TC, BL); Talbot to John Frederick William Herschel, May 21, 1827, Doc. No. 1560, Correspondence (original: Royal Society, London); and Herschel to Talbot, May 22, 1827, Doc. No. 1561, Correspondence (original: TC, BL). This work was published in Talbot, "Facts Relating to Optical Science, No. 1," *London and Edinburgh Philosophical Magazine and Journal of Science* (hereafter *Philosophical Magazine*) 4 (1834): 114.

32. William Henry Fox Talbot, "On the Nature of Light," *Philosophical Magazine* 7 (1835): 117–18; Frank A. J. L. James, "The Debate on the Nature of the Absorption of Light, 1830–1835," *History of Science* 21 (1983): 348–49; and Ware, "Luminescence," 13.

33. James, "Creation of a Victorian Myth," 11–12. James notes that Crookes's republications of Talbot are in *Chemical News* 3 (April 1861): 261–63. Talbot's own version is his "Note on the Early History of Spectrum Analysis," *Transactions of the Royal Society of Edinburgh* 7 (1872): 461–65, followed by "On a New Mode of

Observing Certain Spectra," *Transactions of the Royal Society of Edinburgh* 7 (1872): 466–68, reporting his experiments at Edinburgh University in 1871. See also Talbot to Charles Talbot, March 24, 1871, Doc. No. 4907, Correspondence (original: private collection); and Peter Guthrie Tait to Talbot, April 29, 1871, Doc. No. 387, Correspondence (original, TC, BL).

34. Augustine Brannigan, *The Social Basis of Scientific Discoveries* (Cambridge: Cambridge Univ. Press, 1981), 129–42; and Miller, *Discovering Water*, 11–26.

35. David Edgerton, "From Innovation to Use: Ten (Eclectic) Theses on the History of Technology," *History and Technology* 16 (1999): 1–26.

36. For spark spectra and the velocity of electric discharge, see Frank A. J. L. James, "The Study of Spark Spectra," *Ambix* 30 (1983): 137–62.

37. See also Maimon, "On the Singularity of Early Photography," 970–71.

38. Arnold, *William Henry Fox Talbot*, 262–63. Richard Drayton describes Whig political and scientific support for Hooker's project at Kew in *Nature's Government: Science, Imperial Britain and the "Improvement of the World"* (New Haven: Yale Univ. Press, 2000), 148–69.

39. William Jackson Hooker to Talbot, June 21, 1839, Doc. No. 3895, Correspondence (original: TC, BL).

40. John Frederick William Herschel to Louis-Jacques-Mandé Daguerre, Aug. 1, 1839, in Derek Wood, "The Daguerreotype Patent, the British Government, and the Royal Society" (1980), www.midley.co.uk; Hooker's reminiscence about photographic apparatus is quoted in Glenn Stein, "Photography Comes to the Polar Regions," www.antarctic-circle.org/stein.htm (accessed Jan. 20, 2012); Robert McCormick to Talbot, July 31, 1839, Doc. No. 3911, Correspondence (original: TC, BL) for the "art of incalculable value"; and Talbot to James Clark Ross, Aug. 22, 1839, Doc. No. 3923, Correspondence (original: Royal Society, London). For the magnetic crusade lobby, see Morrell and Thackray, *Gentlemen of Science*, 353–70.

41. Joseph Hooker to William Jackson Hooker, March 17, 1840, in Leonard Huxley, ed., *Life and Letters of Sir Joseph Dalton Hooker*, 2 vols. (London: John Murray, 1918), 1:57–58; Jim Endersby, *Imperial Nature: Joseph Hooker and the Practices of Victorian Science* (Chicago: Univ. of Chicago Press, 2008), 42–49; William Jackson Hooker to Talbot, Dec. 25, 1847, Doc. No. 6068, Correspondence (original: TC, BL); and Stein, "Photography Comes to the Polar Regions."

42. Derek Wood, "J. B. Reade and the Early History of Photography, part 2: Gallic Acid and Talbot's Calotype Patent," *Annals of Science* 27 (1971): 47–83; Wood, "The Involvement of Sir John Herschel in the Photographic Patent Case, Talbot v. Henderson, 1854," *Annals of Science* 27 (1971): 239–64; Wood, "The Calotype Patent Lawsuit of Talbot v. Laroche 1854" (1975), www.midley.co.uk (accessed Jan. 18, 2012); and Arnold, *William Henry Fox Talbot*, 182–216.

43. The Swing riots involved a widespread agrarian revolt against mechanization and low wages that spread through southern England in the second half of 1830. The Chartist campaign for political reform promoted a meeting of a national convention in Birmingham from May 1839 that led to riots in that city after Parliament rejected the Chartists' petition. For Talbot's fears, see Talbot to John Frederick William Herschel, July 19, 1839, Doc. No. 3908, Correspondence (original: Royal Society, London); and Arnold, *William Henry Fox Talbot*, 55–61. For Wiltshire riots around Lacock, see Adrian Randall and Edwina Newman, "Protest, Proletarians and Paternalists: Social Conflict in Rural Wiltshire, 1830–1850," *Rural History* 6 (1995): 208–13. For the Chartist riots in Birmingham in 1839, see Morrell and Thackray, *Gentlemen of Science*, 251.

44. Arnold, *William Henry Fox Talbot*, 176–81; Morrell and Thackray, *Gentlemen of Science*, 256; Morus, "Manufacturing Nature," 411, 422; Clare Pettitt, *Patent Inventions: Intellectual Property and the Victorian Novel* (Oxford: Oxford Univ. Press, 2004), 41–42, 61–62; Edwards, *Making of English Photography*, 35–36; and Christine MacLeod, *Heroes of Invention: Technology, Liberalism and British Identity, 1750–1914* (Cambridge and New York: Cambridge Univ. Press, 2007), 183–84.

45. Adrian Desmond, "Designing the Dinosaur," *Isis* 70 (1979): 224–34; David Gooding, "'He Who Proves Discovers': Herschel, Pepys and the Faraday Effect," *Notes and Records of the Royal Society* 39 (1985): 224–44; Robert Smith, "The Cambridge Network in Action: The Discovery of Neptune," *Isis* 80 (1989): 395–422; and Miller, *Discovering Water*, 215–41.

46. Lady Elisabeth Feilding to Talbot, Oct. 4, 1839, Doc. No. 3947, Correspondence (original: TC, BL); Talbot to Lady Elisabeth, Oct. 5, 1839, Doc. No. 3949, Correspondence (original: National Media Museum, Royal Photographic Society Collection, Bradford); and Miller, *Discovering Water*, 109, 129–37.

47. François Arago, *Historical Eloge of James Watt* (London: John Murray, 1839), 94; and "Popular Information on Commercial Economy—Patents," *Chambers' Edinburgh Journal* 10 (March 1841): 60. For an attack on patent laws because many discoveries are simultaneous, see Isambard Brunel, *Life of Isambard Kingdom Brunel* (London: Longmans, Green, 1870), 492, quoted in MacLeod, *Heroes of Invention*, 267. For an attack on patent laws because they martyr inventors, see [David Brewster], "The Decline of Science in England," *Quarterly Review* 43 (1830): 330–41.

48. For experts in "the useful arts" see [Brewster], "Decline of Science in England," 341. For another denunciation of the patent system, see [David Brewster], "Photogenic Drawing, or Drawing by the Agency of Light," *Edinburgh Review* 86, no. 154 (Jan. 1843): 344. For Brewster's remarks on "a gentleman with an independent fortune," see David Brewster to Talbot, Feb. 4, 1839, Doc. No. 3789, Correspondence (original: TC, BL).

49. Talbot to William Read King, Jan. 1843, Doc. No. 4686, Correspondence (original: National Media Museum, Bradford); Robert Davidson to Talbot, Jan. 11, 1843, Doc. No. 4694, Correspondence (original: TC, BL); Talbot to William Robert Grove, Jan. 12, 1843, Doc. No. 4697, Correspondence (original: Royal Society, London); Grove to Talbot, Jan. 13, 1843, Doc. No. 4698, Correspondence (original: TC, BL); [William Robert Grove], "Physical Science in England," *Blackwood's Edinburgh Magazine* 54 (Oct. 1843): 521; and Arnold, *William Henry Fox Talbot*, 227. See also Morus, "Correlation and Control," 617–18, for the politics of Grove's campaign.

50. Wood, "Calotype Patent Lawsuit"; and Pettitt, *Patent Inventions*, 100–110, 124–30. For the writer and moralist Samuel Smiles and the heroic genre of biography he cultivated, see MacLeod, *Heroes of Invention*, 280–314.

51. Talbot to Henry Labouchere, Dec. 31, 1851, Doc. No. 6542, Correspondence (original: TC, BL); Arnold, *William Henry Fox Talbot*, 179–80; and MacLeod, *Heroes of Invention*, 251–53.

52. "The Photographic Patents," *Art-Journal* 6 (Aug. 1854): 236–38; and Arnold, *William Henry Fox Talbot*, 175. Brewster identified Hunt as the author in David Brewster to Talbot, Aug. 31, 1854, Doc. No. 7037, Correspondence (original: National Media Museum, Bradford). Thanks to Larry Schaaf for drawing my attention to Robert Hunt to Talbot, Dec. 23, 1851, Doc. No. 6539, Correspondence (original: private collection), and Talbot to Hunt, Dec. 25, 1851, Doc. No. 8749, Correspondence (original: private collection), which show the men's very cordial relations a couple of years before the article.

Key Dates in Talbot's Scholarly and Scientific Life

1800 Born at Melbury House, Dorset.
1811 Accepted at Harrow School.
1816 Correspondence with the botanist William Jackson Hooker begins.
1817 Enters Trinity College, Cambridge.
1820 Porson Prize for Greek Verse, Cambridge.
1821 BA, twelfth wrangler, Chancellor's Classical Medal, Cambridge.
1822 First mathematical publication.
1824 Meets John Frederick William Herschel.
1826 Meets David Brewster. Botanical expedition to Corfu.
1827 Lacock Abbey becomes principal residence.
1830 Publication of *Legendary Tales, in Verse and Prose.*
1832 Elected to Parliament as a Whig. Marries Constance Mundy.
1833 Visual persistence experiments.
1834 First period of photochemical experiments.
1836 Invitation to give Bakerian Lecture to the Royal Society for his investigations of crystals.
1838 Royal Society's Royal Medal for work in mathematics. Publication of *Hermes, or Classical and Antiquarian Researches*, volume 1.
1839 Reads "Some Account of the Art of Photogenic Drawing" to the Royal Society and presents his photographic invention to the public. Publication of *The Antiquity of the Book of Genesis: Illustrated by Some New Arguments* and volume 2 of *Hermes, or Classical and Antiquarian Researches.*
1841 First patent for calotype process.
1842 Royal Society's Rumford Medal for discoveries and improvements in photography.
1843 Supports founding of first photographic mass-production facility, at Reading, by Nicolaas Henneman. First photographic experimental demonstration at the British Museum.

1844 Publication of first installments of *The Pencil of Nature*. Journey to Scotland.
1845 Publication of *Sun Pictures in Scotland*.
1847 Publication of *English Etymologies*.
1851 Spark illumination experiment for instantaneous photography.
1852 Patent for photographic engraving. Another photography trial at the British Museum, by invitation of Lord Rosse.
1853 First notebooks on Assyriology.
1857 Cuneiform "decipherment event."
1858 Patent for photoglyphic engraving.
1867 First installment of "Contributions towards a Glossary of the Assyrian Language."
1875 Private printing of "Commentary on the Deluge Tablet, Together with a Second Tablet in the British Museum Relating Apparently to the Deluge," three years after George Smith's discovery of the tablet in the British Museum.
1876 Publication of "The Fight Between Bel and the Dragon, and the Flaming Sword Which Turned Every Way."
1877 Dies at Lacock Abbey.

Notes on Contributors

June Barrow-Green is a Senior Lecturer in the History of Mathematics at the Open University. She is associate editor of *The Princeton Companion to Mathematics* (2008); author of *Poincaré and the Three Body Problem* (1997) and "Wranglers in Exile: Mathematics in the British Empire," in *Mathematics in Victorian Britain* (2011); and coauthor of "Geometry at Cambridge, 1863–1940," in *Historia Mathematica* (2006).

Mirjam Brusius is a Postdoctoral Researcher at the Mahindra Humanities Center at Harvard University. She is the author of "Inscriptions in a Double Sense: An Early Scientific Photograph of Script," in *Nuncius: Journal of the History of Science* (2009); "Impreciseness in Julia Margaret Cameron's Portrait Photographs," in *History of Photography* (2010); and "Misfit Objects: Layard's Excavations in Ancient Mesopotamia and the Biblical Imagination in Mid-Nineteenth Century Britain," in the *Journal of Literature and Science* (2012).

Katrina Dean is University Archivist at the University of Melbourne. She is the author of "Demonstrating the Melbourne University Respirator," in the *Australian Journal of Politics and History* (2007), and coauthor of "Data in Antarctic Science and Politics," in *Social Studies of Science* (2008), and "Digital Lives: Report of Interviews with the Creators of Personal Digital Collections," in *Ariadne* (2008).

Vered Maimon is a Senior Lecturer in the Department of Art History at Tel Aviv University. She is coeditor of *Communities of Sense: Rethinking Aesthetics and Politics* (2009) and author of, among others, "Displaced 'Origins': William Henry Fox Talbot's *The Pencil of Nature*," in *History of Photography* (2008); "The Third Citizen: On Models of Criticality in Contemporary Artistic Practices," in *October* (2009); and "On the Singularity of Early Photography: William Henry Fox Talbot's Botanical Images," in *Art History* (2011).

Chitra Ramalingam is a British Academy Postdoctoral Research Fellow in the Department of History and Philosophy of Science at the University of Cambridge. She is the author of "Natural History in the Dark: Seriality and the Electric Discharge in Victorian Physics," in *History of Science* (2010); "Fixing Transience: Photography and Other Images of Time in 1830s London," in *Time and Photography* (2010); and the book "To See a Spark: Experiment and Visual Experience in Victorian Science" (in progress).

Eleanor Robson is a Reader in Ancient Middle Eastern Science in the Department of History and Philosophy of Science at the University of Cambridge. She is the author of *Mesopotamian Mathematics, 2100–1600 BC: Technical Constants in Bureaucracy and Education* (1999); "Words and Pictures: New Light on Plimpton 322," in the *American Mathematical Monthly* (2002); and *Mathematics in Ancient Iraq: A Social History* (2008).

Larry J. Schaaf is an independent historian and consultant based in Baltimore. He is the founder and Director of the online Correspondence of William Henry Fox Talbot Project and author of *Out of the Shadows: Herschel, Talbot, & the Invention of Photography* (1992) and *The Photographic Art of William Henry Fox Talbot* (2000).

Simon Schaffer is Professor of History of Science at the University of Cambridge. He is a member of the Advisory Board of the Science Museum and a Fellow of the British Academy. His most recent publications include essays on disciplines and institutions in nineteenth-century sciences.

Anne Secord is an Affiliated Research Scholar in the Department of History and Philosophy of Science at the University of Cambridge, and an editor of *The Correspondence of Charles Darwin*. She is the author of "Pressed into Service: Specimens, Space, and Seeing in Botanical Practice," in *Geographies of Nineteenth-Century Science* (2011); a new Oxford World's Classics edition of Gilbert White, *The Natural History of Selborne* (2013); and the book "Artisan Naturalists" (in progress).

Graham Smith is Professor Emeritus in Art History at the University of St. Andrews and a Fellow of the Royal Society of Edinburgh. He is the author of *"Light that Dances in the Mind": Photographs and Memory in the Writings of E. M. Forster and His Contemporaries* (2007) and *Photography and Travel* (2012).

Herta Wolf is Professor in the History and Theory of Photography at the Institute for the History of Art at the University of Cologne. She is the author of "The Tears of Photography," in *Grey Room* (2007); "Collections of All Kinds Will Be Formed," in *Photoresearcher* (2010); and "Wolken: Zum Beispiel," in *Wolken: Welt des Flüchtigen* (2013); and the editor of two anthologies on photography, *Diskurse der Fotografie* (2002) and *Paradigma Fotografie: Fotokritik am Ende des fotografischen Zeitalters* (2003).

Credits and Sources

Every effort has been made to credit the photographers and sources of all illustrations in this volume; if there are any errors or omissions, please contact Yale University Press so that corrections can be made in any subsequent edition.

Beinecke Rare Book and Manuscript Library, Yale University: fig. 107

© The British Library Board: figs. 1, 2, 3, 4, 6, 7, 8, 9, 10, 13, 14, 16, 21, 25, 26, 27, 28, 30, 31, 32, 33, 35, 36, 37, 38, 40, 47, 59, 75, 78, 82, 89, 90, 97, 102, 105, 106, 109

© Trustees of the British Museum: figs. 42, 77, 80, 87, 91, 94, 99

Courtesy of Mirjam Brusius: fig. 12

Reproduced by kind permission of the Syndics of Cambridge University Library: fig. 15

Detroit Institute of Arts, USA / Founders Society Purchase, General Membership Fund / The Bridgeman Art Library: fig. 84

Courtesy of George Eastman House, International Museum of Photography and Film: fig. 63

The J. Paul Getty Museum, Los Angeles: fig. 70

Farlow Library of Cryptogamic Botany, Harvard University, Cambridge, Massachusetts, USA: figs. 18, 19, 24

Houghton Library, Harvard University: fig. 98

Courtesy of HathiTrust: figs. 85, 86

Courtesy of JSTOR: fig. 34

Courtesy of Hans P. Kraus, Jr., New York: figs. 61, 65, 71

The Metropolitan Museum of Art, Gift of Jean Horblit, in memory of Harrison D. Horblit, 1994: fig. 92

The Metropolitan Museum of Art, The Rubel Collection, Purchase, Ann Tenenbaum and Thomas H. Lee and Anonymous Gifts, 1997 (1997.382.1): fig. 58

© Musée des arts et métiers-CNAM, Paris / photo P. Faligot: figs. 50, 51

National Gallery of Canada, Ottawa: fig. 72

Reproduced by permission of the Trustees of the National Library of Scotland: fig. 39

© National Media Museum / Science & Society Picture Library: figs. 49, 52, 54, 55, 56, 57, 62, 64, 69

Photographic History Collection, National Museum of American History, Smithsonian Institution: figs. 60, 100

© National Museums Scotland: fig. 108

© National Trust Images: fig. 11

© NMPFT / Royal Photographic Society / Science & Society Picture: fig. 53

Courtesy of Eleanor Robson: fig. 79

© The Royal Institution, London, UK / The Bridgeman Art Library: fig. 48

Courtesy of Anne Secord: fig. 22

By kind permission of the Master and Fellows of Selwyn College, Cambridge: figs. 20, 23

State Library of Victoria: fig. 66

Courtesy of the Bill Douglas Centre for the History of Cinema and Popular Culture, University of Exeter: fig. 103

© The Hunterian, University of Glasgow 2012: fig. 101

Harry Ransom Center, The University of Texas at Austin: figs. 43, 44, 45, 46, 68

© Victoria and Albert Museum, London: figs. 93, 95, 96

Courtesy of the Yale Babylonian Collection: fig. 76

Yale Center for British Art, Paul Mellon Collection: fig. 41

Courtesy of the Yale University Library: figs. 29, 81, 83, 104

Manuscripts and Archives, Yale University Library: fig. 17

Index

Note: Page numbers in *italics* refer to illustrations.